PERMEABILITY PROPERTIES OF PLASTICS AND ELASTOMERS

To my daughter Lindsey McKeen Polizzotti, who continues to amaze me

PLASTICS DESIGN LIBRARY (PDL)

PDL HANDBOOK SERIES

Series Editor: Sina Ebnesajjad, PhD
President, FluoroConsultants Group, LLC
Chadds Ford, PA, USA
www.FluoroConsultants.com

The **PDL Handbook Series** is aimed at a wide range of engineers and other professionals working in the plastics industry, and related sectors using plastics and adhesives.
PDL is a series of data books, reference works and practical guides covering plastics engineering, applications, processing, and manufacturing, and applied aspects of polymer science, elastomers and adhesives.

Recent titles in the series

Sastri, *Plastics in Medical Devices*
ISBN: 9780815520276

McKeen, *Fatigue and Tribological Properties of Plastics and Elastomers*, Second Edition
ISBN: 9780080964508

Wagner, *Multilayer Flexible Packaging*
ISBN: 9780815520214

Chandrasekaran, *Rubber Seals for Fluid and Hydraulic Systems*
ISBN: 9780815520757

Tolinski, *Additives for Polyolefins*
ISBN: 9780815520511

McKeen, *The Effect of Creep and Other Time Related Factors on Plastics and Elastomers*, Second Edition
ISBN: 9780815515852

Ebnesajjad, Handbook of Adhesives and Surface Preparation
ISBN: 9781437744613

Grot, Fluorinated Ionomers, Second Edition
ISBN: 9781437744576

To submit a new book proposal for the series, please contact
Sina Ebnesajjad, Series Editor
sina@FluoroConsultants.com
or
Matthew Deans, Senior Publisher
m.deans@elsevier.com

PERMEABILITY PROPERTIES OF PLASTICS AND ELASTOMERS

Third Edition

Laurence W. McKeen

Amsterdam • Boston • Heidelberg • London • New York • Oxford
Paris • San Diego • San Francisco • Singapore • Sydney • Tokyo
William Andrew is an imprint of Elsevier

William Andrew is an imprint of Elsevier
225 Wyman Street, Waltham, 02451, USA
The Boulevard, Langford Lane, Kidlington, Oxford OX5 1GB, UK

First edition 1995
Second edition 2003
Third edition 2012

Copyright © 2012 Elsevier Inc. All rights reserved.

No part of this publication may be reproduced or transmitted in any form or by any means, electronic or mechanical, including photocopying, recording, or any information storage and retrieval system, without permission in writing from the publisher. Details on how to seek permission, further information about the Publisher's permissions policies and arrangements with organizations such as the Copyright Clearance Center and the Copyright Licensing Agency, can be found at our website: www.elsevier.com/permissions.

This book and the individual contributions contained in it are protected under copyright by the Publisher (other than as may be noted herein).

Notice

Knowledge and best practice in this field are constantly changing. As new research and experience broaden our understanding, changes in research methods, professional practices, or medical treatment may become necessary.

Practitioners and researchers must always rely on their own experience and knowledge in evaluating and using any information, methods, compounds, or experiments described herein. In using such information or methods they should be mindful of their own safety and the safety of others, including parties for whom they have a professional responsibility.

To the fullest extent of the law, neither the Publisher nor the authors, contributors, or editors, assume any liability for any injury and/or damage to persons or property as a matter of products liability, negligence or otherwise, or from any use operation of any methods, products, instructions, or ideas contained in the material herein.

Library of Congress Cataloging-in-Publication Data
A catalog record for this book is available from the Library of Congress

British Library Cataloguing in Publication Data
A catalogue record for this book is available from the British Library

ISBN: 978-1-4377-3469-0

For information on all Elsevier publications
visit our website at elsevierdirect.com

Transferred to Digital Printing in 2012

Working together to grow
libraries in developing countries

www.elsevier.com | www.bookaid.org | www.sabre.org

ELSEVIER   BOOK AID International   Sabre Foundation

Contents

Preface xi

1. Introduction to Permeation of Plastics and Elastomers 1
    1.1. History 1
    1.2. Transport of Gases and Vapors through Solid Materials 2
        1.2.1. Effusion 2
        1.2.2. Solution-Diffusion and Pore-Flow Models 3
            1.2.2.1. Non-Fickian Diffusion 5
            1.2.2.2. Dependence of Permeability, Diffusion, and Solubility Pressure 5
            1.2.2.3. Dependence of Permeability, Diffusion, and Solubility on Temperature—The Arrhenius Equation 8
    1.3. Multiple-Layered Films 9
    1.4. Permeation of Coatings 10
    1.5. Permeation and Vapor Transmission Testing 10
        1.5.1. Units of Measurement 14
        1.5.2. Gas Permeation Test Cells 14
        1.5.3. Vapor Permeation Cup Testing 14
        1.5.4. Permeation Testing of Coatings 15
            1.5.4.1. Electro-Impedance Spectroscopy 16
            1.5.4.2. Salt Spray/Humidity 16
            1.5.4.3. Atlas Cell 17
            1.5.4.4. Adhesion 17
    1.6. Summary 19
    References 19

2. Introduction to Plastics and Polymers 21
    2.1. Polymerization 21
    2.2. Copolymers 21
    2.3. Linear, Branched, and Cross-Linked Polymers 22
    2.4. Polarity 23
    2.5. Unsaturation 23
    2.6. Steric Hindrance 24
    2.7. Isomers 25
        2.7.1. Structural Isomers 25
        2.7.2. Geometric Isomers 25
        2.7.3. Stereoisomers—Syndiotactic, Isotactic, and Atactic 26
    2.8. Inter- and Intramolecular Attractions in Polymers 27
        2.8.1. Hydrogen Bonding 27
        2.8.2. Van der Waals Forces 27
        2.8.3. Chain Entanglement 27
    2.9. General Classifications 28
        2.9.1. Molecular Weight 28
        2.9.2. Thermosets vs. Thermoplastics 29
        2.9.3. Crystalline vs. Amorphous 29
        2.9.4. Orientation 30

2.10. Plastic Compositions ... 30
    2.10.1. Polymer Blends ... 30
    2.10.2. Elastomers ... 31
    2.10.3. Additives ... 31
        2.10.3.1. Fillers, Reinforcement, and Composites ... 31
        2.10.3.2. Combustion Modifiers, Fire, Flame Retardants, and Smoke Suppressants ... 33
        2.10.3.3. Release Agents ... 33
        2.10.3.4. Slip Additives/Internal Lubricants ... 34
        2.10.3.5. Antiblock Additives ... 34
        2.10.3.6. Catalysts ... 35
        2.10.3.7. Impact Modifiers and Tougheners ... 35
        2.10.3.8. UV Stabilizers ... 35
        2.10.3.9. Optical Brighteners ... 35
        2.10.3.10. Plasticizers ... 35
        2.10.3.11. Pigments, Extenders, Dyes, and Mica ... 36
        2.10.3.12. Coupling Agents ... 36
        2.10.3.13. Thermal Stabilizers ... 36
        2.10.3.14. Antistats ... 36
2.11. Summary ... 37
References ... 37

3. Production of Films, Containers, and Membranes ... 39
3.1. Extrusion ... 39
3.2. Blown Film ... 39
3.3. Calendering ... 41
3.4. Casting Film Lines ... 42
3.5. Post Film Formation Processing ... 44
3.6. Web Coating ... 45
    3.6.1. Gravure Coating ... 45
    3.6.2. Reverse Roll Coating ... 46
    3.6.3. Knife on Roll Coating ... 46
    3.6.4. Metering Rod (Meyer Rod) Coating ... 47
    3.6.5. Slot Die (Slot, Extrusion) Coating ... 47
    3.6.6. Immersion (Dip) Coating ... 47
    3.6.7. Vacuum Deposition ... 47
    3.6.8. Web Coating Process Summary ... 48
3.7. Lamination ... 49
    3.7.1. Hot Roll/Belt Lamination ... 49
    3.7.2. Flame Lamination ... 49
3.8. Orientation ... 50
    3.8.1. Machine Direction Orientation ... 51
    3.8.2. Biaxial Orientation ... 51
    3.8.3. Blown Film Orientation ... 52
3.9. Membrane Production ... 52
3.10. Molding of Containers ... 53
    3.10.1. Blow Molding ... 54
    3.10.2. Extrusion Blow Molding ... 54
    3.10.3. Injection Blow Molding ... 54
    3.10.4. Stretch Blow Molding ... 54
    3.10.5. Rotational Molding/Rotomolding ... 55
3.11. Fluorination ... 56
    3.11.1. Permeation Testing of Fluorinated Containers ... 56
3.12. Coatings ... 57
3.13. Summary ... 57
References ... 57

4. Markets and Applications for Films, Containers, and Membranes 59
4.1. Barrier Films in Packaging 59
4.1.1. Water Vapor 59
4.1.2. Atmospheric Gases 59
4.1.3. Odors and Flavors 60
4.1.4. Markets and Applications of Barrier Films 60
4.1.5. Some Illustrated Applications of Multilayered Films 65
4.2. Containers 65
4.3. Automotive Fuel Tanks and Hoses 67
4.4. Coatings 68
4.5. Gloves 68
4.6. Membranes 69
4.6.1. Dialysis 69
4.6.2. Reverse Osmosis 70
4.6.3. Pervaporation 70
4.6.4. Gas Separation 71
4.6.5. Membrane Structures 72
4.6.5.1. Plate and Frame Modules 72
4.6.5.2. Hollow Fiber Modules 72
4.6.5.3. Tubular Membrane Modules 73
4.6.5.4. Spiral Wound modules 73
References 75

5. Styrenic Plastics 77
5.1. Acrylonitrile–Butadiene–Styrene Copolymer 77
5.2. Acrylonitrile–Styrene–Acrylate 80
5.3. Polystyrene 82
5.4. Styrene–Acrylonitrile Copolymer 86
References 87

6. Polyesters 89
6.1. Liquid Crystalline Polymers 89
6.2. Polybutylene Terephthalate (PBT) 92
6.3. Polycarbonate (PC) 93
6.4. Polycyclohexylene-dimethylene Terephthalate 99
6.5. Polyethylene Naphthalate 100
6.6. Polyethylene Terephthalate (PET) 100
References 106

7. Polyimides 107
7.1. Polyamide–Imide 107
7.2. Polyetherimide 113
7.3. Polyimide 118
References 120

8. Polyamides (Nylons) 121
8.1. Amorphous Polyamide (Nylon) 122
8.2. Polyamide 6 (Nylon 6) 126
8.3. Polyamide 11 (Nylon 11) 131
8.4. Polyamide 12 (Nylon 12) 133
8.5. Polyamide 66 (Nylon 66) 135
8.6. Polyamide 66/610 (Nylon 66/610) 136
8.7. Polyamide 6/12 (Nylon 6/12) 137
8.8. Polyamide 666 (Nylon 666 or 6/66) 139
8.9. Polyamide 6/69 (Nylon 6/6.9) 140

8.10. Polyarylamide	141
8.11. Polyphthalamide/High Performance Polyamide	142
References	144

9. Polyolefins, Polyvinyls, and Acrylics ... 145
- 9.1. Polyethylene ... 145
    - 9.1.1. Unclassified Polyethylene ... 146
    - 9.1.2. Ultra-low-Density Polyethylene ... 150
    - 9.1.3. Linear Low-Density Polyethylene ... 151
    - 9.1.4. Low-Density Polyethylene ... 151
    - 9.1.5. Medium-Density Polyethylene ... 153
    - 9.1.6. High-Density Polyethylene ... 153
- 9.2. Polypropylene ... 156
- 9.3. Polybutadiene ... 158
- 9.4. Polymethylpentene ... 159
- 9.5. Cyclic Olefin Copolymer ... 160
- 9.6. Ethylene–Vinyl Acetate Copolymer ... 161
- 9.7. Ethylene–Vinyl Alcohol Copolymer ... 163
- 9.8. Polyvinyl Butyral ... 179
- 9.9. Polyvinyl Chloride ... 179
- 9.10. Polyvinylidene Chloride ... 182
- 9.11. Polyacrylics ... 186
- 9.12. Acrylonitrile–Methyl Acrylate Copolymer ... 188
- 9.13. Ionomers ... 190
    - 9.13.1. Ionomer—Ethylene Acrylic Acid Copolymer (EAA) ... 190
    - 9.13.2. Ionomer—Perfluorosulfonic Acid (PFSA) ... 192
- References ... 192

10. Fluoropolymers ... 195
- 10.1. Polytetrafluoroethylene (PTFE) ... 196
    - 10.1.1. PTFE Homopolymer ... 197
    - 10.1.2. Modified PTFE ... 202
- 10.2. Fluorinated Ethylene Propylene (FEP) ... 203
- 10.3. Perfluoroalkoxy (PFA) ... 207
    - 10.3.1. PFA ... 208
    - 10.3.2. MFA ... 210
- 10.4. Hexafluoropropylene, Tetrafluoroethylene, Ethylene Terpolymer (HTE) ... 212
- 10.5. Tetrafluoroethylene, Hexafluoropropylene, Vinylidene Fluoride Terpolymer (THV™) ... 212
- 10.6. Amorphous Fluoropolymer (AF)—Teflon AF® ... 214
- 10.7. Polyvinyl Fluoride (PVF) ... 215
- 10.8. Polychlorotrifluoroethylene (PCTFE) ... 217
- 10.9. Polyvinylidene Fluoride (PVDF) ... 219
- 10.10. Ethylene–Tetrafluoroethylene Copolymer (ETFE) ... 223
- 10.11. Ethylene–Chlorotrifluoroethylene Copolymer (ECTFE) ... 226
- References ... 231

11. High-Temperature and High-Performance Polymers ... 233
- 11.1. Polyether Ether Ketone ... 233
- 11.2. Polysiloxane ... 233
- 11.3. Polyphenylene Sulfide ... 234
- 11.4. Polysulfone ... 237
- 11.5. Polyethersulfone ... 241
- 11.6. Polybenzimidazole ... 243
- 11.7. Parylene (poly(*p*-xylylene)) ... 245

11.8. Polyoxymethylene (POM or Acetal Homopolymer)/Polyoxymethylene Copolymer
(POM-Co or Acetal Copolymer) 247
References 250

12. Elastomers and Rubbers 251
12.1. Thermoplastic Polyurethane Elastomers (TPU) 251
12.2. Olefinic TPEs (TPO) 256
12.3. Thermoplastic Copolyester Elastomers (TPE-E or COPE) 257
12.4. Thermoplastic Polyether Block Polyamide Elastomers (PEBA) 259
12.5. Styrenic Block Copolymer (SBC) TPEs 262
12.6. Ethylene Acrylic Elastomers (AEM) 263
12.7. Bromobutyl Rubber 264
12.8. Butyl Rubber 266
12.9. Chlorobutyl Rubber (Polychloroprene) 268
12.10. Ethylene–Propylene Rubbers (EPM, EPDM) 270
12.11. Epichlorohydrin Rubber (CO, ECO) 272
12.12. Fluoroelastomers (FKM) 273
12.13. Natural Rubber 278
12.14. Acrylonitrile–Butadiene Copolymer (NBR) 281
12.15. Styrene–Butadiene Rubber (SBR) 283
References 284

13. Environmentally Friendly Polymers 287
13.1. Cellophane™ 291
13.2. Nitrocellulose 293
13.3. Cellulose Acetate 294
13.4. Ethyl Cellulose 295
13.5. Polycaprolactone 299
13.6. Poly(Lactic Acid) 301
13.7. Poly-3-Hydroxybutyrate 303
References 304

14. Multilayered Films 305
14.1. Metalized Films 305
14.2. Silicon Oxide Coating Technology 305
14.3. Co-continuous Lamellar Structures 306
14.4. Multilayered Films 310
References 318

Appendix A 319
Appendix B 321
Appendix C 323
Index 329

Preface

This book has is an extensive update and extension to the second edition by the same title. The second edition was published in 2002, and a lot has changed in the field since then. There are new plastic materials. There has been an expanded interest in green materials, those made from renewable resources or those that decompose relatively quickly in the environment. There has been a turnover in ownership of the plastic producing companies. There has been a lot of consolidation, which of course means discontinued products. This update is much more extensive than the usual "next edition."

It has been reorganized from a polymer chemistry point of view. Plastics of mostly similar polymer types are grouped into 10 data chapters. A brief explanation of the chemistry of the polymers used in the plastics is discussed at the start of each plastic section.

An extensive introduction has been added as four chapters. The initial chapter focuses on permeation, what it is, how it occurs, is measured and data are presented. The second chapter covers polymer chemistry and plastics composition and how it relates to permeation. The third chapter focuses on production of films, containers, and membranes. The fourth chapter focuses on the uses of barrier films and membranes. Membranes were largely ignored in the previous editions, but the function of membranes is critically affected by their permeation properties.

Chapters 5 through 14 are a databank that serves as an evaluation of permeation performance of plastics. Each of these chapters starts with a brief outline of the chemistry of the polymer in that section. There are hundreds of uniform graphs and tables for more than 60 generic families of plastics that are contained in these chapters.

The data in each chapter are generally organized with chemistry, a manufacturer and trade name list, an applications and uses list followed by the data. Tabular data are first, followed by graphical data. The tabular data from the second edition has been verified and reformatted to take up much less space, whereas this new edition does not have many more pages, there is far more information contained. A list of conversion factors for gas permeation and vapor transmission measures is also included. There is an appendix that covers the important application of gloves. There is also an appendix of standard fuels used in testing plastic for auto fuel systems applications, an important use as the industry moves away from metal to plastics to save weight.

Numerous references are included. Some data from the earlier edition have been removed or replaced with updated data. Removed data include discontinued products. Product names and manufacturers have been updated.

I am especially appreciative of the confidence, support, and patience of my friend Sina Ebnesajjad. He was also the primary proofreader of the manuscript. I would not have been given the opportunity to do this work had it not been for the support of Matthew Deans, Senior Publisher at Elsevier. I also wish to acknowledge Bill Bennett, Marketing Manager, Americas Region of Ansell Limited and Joe Yachanin of Force 12 Design Ltd for their work on the permeation of gloves index. My family has been particularly supportive through the long hours of writing and research from my home office.

Laurence McKeen
2011

1 Introduction to Permeation of Plastics and Elastomers

This book is about the passage of liquids, vapors, or gases through plastic or polymeric materials, such as films, membranes, and containers. The passage of small molecules through solid materials is called *permeation*. *Permeability* properties are important in many applications and are important in everyday life. The small molecule that passes through the solid is called the *permeant*. The permeability of the packaging materials (wrapping films, containers seals, closures, etc.) needs to be matched with the sensitivity of the packaged contents to that permeant and the specified shelf life. Some packages must have nearly hermetic seals, while others can (and sometimes must) be selectively permeable. Knowledge about the exact permeation rates is therefore essential. There are many examples for the importance of permeation properties, some are as follows:

- The air pressure in tires should decrease as slowly as possible, so it is good to know which gas permeates slowest through the rubber wall and at what rate.
- Packaged meat needs to retain its moisture within the package to prevent it from drying out, but it needs to keep oxygen out too, which slows down spoilage.
- The water vapor permeation of insulating material is important to protect underlying from corrosion.
- To meet legal regulations, e.g., California Air Resource Board (CARB) for low-emission vehicles, it is essential to use barrier materials for fuel hoses and tanks.

This first chapter focuses on permeation, what it is, how it occurs, how it is measured, and how its data are presented. The second chapter covers polymer chemistry and plastics composition and how composition affects permeation. The third chapter concentrates on production of films, containers, and membranes. The fourth chapter focuses on the uses of barrier films, containers, and membranes.

Chapters 5 to 14 are a databank that serves as an evaluation of permeation performance of plastics. Each of these chapters starts with a brief outline of the chemistry of the polymer in that section. There are hundreds of uniform graphs and tables for more than 60 generic families of plastics that are contained in these chapters. The data in each chapter are generally organized with chemistry, a manufacturer and trade name list, and an applications and uses list followed by the data.

1.1 History

It appears that the first scientific mention of water permeation was made by a physicist, Abbé Jean-Antoine Nollet (1700−1770).[1] Nollet sealed wine containers with a pig's bladder and stored them under water. After a while, he noticed that the bladder bulged outward. Because of his scientific curiosity, he did the experiment the other way round: he filled the container with water and stored it in wine. The result was a shrinking inward of the bladder. This was an evidence of water permeation through the pig's bladder from an area of high concentration to an area of lower concentration.

The first study of gas permeation through a polymer was conducted by Thomas Graham in 1826.[2] Graham observed a loss in volume of a wet pig bladder inflated with carbon dioxide. In 1831, John Kearsley Mitchell, professor of medicine and physiology at the Philadelphia Medical Institute, observed that balloons shrunk at different rates when they were filled with different gases.

In 1856, Henry Gaspard Philibert Darcy (born 1803) formulated his homogeneous linear law of water transport through a porous medium (essentially sand and gravel beds). Darcy's law is an important basic relationship in hydrogeology. Examples of its application include the flow of water through an aquifer and the flow of crude oil, water, and gas through the rock in petroleum reservoirs. Darcy's law is a relationship between the instantaneous discharge rate through a porous medium, the viscosity of the fluid, and the pressure drop over a given distance. Darcy's law in equation form is given in Eqn (1).

$$Q = \frac{-\kappa A(P_b - P_a)}{\mu L} \quad (1)$$

where (refer Fig. 1.1)
Q = total discharge rate (units of volume per time, such as m^3/s)
κ = permeability of the medium (units of area, such as m^2)
A = cross-sectional area to flow (units of area, such as m^2)
$(P_b - P_a)$ = pressure drop [units of pressure, such as Pa (Pascal)]
μ = dynamic viscosity (units such as Pa s)
L = the length, where the pressure drop is taking place (units such as m).

The permeability κ from Darcy's law should not be confused with the permeability coefficient P, described later in this chapter, used when describing permeation through plastic materials.

A theory for gas permeation through polymeric materials was not developed until 1866, when Thomas Graham proposed the solution-diffusion process, where he postulated that the permeation process involved the dissolution of penetrant, followed by transmission of the dissolved species through the membrane. The other important observations Graham made at the time were as follows:

1) Permeation was independent of pressure.
2) Increase in temperature leads to decrease in penetrant solubility, but made the membrane more permeable.
3) Prolonged exposure to elevated temperature affected the retention capacity of the membrane.
4) Differences in the permeability could be exploited for application in gas separations.
5) Variation in membrane thickness altered the permeation rate but not the separation characteristics of the polymer.

1.2 Transport of Gases and Vapors through Solid Materials

There are two ways for small molecules to pass through a solid material. One is by passing through a small hole or leak. The second is for the small molecule to work its way through the solid between the small spaces between the molecules (in the case of polymers or plastics) or atoms in the crystal structure of inorganic solids or metals.

1.2.1 Effusion

Passage of a gas or a liquid may occur through a defect such as a pinhole in a film. Although the end result might be the same, the passage of the gas or liquid, this is not permeation. This type of transport through a film is more appropriately called *effusion*. Effusion is governed by *Graham's law*, also known as *Graham's law of effusion*. It was formulated by Scottish physical chemist Thomas Graham. Graham found through experimentation that the rate of effusion of a gas is inversely proportional to the square root of the mass of its particles as given in Eqn (2):

$$\frac{\text{Rate}_1}{\text{Rate}_2} = \sqrt{\frac{M_2}{M_1}} \quad (2)$$

where
Rate$_1$ is the rate of effusion for gas 1 (volume or number of moles per unit time)
Rate$_2$ is the rate of effusion for gas 2
M_1 is the molar mass of gas 1
M_2 is the molar mass of gas 2.

When one plots the time, it takes 25 mL of gas to be transported through a pinhole into vacuum against the molecular mass of the gas, a linear relationship is obtained as shown in Fig. 1.2.

A complete theoretical explanation of Graham's law was provided years later by the kinetic theory of

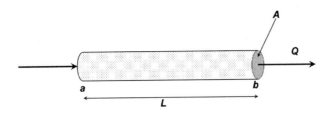

Figure 1.1 Parameter descriptions for Darcy's law.

1: Introduction to Permeation of Plastics and Elastomers

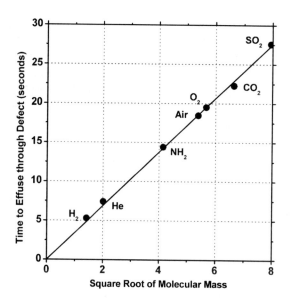

Figure 1.2 The time required for 25 mL samples of different gases to diffuse through a pinhole into a vacuum.

gases. Graham's law is most accurate for molecular effusion, which involves the movement of one gas at a time through a hole. It is only approximate for diffusion of one gas into another.

An example where effusion would be appropriate is given below: oxygen and nitrogen molecules were passing through a relatively large passage way through the tire wall, such as a leak. Graham's law for "effusion" applies only if the exit pinhole through which the molecules pass is relatively large compared with the size of the molecules and does not obstruct or constrain one molecule from passing through relative to the other molecule.

If there are no pinholes in the material then for a molecule to get through that material, it must find its way through a tangle of molecular chains. Since all matter are in continuous motion when it is above absolute zero, a small molecule may be able to work its way through that entanglement of polymer chains as they bend forming a pocket in the material through which a small molecule can fit. Those little pockets may continuously close and form again allowing transport of a small molecule through the material. One would expect that smaller molecules would have an easier time diffusing through that molecular "jungle."

The *kinetic diameter* is a reflection of the smallest effective dimension of a given molecule. Another way to characterize kinetic diameter is to determine the diameter of a molecule, assuming it to be

Table 1.1 Kinetic Diameters of Various Permeants[3]

Molecule	Diameter (nm)
Helium (He)	0.26
Hydrogen (H_2)	0.289
Nitric oxide (NO)	0.317
Carbon dioxide (CO_2)	0.33
Argon (Ar)	0.34
Oxygen (O_2)	0.346
Nitrogen (N_2)	0.364
Carbon monoxide (CO)	0.376
Methane (CH_4)	0.38
Ethylene (C_2H_4)	0.39
Xenon (Xe)	0.396
Propane (C_3H_8)	0.43
n-Butane (C_4H_{10})	0.43
Difluorodichloromethane (CF_2Cl_2)	0.44
Propene (C_3H_6)	0.45
Tetrafluoromethane (CF_4)	0.47
i-Butane (C_4H_{10})	0.50

spherical. Table 1.1 lists the kinetic diameters of various permeants.

1.2.2 Solution-Diffusion and Pore-Flow Models

A relationship that governs "permeation" is based on Fick's (First) law of diffusion and Henry's law of solubilities. Fick's laws of diffusion were derived by Adolf Fick in 1855. His laws described diffusion and can be used to solve for the diffusion coefficient, D. Henry's law is one of the gas laws, formulated by William Henry in 1803. It states that: *At a constant temperature, the amount of a given gas dissolved in a given type and volume of liquid is directly proportional to the partial pressure of that gas in equilibrium with that liquid.* These laws take into account the relative sizes of the molecules and their sizes compared to the very small passage way dimensions in the solid material (such as a rubber) through which the molecules "permeate." It is often mistakenly assumed that "molecular size" correlates directly with "molecular weight." Oxygen does have a greater molecular weight (32) than nitrogen

(28), but the oxygen molecule is actually smaller in size.

Combining Fick's and Henry's laws yields the overall equation governing permeation of small molecules, such as gases, into material such as rubbers and other plastics.

The solution-diffusion and pore-flow models are commonly used to understand and predict permeation through films. Both the models are similar and suggest permeant transfer through polymer films or membranes progresses through five consecutive steps as follows (refer Fig. 1.3):

1. Permeant diffusion to the polymer film from the upstream atmosphere.
2. Adsorption of the permeant by the polymer film at the interface with the upstream atmosphere.
3. Diffusion of the permeant inside and through the polymer film. The diffusion step is the slowest and becomes the rate-determining step in gas permeation.
4. Desorption of the permeant at the interface of the downstream side of the film.
5. Diffusion of the permeant away from the polymer film into the downstream atmosphere.

The primary difference between the solution-diffusion model and the pore-flow model is the assumption of the pressure differential through the membrane.[4] For the solution-diffusion model, the assumption is that the pressure is uniform through the membrane and equal to the upstream pressure. For the pore-flow model, the pressure is assumed to drop uniformly across the membrane. This difference affects the mathematics of modeling the membrane processes of dialysis, gas separation, reverse osmosis, and pervaporation.

Although the difference is small, gas permeation through a nonporous dense polymer membrane is usually described using the solution-diffusion model, the basis of which is described by the fairly simple Eqn (3).

$$P = DS \qquad (3)$$

where
P = permeability
D = diffusivity
S = solubility.

The units of permeability are (amount of gas × thickness of membrane)/(area of membrane × time × pressure).

The standard units used in this book are $(cm^3\ mm)/(m^2\ day\ atm)$ for gas volumes or $(g\ mm)/(m^2\ day\ atm)$ for gas mass. When gas volumes are used, they are usually for gas at standard temperature and pressure (STP) conditions. The current version of IUPAC's STP is a temperature of 0 °C (273.15 K, 32 °F) and an absolute pressure of 100 kPa [14.504 pounds per square inch (psi), 0.986 atm], while NIST's version is a temperature of 20 °C (293.15 K, 68 °F) and an absolute pressure of 101.325 kPa (14.696 psi, 1 atm). International Standard Metric Conditions for natural gas and similar fluids is 288.15 K and 101.325 kPa. Appendix A contains a large list of other units and their conversion factors to the standard used in this book.

The units of diffusivity are (area of membrane)/time.

The units of solubility are (amount of gas)/(volume of polymer × pressure).

All three parameters can be experimentally determined from a time-lag experiment, assuming that a membrane is totally free of permeant and the diffusion coefficient is assumed to be constant. The time-lag technique was originally conceived by Daynes in 1920.[5] The basic experiment measures the amount of vapor that crosses through a film versus time shown in Fig. 1.4 (further details are given in Section 1.5). Before the actual experiment starts, valve A is closed and a vacuum is drawn through open valves B and C. Valve B is closed and valve A is opened at the beginning of the experiment ($t = 0$) and

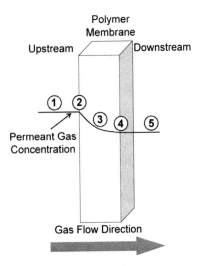

Figure 1.3 Solution-diffusion model of permeation through polymer films or membranes.

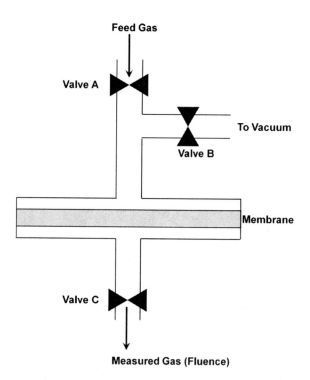

Figure 1.4 Schematic representation of the setup for gathering the data for the time-lag measurement plot.

the amount of gas passing through the membrane is measured.

One plots the amount of vapor that crosses through a film of l thickness, a area, and Δp pressure differential versus time as shown in Fig. 1.5.

A straight line is drawn through the linear portion of the plot as shown. The x-axis intercept is the *time lag*, L. The time lag is used to determine the diffusivity D using the following equation (see Ref. 6 for derivation).

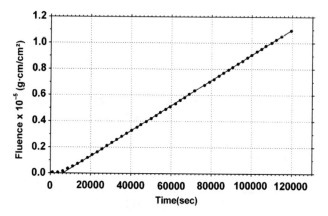

Figure 1.5 Typical time-dependent permeation data (time-lag plot) for a single-layer film.

$$D = \frac{l^2}{6L} \quad (4)$$

where l is the film thickness and L is the time lag determined from the plot.

In this particular example, the film thickness, l, was 0.0177 cm and the time lag was 6264 s. This results in the diffusivity calculated to be 8.62×10^{-9} cm^2/s. Note that this formula for time lag can apply when the diffusion coefficient is constant and not a function of concentration. Concentration may affect the diffusion coefficient if the film swells.

In this example, the Δp pressure differential was 1. The slope of the line through the linear portion of the plot is the permeability, P, which is 1.4×10^{-9} g cm/cm^2 s atm (which reduces to g/cm s atm).

By rearranging Eqn (5), the solubility can be calculated:

$$S = \frac{P}{D} = \frac{1.4 \times 10^{-9} \; (\text{g} \times \text{cm/cm}^2 \times \text{s} \times \text{atm})}{8.62 \times 10^{-9} \; (\text{cm}^3/\text{s})}$$
$$= 0.16 \; (\text{g/cm}^3 \times \text{atm})$$
$$(5)$$

Solubility can also be determined directly using a microbalance to determine the weight gain before and after exposure of a given amount of plastic/polymer to a permeant at a given pressure.[7]

1.2.2.1 Non-Fickian Diffusion

Fick's first law of diffusion works when the diffusing penetrant is nonswelling. When the penetrant swells, the plastic structural changes are introduced (the dimensions are changed) and that may introduce local deformation and stress. In some cases, the stresses are so large that cracks (often called crazing) may form. The stresses generated may affect diffusion and solubility, which in turn affect permeation per Eqn (3). The mathematics of non-Fickian transport is dependent on the model chosen and is mathematically complex.[8-10] Furthermore, the temperature, dimensions, and material history may have an effect on transport.

1.2.2.2 Dependence of Permeability, Diffusion, and Solubility Pressure

The solubility coefficient depends on pressure. Pressure is especially important when using membranes for separations. It also depends on gas

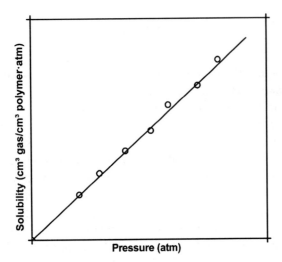

Figure 1.6 Solubility coefficient depending on pressure under Henry's law of sorption.

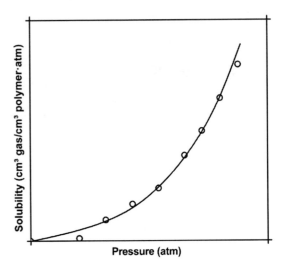

Figure 1.8 Solubility coefficient depending on pressure under Flory–Huggins sorption.

compressibility and interactions between the penetrant and the polymer. In the simplest case, Henry's law of sorption, the solubility is directly proportional to the pressure as shown in Fig. 1.6.

There are other interactions, however, that cause this relationship to deviate from linearity. A common model is the Langmuir model for absorption of a gas on a surface. It assumes that the coverage of the surface by a monomolecular layer. The model approach also assumes that only one gas is being adsorbed. At a given temperature, only a part of the surface will be covered with the adsorbed molecule and another part will not be covered. A dynamic equilibrium will exist between the free gas and the adsorbed gas where there will be as many molecules adsorbing as there will be desorbing. Using this model, the solubility coefficient versus pressure will be in the form shown in Fig. 1.7.

Flory–Huggins is often used when organic vapors are adsorbed at high-pressures swelling by the permeant in rubbery polymers and the relationship is shown in Fig. 1.8.

The Langmuir model assumes monomolecular layer adsorption. However, molecules are often adsorbed on already adsorbed molecules and the Langmuir isotherm is not valid. In 1938, a model was developed by Brunauer, Emmett, and Teller (BET) that takes multilayer adsorption into account. Their

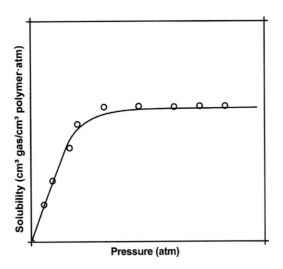

Figure 1.7 Solubility coefficient depending on pressure under Langmuir sorption.

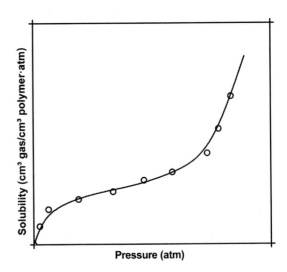

Figure 1.9 Solubility coefficient depending on pressure under BET sorption.

1: Introduction to Permeation of Plastics and Elastomers

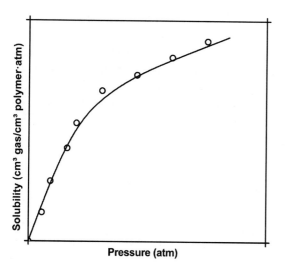

Figure 1.10 Solubility coefficient depending on pressure under dual-mode sorption.

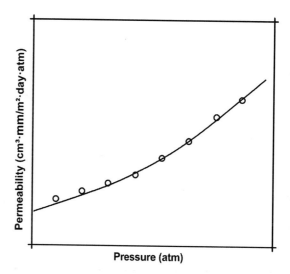

Figure 1.12 Permeability coefficient depending on pressure when the penetrant plasticizes the polymer.

theory is called BET theory, and Fig. 1.9 shows the predicted solubility versus pressure relationship. It is often applied in situations where the permeant is highly soluble in a glassy polymer.

Dual-mode sorption is used most often when gases are absorbed by glassy polymers (Fig. 1.10).

The permeability coefficient may or may not also vary with pressure. In the ideal case, diffusion and solubility are not dependent on pressure and neither is the permeability coefficient as shown in Fig. 1.11.

When the penetrant acts like a plasticizer to the polymer such as with the adsorption of organic vapors by a rubbery polymer, the relationship would be more similar to that shown in Fig. 1.12.

High-penetrant solubility in a glassy polymer might give a relationship that looks similar to that shown in Fig. 1.13.

A combination of the two previous responses can lead to permeability versus pressure relationship similar to that shown in Fig. 1.14.

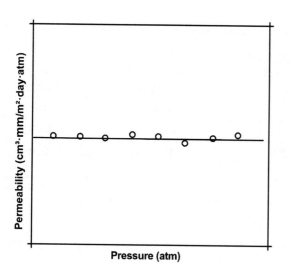

Figure 1.11 Permeability coefficient depending on pressure in ideal case.

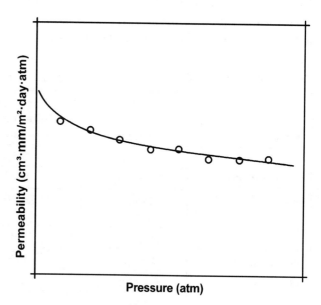

Figure 1.13 Permeability coefficient depending on pressure when the penetrant plasticizes a rubbery polymer.

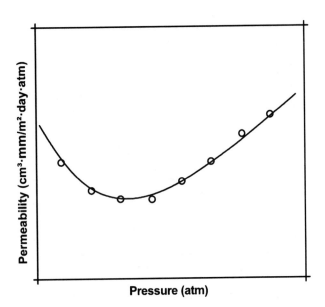

Figure 1.14 Permeability coefficient depending on pressure when the penetrant plasticizes a glassy polymer.

1.2.2.3 Dependence of Permeability, Diffusion, and Solubility on Temperature—The Arrhenius Equation

The Arrhenius equation is a simple, but remarkably accurate, formula for the temperature dependence of the chemical reaction rate constant. The equation was first proposed by the Dutch chemist Jacobus Hendricus van't Hoff in 1884 but 5 years later in 1889, the Swedish Chemist Svante Arrhenius provided a physical justification and interpretation for it. Today, it is generally viewed as an empirical relationship that works well, over moderate temperature ranges, to model the temperature–variance of permeation, diffusion and solubility coefficients, and other chemical processes.

The equation used to forecast or fit the permeation coefficient temperature dependence data is given below.

$$P = P_0 \exp\left(\frac{-\Delta E_p}{RT}\right) \quad (6)$$

P_0 and ΔE_p are characteristics of a particular material and permeant pair. The units of P_0 are the same as that of permeability. The units of ΔE are commonly energy per mole, kJ/mol. The units of the gas constant, R, are of energy (kJ/mol) per degree Kelvin (K) per mole. The temperature, T, is K, which is 273.15+ °C. Table 1.2 shows some of the data that are used in the Arrhenius equation.

There are similar Arrhenius equations that model the diffusion and solubility coefficients. Equation (7) applies for diffusion.

$$D = D_0 \exp\left(\frac{-\Delta E_d}{RT}\right) \quad (7)$$

Equation 8 applies for solubility.

$$S = PH_0 \exp\left(\frac{-\Delta H_s}{RT}\right) \quad (8)$$

Table 1.2 Arrhenius Equation Parameters for the Permeability of Some Polymer/Permeant Pairs[11]

Polymer	Permeant	P_0, Source Document Units [cm³(STP) cm/ cm² s Pa]	P_0, Normalized Units (cm³ mm/ m² day atm)	ΔE_p (kJ/mol)
Low density polyethylene	Oxygen	6.65×10^{-6}	5.82×10^9	42.7
	Nitrogen	3.29×10^{-5}	2.88×10^{10}	49.9
	Carbon dioxide	6.20×10^{-6}	5.43×10^9	38.9
	Water	4.88×10^{-6}	4.27×10^9	33.5
Poly(vinylidene chloride)	Nitrogen	9.00×10^{-5}	7.88×10^{10}	70.3
	Oxygen	8.25×10^{-5}	7.22×10^{10}	66.6
	Carbon dioxide	2.47×10^{-6}	2.16×10^9	51.5
	Water	8.63×10^{-6}	7.55×10^{10}	46.1

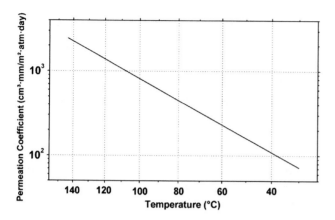

Figure 1.15 Arrhenius plot of the permeability coefficient versus temperature for carbon dioxide through polyvinylidene fluoride.[21]

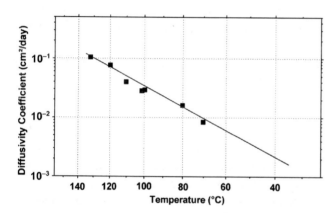

Figure 1.16 Arrhenius plot of the diffusion coefficient versus temperature for carbon dioxide through polyvinylidene fluoride.[7]

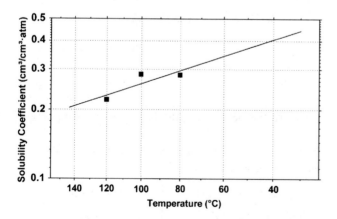

Figure 1.17 Arrhenius plot of the solubility coefficient versus temperature for carbon dioxide through polyvinylidene fluoride.[7]

The parameters (P_0, D_0, H_0) and (ΔE_p, ΔE_d, ΔH_s) are determined from the corresponding Arrhenius plots of experimentally measured data. An Arrhenius graph plots the log or natural log of the measured parameter (P, D, or S) against the inverse of absolute temperature (1/K). When plotted in this manner, the value of the "y-intercept" will correspond to (P_0, D_0, H_0), and the slope of the line will be equal to (ΔE_p, ΔE_d, ΔH_s)/R.

Examples of Arrhenius plots for permeability, diffusion, and solubility are given in Figs. 1.15, 1.16, and 1.17, respectively.

One should use caution when extrapolating these plots too far past the experimental data temperature range. It is interesting to note that for both permeation and diffusion, the parameters increase with increasing temperature, but the solubility relationship is the opposite. The reduced solubility with increasing temperature follows the solubility of oxygen in water. Excessively warm water temperatures have been known to lead to fish kills caused by low-oxygen concentration in warm water.

1.3 Multiple-Layered Films

Many barrier films are multilayered as discussed in the later section of this chapter. Multilayer films can achieve performance levels that single-layered films cannot. Depending on the choice of the composition of each layer, one may be able to use one layer for water permeation barrier properties and another layer for oxygen barrier properties giving a total film that is a barrier to both. A single-layer barrier film may not be able to achieve that. If the permeation and the thickness of each layer are known, one can estimate the permeability of the multilayered structure by using Eqn (9).

$$\frac{1}{P} = \sum_{i=1}^{n} \frac{E_i}{P_i} \qquad (9)$$

where

E_i = thickness of layer i
P = permeability of the multilayered structure
P_i = permeability of each layer
n = number of layers.

Of course, one may test for the permeation coefficients as described in Section 1.5 on testing. Later chapters in this book discuss commercial examples of multilayered structures and methods to produce them.

1.4 Permeation of Coatings

Coatings are not usually discussed in books and articles about permeation. Coatings are frequently intended to function as adhered barrier films, protecting the substrate to which they are applied from the environment, particularly from oxidation/corrosion. Permeation to oxygen and water vapor is certainly of interest as these are known factors in the rusting of ferrous metals. Permeation of other chemicals, such as the corrosive gases carbon dioxide and hydrogen sulfide, are often of interest.

Barrier coatings for corrosion protection are designed to block or at least slowdown the passage of oxygen and water to the metal surface being protected. These are essential components of the corrosion reactions, which are described below. Corrosion is an electrochemical reaction, so there is an oxidation reaction and a reduction reaction:

Iron is oxidized: $Fe \rightarrow Fe^{2+} + 2e^-$
Oxygen is reduced: $O_2 + H_2O + 2e^- \rightarrow 2(OH)^-$

The ferrous (Fe_2^+) and hydroxyl ions (OH^-) react to form ferrous hydroxide: $Fe + 2(OH)^- \rightarrow Fe(OH)_2$

The ferrous hydroxide ($Fe(OH)_2$) is oxidized further to ferric oxide (Fe_2O_3, red rust):

$4Fe(OH)_2 + O_2 \rightarrow 2Fe_2O_3 \cdot H_2O + 2H_2O$

Before the early 1950s, coatings were generally believed to protect steel by acting as a barrier to keep water and oxygen away from the steel surface. However, Mayne[12] found that the permeability of paint films was generally so high that the concentration of water and oxygen coming through the films would be higher than the rate of consumption of water and oxygen in the corrosion of the steel if it were uncoated steel. He concluded that the barrier properties did not explain the effectiveness of the corrosion protection of coatings. At that time, he proposed that the electrical conductivity of coating films is the variable that controls the degree of corrosion protection since corrosion is an electrochemical process. He confirmed experimentally that coatings with very high conductivity provide poor corrosion protection. However, in comparisons of films with relatively low conductivity, little correlation between conductivity and corrosion protection was found.

According to Funke, the corrosion protection of coating is now considered to be dependent on three mechanisms:[13]

1. Electrochemical mechanism—Conductivity and active anticorrosion pigmentation
2. Physiochemical mechanism—Diffusion and permeation
3. Adhesional mechanism—Adhesion to substrate under wet conditions

The electrochemical process was described above. Anticorrosion additives need to be able to act at the substrate–coating interface where the corrosion occurs.

The physiochemical mechanism is focused on permeation. It is not only permeation of water and oxygen but also ionic species such as chloride ion (Cl^-), sulfate ion (SO_4^{-2}), and nitrate ion (NO_3^-) that act as a corrosion stimulant.

The adhesion mechanism applies to the adhesion of the coating under wet conditions, *wet adhesion*. Water permeating through an intact film could displace areas of film from steel if the film has poor wet adhesion. Water and oxygen dissolved in the water would then be in direct contact with the steel surface leading to the start of corrosion. As corrosion proceeds, ferrous and hydroxide ions are generated, leading to formation of an osmotic cell under the coating film, the start of a small blister (as shown in Fig. 1.23). Osmotic pressure (see Section 4.5.2) can provide a force to remove more coating from the substrate, making the blister larger. Osmotic pressure once established is expected to far exceed adhesion forces. Thus, blisters form and expand, exposing more unprotected steel surface to water and oxygen.[14]

1.5 Permeation and Vapor Transmission Testing

There are many permeation and vapor transmission tests that are listed in Tables 1.4–1.7. Although there are specialized tests that deal with specific materials with specific ways to detect and measure the permeants, the tests operate similarly.

Table 1.3 Test Film Surface Areas of Elcometer Payne Permeability Cups

Cup Identification	Film Area (cm³)
Elcometer 5100/1	10
Elcometer 5100/2	30
Elcometer 5100/3	50

Table 1.4 List of Some ASTM (ASTM International, formerly known as the American Society for Testing and Materials) Permeation; Vapor Transmission; and Related Film, Packaging, and Sheeting Standards

ASTM Standard	Description
D 1434-82(2009)e1	Standard test method for determining gas permeability characteristics of plastic film and sheeting
D 1653-03(2008)	Standard test methods for water vapor transmission of organic coating films
D 2103-10	Standard specification for polyethylene film and sheeting
D 2684-10	Standard test method for permeability of thermoplastic containers to packaged reagents or proprietary products
D 3079-94(2009)e1	Standard test method for water vapor transmission of flexible heat-sealed packages for dry products
D 3981-09a	Standard specification for polyethylene films made from medium-density polyethylene for general use and packaging applications
D 3985-05(2010)e1	Standard test method for oxygen gas transmission rate through plastic film and sheeting using a coulometric sensor
D 4491-99a(2009)	Standard test methods for water permeability of geotextiles by permittivity
D 4635-08a	Standard specification for polyethylene films made from low-density polyethylene for general use and packaging applications
D 5047-09	Standard specification for polyethylene terephthalate film and sheeting
D 3816/D3816M-96(2007)	Standard test method for water penetration rate of pressure-sensitive tapes
D 3833/D3833M-96(2006)	Standard test method for water vapor transmission of pressure-sensitive tapes
D 4279-95(2009)	Standard test methods for water vapor transmission of shipping containers constant and cycle methods
D 6701-01 (withdrawn)	Standard test method for determining water vapor transmission rates through nonwoven and plastic barriers
D 726-94(1999)	Standard test method for resistance of nonporous paper to passage of air
E 104-02(2007)	Standard practice for maintaining constant relative humidity by means of aqueous solutions
E 171-94(2007)	Standard specification for standard atmospheres for conditioning and testing flexible barrier materials
E 252-06	Standard test method for thickness of thin foil, sheet, and film by mass measurement
E 398-03(2009)e1	Standard test method for water vapor transmission rate of sheet materials using dynamic relative humidity measurement
E 96/E96M-10	Standard test methods for water vapor transmission of materials
F 1115-95(2008)e1	Standard test method for determining the carbon dioxide loss of beverage containers
F 1249-06	Standard test method for water vapor transmission rate through plastic film and sheeting using a modulated infrared sensor
F 1306-90(2008)e1	Standard test method for slow rate penetration resistance of flexible barrier films and laminates

(Continued)

Table 1.4 (*Continued*)

ASTM Standard	Description
F 1307-02(2007)	Standard test method for oxygen transmission rate through dry packages using a coulometric sensor
F 1769-97 (withdrawn)	Standard test method for measurement of diffusivity, solubility, and permeability of organic vapor barriers using a flame ionization detector
F 1770-97 (withdrawn)	Standard test method for evaluation of solubility, diffusivity, and permeability of flexible barrier materials to water vapor
F 1927-07	Standard test method for determination of oxygen gas transmission rate, permeability, and permeance at controlled relative humidity through barrier materials using a coulometric detector
F 372-99(2003) (withdrawn)	Standard test method for water vapor transmission rate of flexible barrier materials using an infrared detection technique
F 88/F 88M-09	Standard test method for seal strength of flexible barrier materials
F 119-82(2008)	Standard test method for rate of grease penetration of flexible barrier materials (rapid method)
F 17-08	Standard terminology relating to flexible barrier materials
F 1927-07	Standard test method for determination of oxygen gas transmission rate, permeability, and permeance at controlled relative humidity through barrier materials using a coulometric detector
F 2338-09	Standard test method for nondestructive detection of leaks in packages by vacuum decay method
F 2391-05	Standard test method for measuring package and seal integrity using helium as the tracer gas
F 2475-05	Standard guide for biocompatibility evaluation of medical device packaging materials
F 2476-05	Test method for the determination of carbon dioxide gas transmission rate (CO_2TR) through barrier materials using an infrared detector
F 2559-06/F2559M-06 (2010)e1	Standard guide for writing a specification for sterilizable peel pouches
STP 912	Current technologies in flexible packaging

Table 1.5 List of Some ISO Permeation; Vapor Transmission; and Related Film, Packaging, and Sheeting Standards

ISO Standard	Description
ISO 15105-1	Plastics—Film and sheeting—Determination of gas-transmission rate—Part 1: Differential-pressure method
ISO 15105-2	Plastics—Film and sheeting—Determination of gas-transmission rate—Part 2: Equal-pressure method
ISO 15106-1	Plastics—Film and sheeting—Determination of water vapor-transmission rate—Part 1: Humidity detection sensor method
ISO 15106-2	Plastics—Film and sheeting—Determination of water vapor-transmission rate—Part 2: Infrared detection sensor method
ISO 15106-3	Plastics—Film and sheeting—Determination of water vapor-transmission rate—Part 3: Electrolytic detection sensor method

1: Introduction to Permeation of Plastics and Elastomers

Table 1.5 (*Continued*)

ISO Standard	Description
ISO 15106-4	Plastics—Film and sheeting—Determination of water vapor-transmission rate—Part 4: Gas-chromatographic detection sensor method
ISO 1663:2007	Rigid cellular plastics—Determination of water vapor-transmission properties
ISO 2528	Sheet materials—Determination of water vapor-transmission rate—Gravimetric (dish) method
ISO 2556-2001	Plastics—Determination of the gas-transmission rate of films and thin sheets under atmospheric pressure—Manometric method
ISO 7229:1997	Rubber- or plastics-coated fabrics—Measurement of gas permeability
ISO 9913-1:1996 (withdrawn)	Optics and optical instruments—Contact lenses—Part 1: Determination of oxygen permeability and transmissibility with the FATT method
ISO 9913-2:2000 (withdrawn)	Optics and optical instruments—Contact lenses—Part 2: Determination of oxygen permeability and transmissibility by the coulometric method
ISO 9932:1990	Paper and board—Determination of water vapor-transmission rate of sheet materials—Dynamic sweep and static gas methods

Table 1.6 List of DIN (Deutsches Institut für Normung e.V. or in English, the German Institute for Standardization) Permeation; Vapor Transmission; and Related Film, Packaging, and Sheeting Standards

Standard	Description
DIN 16906	Testing of plastic sheeting and films, sample and specimen preparation, and conditioning
DIN 16995	Films for packaging, plastic films, properties, and testing
DIN 53122-1	Determination of the water vapor-transmission rate of plastic film, rubber sheeting, paper, board, and other sheet materials by gravimetry
DIN 53122.2	Testing of plastic films, rubber films, paper, board, and other sheet materials; determination of water vapor transmission, electrolysis method
DIN 53380-1	Determining the gas-transmission rate of plastic film by the volumetric method
DIN 53380-3	Determining the gas-transmission rate of plastic film, sheeting, and moldings by the carrier gas method

Table 1.7 List of Other Permeation; Vapor Transmission; and Related Film, Packaging, and Sheeting Standards (JIS—Japanese Industrial Standards, TAPPI—Technical Association of the Pulp and Paper Industry)

Standard	Description
JIS K 7126	Testing method for gas-transmission rate through plastic film and sheeting
JIS K 7126-1:2006	Plastics—Film and sheeting—Determination of gas-transmission rate—Part 1: Differential-pressure method
JIS K 7129-2008	Testing methods for water vapor-transmission rate of plastic film and sheeting (instrument method)
JIS Z 0208:1976	Testing methods for determination of the water vapor-transmission rate of moisture-proof packaging materials (dish method)
TAPPI T 557 pm-95	Water vapor-transmission rate through plastic film and sheeting using a modulated infrared sensor

1.5.1 Units of Measurement

Permeability and vapor transport properties are reported in dozens of different units.[15]

The units of gas permeability are

$$P = \frac{\text{amount of gas} \times \text{thickness of membrane}}{\text{area of membrane} \times \text{time} \times \text{pressure}}$$

The units of vapor or liquid permeability are

$$P = \frac{\text{amount of liquid} \times \text{thickness of membrane}}{\text{area of membrane} \times \text{time}}$$

The amount of gas may be reported as volume or mass. When gas volumes are used, they are usually used for gas at STP conditions. The current version of IUPAC's STP is a temperature of 0 °C (273.15 K, 32 °F) and an absolute pressure of 100 kPa (14.504 psi, 0.986 atm), while NIST's version is a temperature of 20 °C (293.15 K, 68 °F) and an absolute pressure of 101.325 kPa (14.696 psi, 1 atm). International Standard Metric Conditions for natural gas and similar fluids are 288.15 K and 101.325 kPa. The standard units used in this book are $(cm^3\,mm)/(m^2\,day\,atm)$ for gas volumes or $(g\,mm)/(m^2\,day\,atm)$ for gas mass.

For liquid or vapor permeation, the amount of liquids is reported in mass units. The standard units used in this book are $(g\,mm)/(m^2\,day)$.

Appendix A contains a large list of other units and their conversion factors to the standard used in this book.

1.5.2 Gas Permeation Test Cells

The gas permeation test cell is used to measure gas permeation and a schematic of one is shown in Fig. 1.18. This type of cell is usually run at atmospheric pressure and is kept at a constant temperature. The test film separates the cell into two halves, the feed side and the permeate side. On the feed side, the gas of interest, such as oxygen, is fed continuously. On the opposite side, a sweep gas such as nitrogen removes the permeant gas as it diffuses through the test film. By sweeping the permeant out of the cell, the partial pressure of the permeant gas is kept nearly zero. The sweep gas and permeate gas stream are analyzed to determine the amount of permeate that has passed through the test film. The method of analysis varies depending on the gas. There are many ways to analyze for permeants.

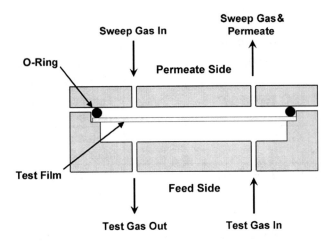

Figure 1.18 Schematic representation of a generic gas permeation test cell.

The data obtained from this experiment are weight or volume of permeant gas versus time. Couple this with the exposed surface area of the test film and the partial pressure difference of the permeant gas between the two sides (usually ~1 atm); one can calculate the permeability coefficient. These calculations were reviewed in Section 1.2.2.

There are many ways to measure the amount of permeate that passes through the film. For oxygen, a coulometric sensor is often used, but modulated infrared sensor, flame ionization detector, electrolytic detection sensor, pressure decay, and other methods may be specified.

Test instruments are available to simplify the measurements. MOCON (http://www.mocon.com) is a company that provides a wide range of instruments used to measure permeation of a number of gases through films. Labthink® Ltd. Produces the *PERME series of permeation test instruments (http://www.labthink.cn). OxySense® (http://www.oxysense.com) is a source of oxygen permeation testing solutions.

1.5.3 Vapor Permeation Cup Testing

One of the simplest tests is the measurement of water vapor transmission through films by the cup method. These cups are the BYK-Gardner Perm Cups or the Elcometer 5100 Payne Permeability Cups, shown in Figs. 1.19 and 1.20, respectively. The

Figure 1.19 Picture of the different sizes of BYK-Gardner permeability cups (photo courtesy of Paul N. Gardner Company, Inc.).

BYK-Gardner Perm Cup is a shallow cylinder with a threaded flange, flat retaining washer, and threaded ring cover. Rubber gaskets are used to tightly seal the specimen between the cup and the ring cover. The cup and cover are knurled for easier handling. Two different size cups are available, a large 25-cm^2 cup meets the requirement of the ASTM standard, and a smaller 10-cm^2 cup, which allows the use of a smaller specimen. The Elcometer permeability cups come in three sizes as shown in Table 1.3. They use small clamps to secure the test film to the cup.

The permeability of a coating or other free film to water vapor is measured by suspending the film of the material across the top of a wide shallow cup that is filled with some water. The cups are sealed so that the only way for water to escape is through the suspended film. Then, the cups are placed in a controlled environment such as in a desiccator. During the test, vapor passes from the cup through the film specimen into the desiccator in a controlled environment. Weight loss of the cup's content over a specified period is used to determine the rate of vapor transmission through the film. Other liquids may be tested besides water, but these are usually put into a vacuum. The two most common standards for this test are ASTM D1653 and ISO 7783-2.

1.5.4 Permeation Testing of Coatings

As described earlier, permeation through coatings is a concern to users and formulators. Often coatings may be tested by the same tests that are used to test films, if a defect-free free-standing film can be prepared and handled. Free films may be prepared in a number of ways. The paint or coating may be applied onto a surface with which it will not stick strongly. Most often this is a sheet of glass, polyethylene, or fluoropolymer. After the paint has dried or cured, it sometimes will peel off these substrates. For those films that will not peel off readily, sometimes soaking in water will permit them to be removed in one piece. Films removed in this fashion must be dried completely before testing. When this does not work, one may coat aluminum foil, cure the coating, and then etch the foil away in an acid such as hydrochloric acid or a base such as sodium hydroxide. If this process is used, one must keep in mind that the acid or base may alter the chemical structure of the paint film.

Figure 1.20 Elcometer Payne permeability cups (photo courtesy of Elcometer Ltd.).

Free paint film preparation is not usually easy to do. Paint films might be too thin or too brittle to handle even if they are prepared. Paint films often have defects such as pinholes or craters in them caused by contaminants or application problems. Films with defects will not give useful free film permeation test results.

Another complication is that coatings are often systems, layers of different coatings on each other, such as a topcoat on a primer.

There are tests that are related to permeation measurements that may be used on coatings as applied to the substrates. Few of these tests are

- Electro-impedance spectroscopy (EIS)
- Salt spray
- Humidity
- Atlas cell.

These tests will be discussed briefly in the following sections.

1.5.4.1 Electro-Impedance Spectroscopy

EIS is a well-established laboratory technique for characterizing electrochemical systems.[16–18] Since corrosion is electrochemical in nature, it is used for evaluating the corrosion protection of organic coatings. The technique uses alternating current (AC) electricity to measure the electrical resistance (i.e., impedance) of a coating. The impedance of the coating is used to assess coating integrity before and after exposure to a corrosive environment and to follow deterioration. As the impedance decreases, the coating is less protective.

The impedance measurement is a nondestructive technique in which measurements can be obtained rapidly. The technique is also highly sensitive, where impedance decreases rapidly as a coating deteriorates due to increased permeability to water and electrolyte.

The impedance of a coating is measured with an electrochemical cell consisting of the coated panel (test electrode), a reference electrode, and an inert counter electrode. A wide variety of cell designs is available. The electronic measuring equipment consists of a potentiostat, either a lock-in amplifier or frequency response analyzer, and a computer with specialty EIS software. Coating impedance is measured as a function of the frequency of an applied AC voltage. The frequency ranges from 0.001 Hz to 100 kHz. The resulting data are normally presented in the form of a Bode plot, consisting of Log Z plotted versus Log f, where Z is impedance in ohms centimeter, and f is frequency in Hertz. Information about coating protection and barrier properties is obtained at the low frequency end of the plot. To facilitate comparison of impedance measurements from different coatings, or to examine the change in impedance as a function of time of exposure, many laboratories report the coating impedance at 0.1 Hz. Selection of Log Z at 0.1 Hz is somewhat arbitrary.

1.5.4.2 Salt Spray/Humidity

The most common corrosion type test for coatings is what is commonly called the salt spray or salt fog test. This test is considered to be most useful for measuring relative corrosion resistance of related coating materials rather than providing absolute performance measures. The relationship of corrosion to permeation of water and oxygen was discussed in Section 1.4.

The test provides a controlled corrosive environment that represents accelerated marine type atmospheric conditions. The equipment setup, conditions, and operational procedures are described in detail by the ASTM standard B117-03 Standard Practice for Operating Salt (see Fig. 1.21). It does not describe the type of test specimen and exposure periods nor does it describe how to interpret the results. There are other ASTM practices to help with evaluation of exposed panels.

The test is typically run by preparing coated test panels under controlled conditions that is controlled surface preparation, film thickness, and bake. The panels are exposed at a somewhat elevated temperature, often 40 °C (105 °F), to a mist of 5% salt water. The panels are examined periodically, typically after 24, 48, 96, 168, 336, 500, and 1000 h. And evaluated for defects such as, degree of rusting, blistering, and chalking. Blistering is most directly related to water permeation.

Some ASTM methods for evaluating panels being tested for corrosion are as follows:

- D714-02 Standard Test Method for Evaluating Degree of Blistering of Paints—Ratings vary depending on size and number of blisters. For

Figure 1.21 Photographs of typical salt spray cabinet.

multilayer coatings, blistering may form between any of the layers, so that they also must be noted.

- D1654-92(2000) Standard Test Method for Evaluation of Painted or Coated Specimens Subjected to Corrosive Environments.

1.5.4.3 Atlas Cell

Atlas cell is a test that allows the estimation of the resistance of a coating in contact with a chemical at a given temperature. The picture in Fig. 1.22 describes the test. The coating is applied on the inside of the panels, which close the glass pipe. They are held in place by clamps with a chemically resistant gasket such as GORE-TEX®, EPDM, or other heat and chemical-resistant elastomer insuring a leak-free seal. The test liquid is heated. The heater can be controlled at a particular temperature or the liquid can be run at reflux with a condenser returning the liquid to the cell. The coating is exposed to both a liquid and a vapor phase. There also seems to be a temperature gradient across the coating and the panel. The effect of the gradient is a very important part of this test. Even many thick coatings will blister within a week when exposed to these conditions.

Most often the liquid is just boiling water. With precautions taken for leaks and spills, other liquids may be used. Sulfuric, nitric, hydrochloric, and hydrofluoric acids and strong bases such as sodium hydroxide may also be studied with Atlas cells. Visual inspection is done daily at the start of the test, but can be done less often as the test progresses. Most of the time, after 720 h, the test is stopped. Often the panels blister as shown in Fig. 1.23. If the panel is not severely blistered then adhesion can be tested.

The Atlas cell test is a key test for chemical equipment applications, such as tanks, mixers, and reactors. The ASTM procedure is C868-02 Standard Test Method for Chemical Resistance of Protective Linings.

Figure 1.22 Atlas cell testing apparatus.

1.5.4.4 Adhesion

As mentioned above, adhesion of a coating to substrate is often measured before and after exposure. Permeation of small molecules, particularly water, affects this important coating property.

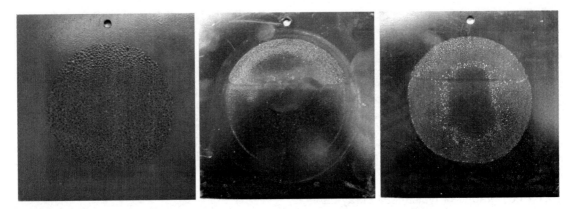

Figure 1.23 Sample atlas cell panel failures. (Left) Blisters uniformly over entire face of panel; (middle) blisters primarily in the vapor phase exposed area; (right) blisters form from outer edge first and move toward middle.

1.5.4.4.1 Instron Peel Test

One quantitative adhesion test is D1876-01 Standard Test Method for Peel Resistance of Adhesives (T-Peel Test). The test requires very thick coatings and an expensive Instron machine, but the data can be very useful. Basically, the Instron pulls the coating and substrate apart and measures the force required to do that.

The test panel or foil must be prepared to allow easy separation of the coating from the substrate for at least an inch (25 mm) at one end. Covering that end with a silicone coating or a piece of Kapton® film; then applying the coating over the entire panel can accomplish this. For this test, the coating must be thick, generally 40 mils or more (1 mm). Thinner coatings may break before the adhesive of the coating to substrate fails. Figure 1.24 shows a sample with the layers separating, while being pulled by the jaws on the Instron. The width of the test strip is measured; usually it is cut to 1 in. (25 mm). The force in pounds per inch (kg/cm) to pull apart the two strips is a measure of adhesion of the coating to the substrate.

1.5.4.4.2 Pull-Off Adhesion Tests

There are a number of adhesion tests that are described as pull-off tests. Most of the tests require

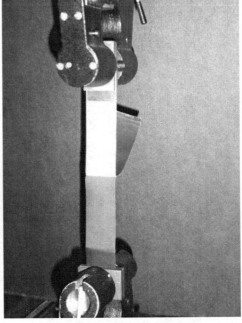

Figure 1.24 Instron peel strength adhesion test.

gluing of a metal probe, often called a dolly or stub, to the surface of the coating. These are described in ASTM standard D4541-02, Standard Test Method for Pull-Off Strength of Coatings Using Portable Adhesion Testers. By the use of a portable pull-off adhesion tester, a load is increasingly applied to the surface until the dolly is pulled off. Testers operate using mechanical (twist by hand), hydraulic (oil), or pneumatic (air) pressure. The force required to pull the dolly off or the force the dolly withstood yields the tensile strength in pounds per square inch or mega Pascals. Failure will occur along the weakest plane within the system comprised of the dolly, adhesive, coating system, and substrate, and will be exposed by the fracture surface. If the failure is at the coating–substrate interface then the measured tensile strength may be called the adhesion strength. This test method maximizes tensile stress as compared with the shear stress applied by the peel adhesion tests. A standard method for the application and performance of this test is available in ASTM D4541 and ISO 4624.

Other adhesion tests include knife, tape, and scrape tests but they are not quantitative and are mainly useful in comparing coating systems.[19,20]

1.6 Summary

This chapter has introduced permeation, diffusion, and solubility. Theories and models have been suggested. Measurement and tests methods have also been discussed. The subject of chemistry of the plastics has been raised, but the effect of chemistry on permeation properties has been deferred. Chapter 2 will delve into the relationship of polymer structure and plastic composition to permeation properties.

References

1. Suloff EC. *Sorption behavior of an aliphatic series of aldehydes in the presence of poly(ethylene terephthalate) blends containing aldehyde scavenging agents (Dissertation)*; 2002: 29–30.
2. Robertson GL. *Food packaging: principles and practice*. 2nd ed. CRC Press; 2005.
3. Wood-Adams P. *Physical chemistry of polymers (Course Notes)*. Concordia University; Fall 2006.
4. Wijmans J. The solution-diffusion model: a review. *J Membr Sci* 1995;**107**:1–21.
5. Rutherford SW, Do DD. Review of time lag permeation technique as a method for characterisation of porous media and membranes. *Adsorption* 1997;**3**:283–312.
6. Graff GL, Williford RE, Burrows PE. Mechanisms of vapor permeation through multilayer barrier films: lag time versus equilibrium permeation. *J Appl Phys* 2004;**96**:1840.
7. Solms N, Zecchin N, Rubin A, Andersen S, Stenby E. Direct measurement of gas solubility and diffusivity in poly(vinylidene fluoride) with a high-pressure microbalance. *Eur Polym J* 2005;**41**:341–8.
8. Edwards DA. Non-fickian diffusion in thin polymer films. *J Polym Sci B: Polym Phys* 1996;**34**:981–97.
9. Bauermeister N, Shaw S. Finite-element approximation of non-Fickian polymer diffusion. *IMA J Numer Anal* 2009;**30**:702–30.
10. Shaw S. Finite element approximation of a non-local problem in non-fickian polymer diffusion. *Int J Num Anal Model* 2010;**8**:226–51.
11. Wood-adams P. *Permeation: essential fac.* Polymer; 2006:1–16.
12. Mayne JEO. The mechanism of the inhibition of the corrosion of iron and steel by means of paint. *Off Dig* 1952;**24**:127–36.
13. Funke W. How organic coating systems protect against corrosion. In: Dickie R, editor. *Polymeric materials for corrosion control*. American Chemical Society; 1986:222–8.
14. Sangaj N. Permeability of polymers in protective organic coatings. *Prog Org Coat* 2004;**50**: 28–39.
15. Yasuda H. Units of gas permeability constants. *J Appl Polym Sci* 1975;**19**:2529–36.
16. Loveday D, Peterson P, Rodgers B, Instruments G. Evaluation of organic coatings with electrochemical impedance spectroscopy, Part 1: Fundamentals of electrochemical impedance spectroscopy. *JCT CoatingsTech*; 2004:46–52.
17. Loveday D, Peterson P, Rodgers B, Instruments G. Evaluation of organic coatings with electrochemical impedance spectroscopy,

Part 2: Application of EIS to coatings. *JCT CoatingsTech*; 2004:88–93.
18. Loveday D, Peterson P, Rodgers B, Instruments G. Evaluation of organic coatings with electrochemical impedance spectroscopy, Part 3: Protocols for testing coatings with EIS. *JCT CoatingsTech*; 2005:22–7.
19. Mittal KL. STP640 adhesion measurement of thin films, thick films, and bulk coatings. *ASTM International*; 1978.
20. Lacombe R. *Adhesion measurement methods: Theory and practice*. CRC Press; 2006.
21. Solef® & Hylar® PVDF, Design and Processing Guide, BR2001C-B-2–1106; 2006.

2 Introduction to Plastics and Polymers

The most basic component of plastic and elastomer materials is polymer. The word polymer is derived from the Greek term for "many parts." Polymers are large molecules that comprised many repeat units called monomers that have been chemically bonded together into long chains. Since World War II, the chemical industry has developed a large number of synthetic polymers to satisfy the needs of the materials for a diverse range of products, including paints, coatings, fibers, films, elastomers, and structural plastics. Literally, thousands of materials can be called "plastics," although the term today is typically reserved for polymeric materials, excluding fibers, which can be molded or formed into solids or semisolid objects. The subject of this section includes polymerization chemistry and the different types of polymers and how they can differ from each other. Since plastics are rarely "neat," reinforcement, fillers and additives are reviewed. A basic understanding of plastic and polymer chemistry will make the discussion of permeation of specific plastics, films, and membranes easier to understand, and it also provides a basis for the introductions of the plastic families in later chapters. This section is taken from *The Effect of Temperature and Other Factors on Plastics* book, but it has been rewritten, expanded, and refocused on permeation properties.

2.1 Polymerization

Polymerization is the process of chemically bonding monomer building blocks to form large molecules. Commercial polymer molecules are usually thousands of repeat units long. Polymerization can proceed by one of the several methods. The two most common methods are called addition and condensation polymerization.

In addition polymerization, a chain reaction adds new monomer units to the growing polymer molecule one at a time through double or triple bonds in the monomer. Each new monomer unit creates an active site for the next attachment. The net result is shown in Fig. 2.1. Many of the plastics discussed in a later chapter of this book are formed in this manner. Some of the plastics made by addition polymerization include polyethylene (PE), polyvinyl chloride (PVC), acrylics, polystyrene, and polyoxymethylene (acetal).

The other common method is condensation polymerization in which the reaction between monomer units and the growing polymer chain end group releases a small molecule, often water as shown in Fig. 2.2. This reversible reaction will reach equilibrium and halt unless this small molecular by-product is removed. Polyesters and polyamides are among the plastics made by this process.

Understanding the polymerization process used to make a particular plastic gives insight into the nature of the plastic. For example, plastics made via condensation polymerization, in which water is released, can degrade when exposed to water at high temperature. Polyesters such as polyethylene terephthalate (PET) can degrade by a process called hydrolysis when exposed to acidic, basic, or even some neutral environments severing the polymer chains. As a result, the properties of the polymer are degraded.

2.2 Copolymers

A copolymer is a polymer formed when two (or more) different types of monomer are linked in

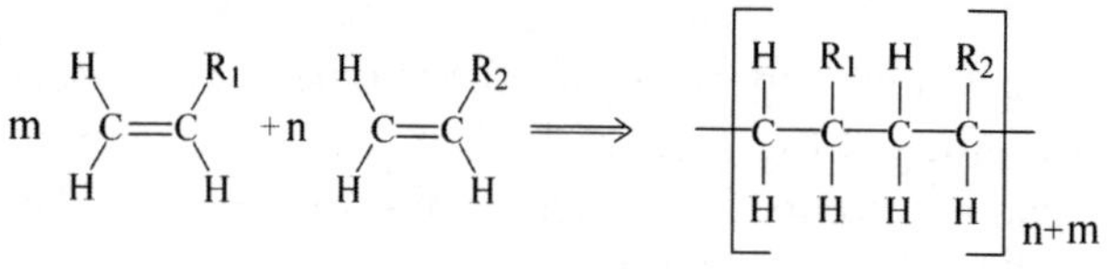

Figure 2.1 Addition polymerization.

Figure 2.2 Condensation polymerization.

$$HO-\overset{\overset{O}{\|}}{C}-R_1-\overset{\overset{O}{\|}}{C}-OH \ + \ HO-R_2-OH \ \Longrightarrow \ \left[\overset{\overset{O}{\|}}{C}-R_1-\overset{\overset{O}{\|}}{C}-O-R_2-O\right]_n + H_2O$$

the same polymer chain, as opposed to a homopolymer where only one monomer is used. If exactly three monomers are used, it is called a terpolymer.

Monomers are only occasionally symmetric; the molecular arrangement is the same no matter which end of the monomer molecule you are looking at. The arrangement of the monomers in a copolymer can be head-to-tail, head-to-head, or tail-to-tail. Since a copolymer consists of at least two types of repeating units, copolymers can be classified based on how these units are arranged along the chain. These classifications include the following:

- Alternating copolymer
- Random copolymer (statistical copolymer)
- Block copolymer
- Graft copolymer.

When the two monomers are arranged in an alternating fashion, the polymer is called, of course, an alternating copolymer:

–A–B–A–B–A–B–A–B–A–B–A–B–A–B–A–B–A–B–
Alternating Copolymer

In the following examples, A and B are different monomers. Keep in mind that A and B do not have to be present in a one-to-one ratio. In a random copolymer, the two monomers may follow any order:

–A–A–B–A–B–B–A–B–A–A–B–B–B–A–B–A–A–
Random Copolymer

In a block copolymer, all of one type of monomers are grouped together, and all of the second monomers are grouped together. A block copolymer can be thought of as two homopolymers joined together at the ends:

–A–A–A–A–A–A–A–A–B–B–B–B–B–B–B–B–
Block Copolymer

A polymer that consists of large grouped blocks of each of the monomers is also considered a block copolymer:

–A–A–A–A–A–A–B–B–B– B–B–B–A–A–A–A–A–
Block Copolymer

When chains of a polymer made of monomer B are connected onto a polymer chain of monomer A, we have a graft copolymer:

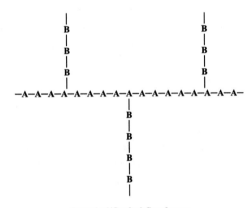

Branched/Grafted Copolymer

High-impact polystyrene (HIPS) is a graft copolymer. It is a polystyrene backbone with chains of polybutadiene grafted onto the backbone. The polystyrene gives the material strength, but the rubbery polybutadiene chains give it resilience to make it less brittle.

2.3 Linear, Branched, and Cross-Linked Polymers

Some polymers are linear, a long chain of connected monomers. PE, PVC, Nylon 66, and polymethyl methacrylate are some linear commercial examples found in this book. Branched polymers can be visualized as a linear polymer with side chains of the same polymer attached to the main chain. While the branches may in turn be branched, they do not connect to another polymer chain. The ends of the branches are not connected to anything. Special

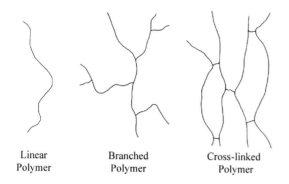

Figure 2.3 Linear, branched, and cross-linked polymers.

types of branched polymers include star polymers, comb polymers, brush polymers, dendronized polymers,[1] ladders, and dendrimers. Cross-linked polymer, sometimes called network polymer, is one in which different chains are connected. Essentially, the branches are connected to different polymer chains on the ends. These three polymer structures are shown in Fig. 2.3.

2.4 Polarity

A molecule is two or more atoms joined by a covalent bond. Basically, the positively charged atom nuclei share the negatively charged electrons. However, if the atoms are different, they may not share the electrons equally. The electrons will be denser around one of the atoms. This would make that end more negatively charged than the other end and that creates a negative pole and a positive pole (a *dipole*), and such a bond is said to be a *polar bond* and the molecule is polar and has a *dipole moment*. A measure of how much an atom attracts electrons is *electronegativity*. The electronegativity of common atoms in the polymers is given as follows:

F > O > Cl and N > Br > C and H.

The polarity of a molecule affects the attraction between molecular chains, which affects the structure of the polymer and the attraction of polar molecules, so one would expect polarity to affect solubility which in turn affects permeability.

How does one predict molecular polarity? When there are no polar bonds in a molecule, there is no permanent charge difference between one part of the molecule and another, so the molecule is nonpolar. For example, the Cl_2 molecule has no polar bonds because the electron charge is identical on both atoms. It is therefore a nonpolar molecule. The C—C and C—H bonds in hydrocarbon molecules, such as ethane, C_2H_6, are not significantly polar, hence hydrocarbons are nonpolar molecular substances and hydrocarbon polymers like PE or polypropylene are also nonpolar.

However, a molecule can possess polar bonds and still be nonpolar. If the polar bonds are evenly (or symmetrically) distributed, the bond dipoles cancel and do not create a molecular dipole. For example, the three bonds in a molecule of CCl_4 are significantly polar, but they are symmetrically arranged around the central carbon atom. No side of the molecule has more negative or positive charge than another side, and hence the molecule is nonpolar (Table 2.1).

Generally, polar polymers are more permeable to water than nonpolar polymers. Figure 2.4 shows a qualitative ranking of some polymer polarities.

2.5 Unsaturation

Up to this point in the discussion of polymer chemistry, the atom-to-atom structure has not been discussed. The covalent bonds between atoms in

Table 2.1 Dipole Moments in Some Small Molecules

Molecule	Dipole Moment	Molecule	Dipole Moment	Molecule	Dipole Moment
H_2	0	HF	1.75	CH_4	0.0
O_2	0	H_2O	1.84	CH_3Cl	1.86
N_2	0	NH_3	1.46	CCl_4	0
Cl_2	0	NF_3	0.24	CO_2	0
Br_2	0	BF_3	0		

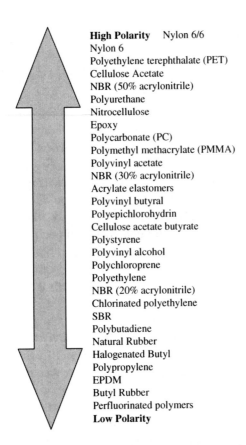

Figure 2.4 Qualitative ranking of polymer polarities.

a polymer can be single, double, triple bonds, or even rings. The presence of bonds other than single bonds generally makes the polymer molecule stiffer and reduces rotation along the polymer chain, and this can affect its properties. It is easier to discuss molecules first and then extend that discussion to polymers. Saturated molecules only contain single bonds with no rings.

Often when talking about molecular unsaturation, the degree of unsaturation (DoU) is noted. The DoU can be calculated from the below equation if the molecular formula is given.

$$DoU = \frac{2C + 2 + N - X - H}{2}, \quad (1)$$

where

C is the number of carbons
N is the number of nitrogens
X is the number of halogens (F, Cl, Br, I)
H is the number of hydrogens.

Oxygen and sulfur are not included in the formula because saturation is unaffected by these elements.

Examples:
Ethylene: C_2H_4

$$DoU = \frac{2C + 2 + N - X - H}{2}$$
$$= \frac{2 \times 2 + 2 + 0 - 0 - 4}{2} = 1 \quad (2)$$

Benzene: C_6H_6

$$DoU = \frac{2C + 2 + N - X - H}{2}$$
$$= \frac{2 \times 6 + 2 + 0 - 0 - 6}{2} = 4 \quad (3)$$

When polymers are used the formula shown is often the repeating unit. This will often have two bonds that are shown to which the repeating unit is supposed to attach. When applying a DoU formula to the repeating unit, one would remove the "+2" from it.

Examples:
PE: $-(CH_2-CH_2)_n-$

$$DoU = \frac{2C + N - X - H}{2}$$
$$= \frac{2 \times 2 + 0 - 0 - 4}{2} = 0 \quad (4)$$

Polyphenylene sulfone (PPS): $-(C_6H_4-S)_n-$

$$DoU = \frac{2C + N - X - H}{2}$$
$$= \frac{2 \times 6 + 0 - 0 - 4}{2} = 4 \quad (5)$$

2.6 Steric Hindrance

As described earlier in this chapter, polymers are long chains of atoms linked together. They may be flexible and bendable. To explain this, one may visualize them as a ball-and-stick model. In chemistry, the ball-and-stick model is a molecular model of a chemical substance, which aims to display both the 3D position of the atoms and the bonds between them. The atoms are typically represented by spheres, connected by rods that represent the bonds. Double and triple bonds are usually represented by two and three curved rods,

2: Introduction to Plastics and Polymers

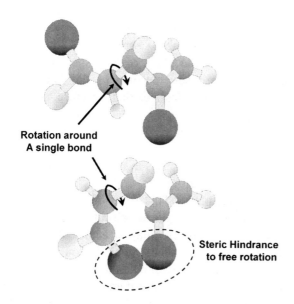

Figure 2.5 Steric hindrance shown with a ball-and-stick molecular model.

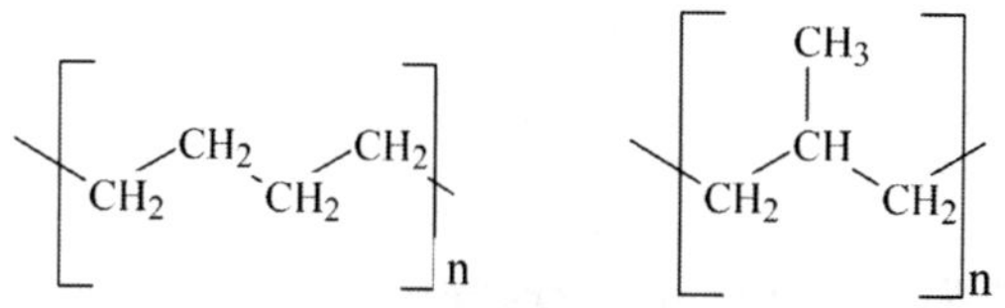

Figure 2.6 Structural isomers.

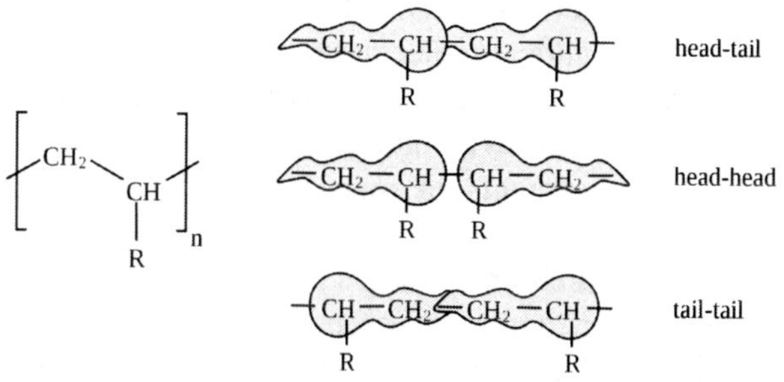

Figure 2.7 Head-to-tail isomers.[2]

respectively. The chemical element of each atom is often indicated by the sphere's color. The top of Fig. 2.5 shows a drawing of a ball-and-stick model of a molecule. Figure 2.5 also indicates that there is free rotation around the single bonds. If there was a double or triple bond, there would not be any rotation possible around those bonds. Similarly, ring structures, while they might flex a little bit, inhibit rotation. In some cases, such as shown in the bottom of Fig. 2.5, large atoms or bulky side groups might bump into each other as the molecule rotates around single bonds. This is called *sterically hindered* or *steric hindrance*. Hindered or inhibited rotation stiffens the polymer molecule and dramatically affects its physical properties.

2.7 Isomers

Isomers (from Greek isomerès; isos = "equal" and méros = "part") are compounds with the same molecular formula but a different arrangement of the atoms in space. There are many kinds of isomers, and the properties can differ widely or almost not at all.

2.7.1 Structural Isomers

Structural isomers have the atoms that are arranged in a completely different order as shown in Fig. 2.6. Here, both polymer repeating groups have the same formula, $-C_4H_8-$, but the atoms are arranged differently. The properties of structural isomers may be very different from each other.

Often the repeating group in a polymer has exactly the same formula, but the repeating group is flipped over as shown in Fig. 2.7. If one views the repeating group as having a head and a tail, then the different ways to connect neighboring repeating units are head–tail, head–head, and tail–tail.

2.7.2 Geometric Isomers

When there is a carbon–carbon double bond in a molecule, there might also be two ways to arrange the groups attached to the double bonds. This is best seen in side-by-side structures such as shown in Fig. 2.8.

These are called *geometric isomers* that owe their existence to hindered rotation about double bonds. If the substituents are on the same side of the double bond then the isomer is referred to as *cis* (Latin: on this side). If the substituents are on

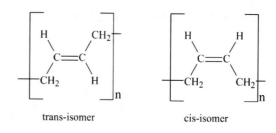

Figure 2.8 The *cis* and *trans* isomers.

the opposite side of the double bond then it is referred to as *trans* (Latin: across).

2.7.3 Stereoisomers— Syndiotactic, Isotactic, and Atactic

Stereoisomerism occurs when two or more molecules have identical molecular formula and the same structural formula (i.e., the atoms are arranged in the same order). However, they differ in their 2D or 3D spatial arrangements of their bonds, which mean different spatial arrangement of the atoms, even though they are bonded in the same order. This may best be understood by an example.

Polypropylenes all have the same simplified structural polymer formula of polypropene as shown in Fig. 2.9.

However, there are subtle differences in the ways to draw this structure. Figure 2.10 shows a longer structure of polypropene and some 3D structures. This structure shows how some bonds (the dotted lines) are behind the plane of the paper and others stick out of the paper (the ones on the ends of the little triangular wedges). In this structure, some of the CH_3 groups are above the paper plane and others are behind the paper plane. This is called *atactic* polypropene.

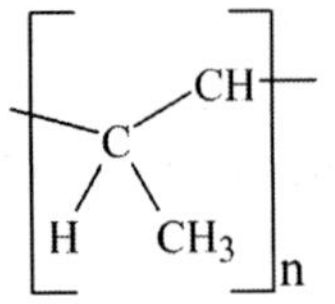

Atactic polypropene has at random about 50% of hydrogen/methyl groups in front/back of C–C–C chain viewing plane. This form of polypropene is amorphous (noncrystalline, discussed in Section 2.9.3) and has an irregular structure due to the random arrangement of the methyl groups attached to the main carbon–carbon chain. It tends to be softer and more flexible than the other forms (described below) and is used for roofing materials, sealants, and other weatherproof coatings.

Isotactic polypropene has all of the methyl groups in front of C–C–C chain viewing plane and all of the H's at back as shown in Fig. 2.11. This stereoregular structure maximizes the molecule–molecule contact and so increases the intermolecular forces compared with the atactic form. This regular structure is much stronger (than the atactic form above) and is used in sheet and film form for packaging and carpet fibers.

Syndiotactic polypropene has a regular alternation of 50% of hydrogen/methyl groups in front/back of C–C–C chain viewing plane as shown in Fig. 2.12.

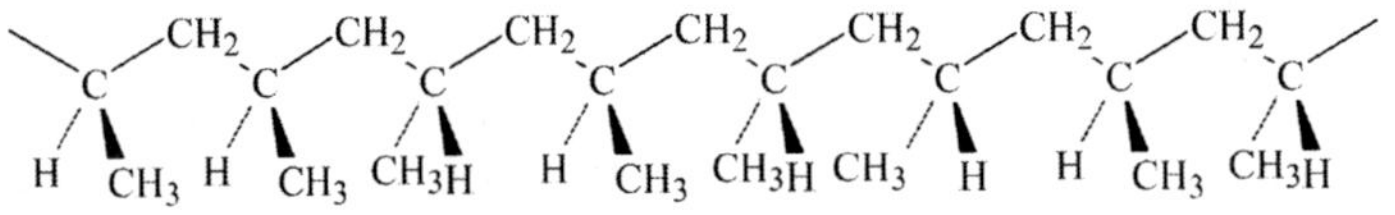

Figure 2.9 The structure of polypropene.

Figure 2.10 The structure of atactic polypropene.

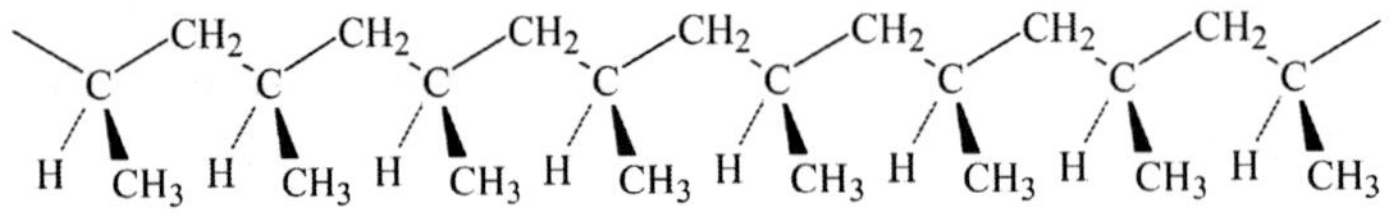

Figure 2.11 The structure of isotactic polypropene.

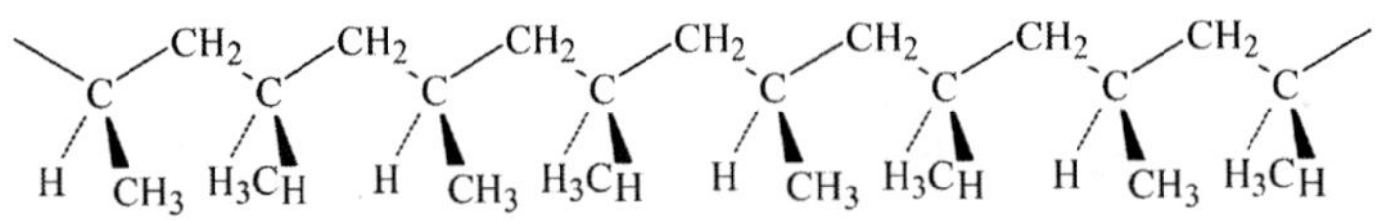

Figure 2.12 The structure of syndiotactic polypropene.

2: Introduction to Plastics and Polymers

Its properties are similar to isotactic polypropene rather than the atactic form, that is, the regular polymer structure produces stronger intermolecular forces and a more crystalline form than the atactic polypropene.

2.8 Inter- and Intramolecular Attractions in Polymers

The attractive forces between different polymer chains or segments within polymer chains play a large part in determining a polymer's properties. As mentioned in Section 2.4, atoms can have polarity or dipole moments. Since negative charges are attracted to the opposite positive charges and repelled by like charges, it is possible to generate attractions that lead to certain structures.

2.8.1 Hydrogen Bonding

One of the strongest dipole interactions is the attraction of some oxygen atoms to hydrogen atoms, even though they are covalently bonded to other atoms. This is called hydrogen bonding, and a schematic representation is shown in Fig. 2.13. The N—H bond provides a dipole when the hydrogen has a slightly positive charge and the nitrogen has a slight negative charge. The carbonyl group, the C=O, likewise is a dipole, where the oxygen has the slight negative charge and the carbon is slightly positive. When polymer chains line up, these *hydrogen bonds* are formed (indicated by the wide gray bars in the figure), bonds that are far weaker than the covalent bonds but bonds of significant strength nonetheless.

Other side groups on the chain polymer can lend the polymer to hydrogen bonding between its own chains. These stronger forces typically result in higher tensile strength and higher crystalline melting points. Polyesters have dipole–dipole bonding between the oxygen atoms in C=O groups and the hydrogen atoms in H—C groups. Dipole bonding is not as strong as hydrogen bonding.

2.8.2 Van der Waals Forces

Many polymers, such as polyethylene, have no permanent dipole. However, attractive forces between polyethylene chains arise from weak forces called Van der Waals forces. Van der Waals forces are much weaker than chemical bonds, and random thermal motion around room temperature can usually overcome or disrupt them.

Molecules can be thought of as being surrounded by a cloud of negative electrons. But the electrons are mobile, and at any one instant, they might find themselves toward one end of the molecule, making that end slightly negative ($\delta-$). The other end will be momentarily short of electrons and so becomes ($\delta+$). Basically, temporary fluctuating dipoles are present in all molecules, and the forces due to these dipoles are the basis for Van der Waals attraction. Van der Waals forces are quite weak, however, so polyethylene can have a lower melting temperature compared to other polymers that have hydrogen bonding.

2.8.3 Chain Entanglement

Polymer molecules are long chains, which can become entangled with one another, much like a bowl of spaghetti. Along with intermolecular forces, chain entanglement is an important factor contributing to the physical properties of polymers. The difficulty in

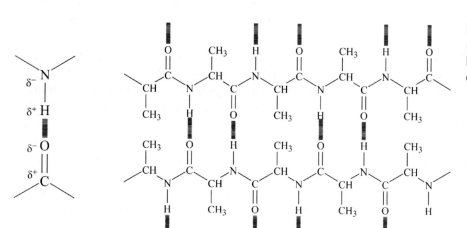

Figure 2.13 Schematic representation of hydrogen bonding in a pair of polymer chains.

untangling their chains makes polymers and the plastic made from them strong and resilient.

2.9 General Classifications

Besides the chemical structures of the polymers in the plastics, there are several other characterizations that are important including molecular weight, thermoplastics vs. thermosets, and crystallinity.

2.9.1 Molecular Weight

The molecular weight of a polymer is the sum of the atomic weights of individual atoms that comprise a molecule. It indicates the *average* length of the bulk resin's polymer chains. All polymer molecules of a particular grade do not have the exact same molecular weight. There is a range or distribution of molecular weights. There are two important but different ways to calculate molecular weight. The most important one is called the number average molecular weight, M_n. For all "i" molecules in a sample, the number average molecular weight is calculated using Eqn (6):

$$M_n = \frac{\sum_i N_i M_i}{\sum_i N_i}, \quad (6)$$

where

i is the number of polymer molecules

N_i is the number of molecules that have the molecular weight M_i.

The weight average molecular weight is a different calculation as in Eqn (7)

$$M_w = \frac{\sum_i N_i M_i^2}{\sum_i N_i M_i} \quad (7)$$

Figure 2.14 shows a molecular weight distribution chart with the two different molecular weight measures noted on it. The ratio M_w/M_n is called the *molar-mass dispersity index*[3] [often called *polydispersity* (PDI)].[3] If all the polymer chains are exactly the same, then the number average and weight average molecular weights are exactly the same, and the PDI is "1." The larger the molar-mass dispersity index, the wider the molecular weight distribution. The molecular weight range can affect many properties of plastic materials.

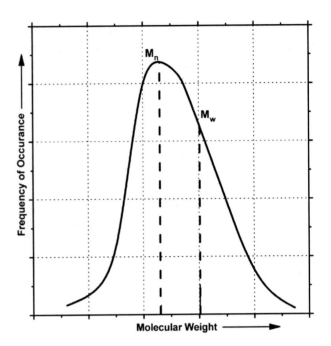

Figure 2.14 Hypothetical molecular weight distribution plot showing number and weight average molecular weights.

Another common means of expressing the length of a polymer chain is the *degree of polymerization*, this quantifies the average number of monomers incorporated into the polymer chain. The average molecular weight can be determined by several means, but this subject is beyond the scope of this book. Low-molecular-weight polyethylene chains have backbones as small as 1000 carbon atoms long. Ultrahigh-molecular-weight polyethylene chains can have 500,000 carbon atoms along their length. Many plastics are available in a variety of chain lengths or different molecular weight grades. These resins can also be classified indirectly by a viscosity value, rather than molecular weight. Within a resin family, such as polycarbonate, high-molecular-weight grades have higher melt viscosities. For example, in the viscosity test for polycarbonate, the melt flow rate ranges from approximately 4 g/10 min for the highest molecular weight, standard grades, to more than 60 g/10 min for lowest molecular weight, high flow, specialty grades.

Selecting the correct molecular weight for an injection molding application generally involves a balance between filling ease and material performance. If the application has thin-walled sections, a lower molecular-weight/lower viscosity grade offers better flow. For normal wall thicknesses, these resins also offer faster mold cycle times and fewer molded in stresses. The stiffer flowing, high-molecular-weight resins offer the ultimate material performance, being tougher and more resistant to chemical and environmental attack. High-molecular-weight films orient better (see Section 3.8 for details). Low-molecular-weight films are often optically clearer.

Molecular weight of the polymers that are used in engineering plastics affects many of the plastics properties. Although it is not always known exactly what the molecular weights are, as mentioned above higher flowing plastics of a given series of products generally are low-molecular-weight polymers. Molecular weight can affect the permeation properties as shown in Fig. 2.15.

Dispersity can also have an effect on permeation rates. For polymers, large molar-mass dispersity index implies that a significant amount of low-molecular-weight polymer is present and that can act like a plasticizer, which increases permeation rates (see Section 2.10.3.10).

2.9.2 Thermosets vs. Thermoplastics

A plastic falls into one of two broad categories depending on its response to heat: thermoplastics or thermosets. Thermoplastics soften and melt when heated and harden when cooled. Because of this behavior, these resins can be injection molded, extruded, or formed via other molding techniques. This behavior also allows production scrap runners and trimmings to be reground and reused. Thermoplastics can often be recycled.

Unlike thermoplastics, thermosets react chemically to form cross-links, as described earlier, that limit chain movement. This network of polymer chains tends to degrade, rather than soften, when exposed to excessive heat. Until recently, thermosets could not be remelted and reused after initial curing. Recent advances in recycling have provided new methods for remelting and reusing thermoset materials.

2.9.3 Crystalline vs. Amorphous

Thermoplastics are further classified by their crystallinity or the degree of order within the overall structure of the polymer. As a crystalline resin cools from the melt, polymer chains fold or align into highly ordered crystalline structures as shown in Fig. 2.16.

Some plastics can be completely amorphous or crystalline. Often specifications in plastics will report what percent of it is crystalline, such as 73% crystallinity. Generally, polymer chains with bulky side groups cannot form crystalline regions. The degree of crystallinity depends on both the polymer and the processing technique. Some polymers such as polyethylene crystallize quickly and reach high levels of crystallinity. Others, such as PET polyester, require slow cooling to crystallize. If cooled

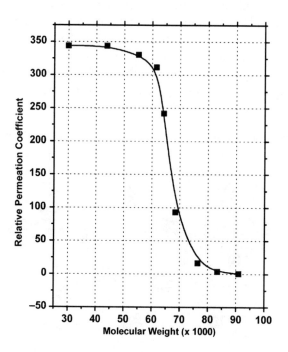

Figure 2.15 Water permeation of ethylene–vinyl alcohol copolymer vs. polymer molecular weight.[4]

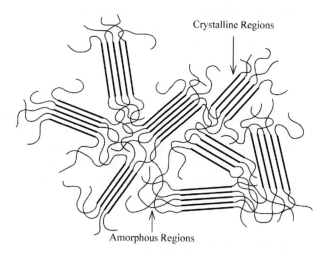

Figure 2.16 Many plastics have crystalline and amorphous regions.

quickly, PET polyester remains amorphous in the final product.

Crystalline and amorphous plastics have several characteristic differences. Amorphous polymers do not have a sharp melting point, but do have what is called a glass transition temperature, T_g. A glass-transition temperature is the temperature at which a polymer changes from hard and brittle to soft and pliable. The force to generate flow in amorphous materials diminishes slowly as the temperature rises above the glass transition temperature. In crystalline resins, the force requirements diminish quickly as the material is heated above its crystalline melt temperature. Because of these easier flow characteristics, crystalline resins have an advantage in filling thin-walled sections of a mold. Crystalline resins generally have superior chemical resistance, greater stability at elevated temperatures, and better creep resistance. Amorphous plastics typically have better impact strength, less mold shrinkage, and less final part warping than crystalline materials. Higher crystallinity usually leads to lower permeation rates. End-use requirements usually dictate whether an amorphous or crystalline resin is preferred.

2.9.4 Orientation

When films are made from plastic polymers, the polymer molecules are randomly intertwined like a bowl of spaghetti. They are amorphous. They are coiled and twisted and have no particular alignment, unless they are crystallized during cooling or aging. However, if the film is drawn or stretched, the amorphous regions of the polymer chains are straightened and aligned to the direction of drawing. The process for doing this is discussed later in this chapter. Oriented films usually have lower oxygen and water permeation rates.

2.10 Plastic Compositions

Plastics are usually formulated products meaning that they are not always neat polymers. They may be blends of polymers and may have any or many additives used to tailor performance properties.

2.10.1 Polymer Blends

Polymers can often be blended. Occasionally, blended polymers have properties that exceed those of either of the constituents. For instance, blends of polycarbonate resin and PET polyester, originally created to improve the chemical resistance of polycarbonate, actually have fatigue resistance and low-temperature impact resistance superior to either of the individual polymers.

Sometimes, a material is needed that has some of the properties of one polymer and some of the properties of another polymer. Instead of going back to the laboratory and trying to synthesize a brand new polymer with all the properties wanted, two polymers can be melted together to form a blend, which will hopefully have some properties of both.

Two polymers that do actually mix well are polystyrene and polyphenylene oxide. A few other examples of polymer pairs that will blend are as follows:

- PET with polybutylene terephthalate
- Polymethyl methacrylate with polyvinylidene fluoride

Phase-separated mixtures are obtained when one tries to mix most polymers. But strangely enough, the phase-separated materials often turn out to be rather useful. They are called immiscible blends.

Polystyrene and polybutadiene are immiscible. When polystyrene is mixed with a small amount of polybutadiene, the two polymers do not blend. The polybutadiene separates from the polystyrene into little spherical blobs. If this mixture is viewed under a high-power microscope, something that looks like the picture in Fig. 2.17 would be seen.

Multiphase polymer blends are of major economic importance in the polymer industry. The most

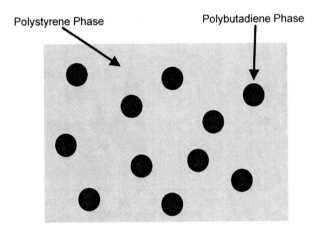

Figure 2.17 Immiscible blend of polystyrene and polybutadiene.

common examples involve the impact modification of a thermoplastic by the microdispersion of a rubber into a brittle polymer matrix. Most commercial blends consist of two polymers combined with small amounts of a third, compatibilizing polymer, typically a block or graft copolymer.

Multiphase polymer blends can be easier to process than a single polymer with similar properties. The possible blends from a given set of polymers offer many more physical properties than do the individual polymers. This approach has shown some success but becomes cumbersome when more than a few components are involved.

Blending two or more polymers offers yet another method of tailoring resins to a specific application. Because blends are only physical mixtures, the resulting polymer usually has physical and mechanical properties that lie somewhere between the values of its constituent materials. For instance, an automotive bumper made from a blend of polycarbonate resin and thermoplastic polyurethane elastomer gains rigidity from the polycarbonate resin and retains most of the flexibility and paintability of the polyurethane elastomer. For business machine housings, a blend of polycarbonate and acrylonitrile-butadiene-styrene copolymer (ABS) resins offers the enhanced performance of polycarbonate flame retardance and ultraviolet (UV) stability at a lower cost.

Additional information on the subject of polymer blends is available in the literature.[5-7]

2.10.2 Elastomers

Elastomers are a class of polymeric materials that can be repeatedly stretched to over twice the original length with little or no permanent deformation. Elastomers can be made of either thermoplastic or thermoset materials and are generally tested and categorized differently than rigid materials. They are commonly selected according to their hardness and energy absorption characteristics, properties rarely considered in rigid thermoplastics. Elastomers are found in numerous applications, such as automotive bumpers and industrial hoses.

2.10.3 Additives

The properties of neat polymers are often not ideal for production or the end use. When this is the case, materials are added to the polymer to improve the performance shortfall. The additives can improve the processing and performance of the plastic. For whatever reason the additive is used, it can affect the permeation, diffusion, and solubility properties.

Additives encompass a wide range of substances that aid processing or add value to the final product.[8,9] Found in virtually all plastics, most additives are incorporated into a resin family by the supplier as part of a proprietary package. For example, you can choose standard polycarbonate resin grades with additives for improved internal mold release, UV stabilization, and flame retardance; or you can choose nylon grades with additives to improve impact performance.

Additives often determine the success or failure of a resin or system in a particular application. Many common additives are discussed in the following sections. Except for reinforcement fillers, most additives are added in very small amounts.

2.10.3.1 Fillers, Reinforcement, and Composites

Reinforcing fillers can be added in large amounts. Some plastics may contain as much as 60% of reinforcing fillers. Often, fibrous materials, such as glass or carbon fibers, are added to resins to create reinforced grades with enhanced properties. For example, adding 30% of short glass fibers by weight to nylon 6 improves creep resistance and increases stiffness by 300%. These glass-reinforced plastics usually suffer some loss of impact strength and ultimate elongation, and these are more prone to warping because of the relatively large difference in mold shrinkage between the flow and cross-flow directions.

Plastics with nonfibrous fillers such as glass spheres or mineral powders generally exhibit higher stiffness characteristics than unfilled resins, but not as high as fiber-reinforced grades. Resins with particulate fillers are less likely to warp and show a decrease in mold shrinkage. Particulate fillers typically reduce shrinkage by a percentage roughly equal to the volume percentage of filler in the polymer, an advantage in tight tolerance molding.

Often reinforced plastics are called *composites*. Often, the plastic material containing the reinforcement is referred to as the matrix. One can envision a number of ways in which different reinforcing materials might be arranged in a composite. Many of these arrangements are shown in Fig. 2.18.

Figure 2.18 Several types of composite materials.

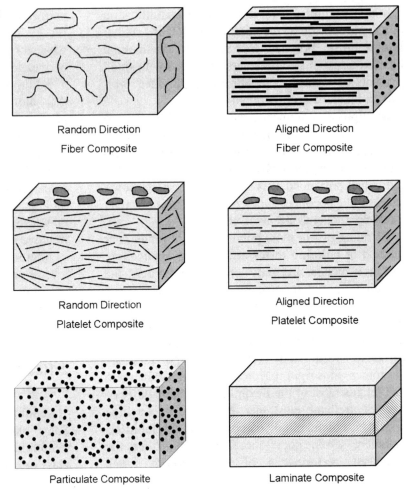

While barrier films and membranes usually do use fiber-reinforcing fillers, platelet and particulate composites are used and laminate composites are very common.

Particulates, in the form of pigments, to impart color may be added. On occasion, particulate, called extender, is added to reduce the amount of relatively expensive polymer used, which reduces overall cost.

Platelet additives may impart color and luster, metallic appearance, or a pearlescent effect, but they also can strongly affect permeation properties. Most of these additives have little or no permeation through themselves so when a film contains particulate additives, the permeating molecule must follow a path around the particulate additive as shown in Fig. 2.19. This is often called a *tortuous path effect*.

Barrier enhancement due to tortuous path through a platelet-filled film may be modeled by Eqn (8).[10]:

$$P_c = P_0 \frac{V_p}{1 + (V_f A_f)/2} \quad (8)$$

where
P_0 is the permeability coefficient of the neat polymer
V_p is the volume fraction of polymer
V_f is the volume fraction of the flake filler
A_f is the aspect ratio of the flake filler, length/thickness (L/W in Fig. 2.19).

These variables assume that the particles are completely separated (exfoliated) and planar in the film. This is relatively easy to do in cast films and coatings, a little less so in extruded and blown films, much more difficult in molded items. Figure 2.20 shows graphically the effect that aspect ratio and concentration of platelet fillers have on the inherent permeability of the polymer matrix.

It has also been modeled for other shaped particulate fillers including rods or cylinders and

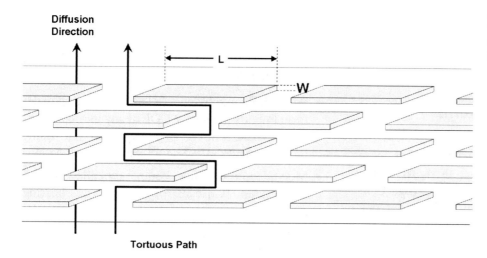

Figure 2.19 Tortuous path of permeant molecule through a particulate containing film.

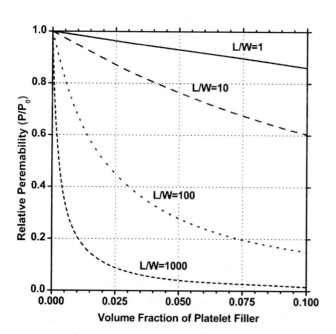

Figure 2.20 Relative permeability vs. volume fraction of platelet fillers of different aspect ratios.

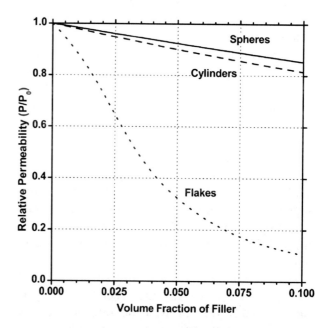

Figure 2.21 Calculated change in the permeation of a film containing particulate of a specified shape vs. the amount of that particulate in the film.

spheres.[11] Figure 2.21 shows the relative effectiveness of particulate in various shapes on the permeation coefficients vs. volume fraction of the particulate.

2.10.3.2 Combustion Modifiers, Fire, Flame Retardants, and Smoke Suppressants

Combustion modifiers are added to polymers to help retard the resulting parts from burning. Generally required for electrical and medical housing applications, combustion modifiers and their amounts vary with the inherent flammability of the base polymer. Polymers designed for these applications often are rated using an Underwriters Laboratories rating system. Use these ratings for comparison purposes only, as they may not accurately represent the hazard present under actual fire conditions.

2.10.3.3 Release Agents

External release agents are lubricants, liquids, or powders, which coat a mold cavity to facilitate part removal. Internal release agents can accomplish the same purpose. The identity of the release agent is rarely disclosed, but frequently they are fine fluoropolymer powders, called micropowders, silicone resins, or waxes.

2.10.3.4 Slip Additives/Internal Lubricants

When polymeric films slide over each other, there is a resistance that is quantified in terms of the coefficient of friction (COF). Films with high COF tend to stick together instead of sliding over one another. Sticking makes the handling, use, and conversion of films difficult. To overcome sticking, slip agents are added to the films.

Slip additives can be divided into two types: migrating and nonmigrating types. Migrating slip additives are the most common class, and they are used above their solubility limit in the polymer. These types of additives are molecules that have two distinct parts, typically pictured as a head and tail as shown in Fig. 2.22. One part of the molecule, usually the head, is designed to be soluble in the polymer (particularly when it is molten during processing) making up the plastic. The other part, the tail, is insoluble. As the plastic cools and solidifies from its molten state, these molecules migrate to the surface, where the insoluble end "sticks up" reducing the COF. This process is shown in Fig. 2.22. These additives are typically fatty acid amides.

There are migrating slip additives that are not of this two-part structure. One additive is perfluoropolyether (PFPE) synthetic oil marketed by DuPont™ under the trademark Fluoroguard®, which is an internal lubricant that imparts improved wear and low-friction properties. Silicone fluids, such as those made by Dow Corning, can also act as a boundary lubricant. Both these materials may migrate to the surface of the plastic over time.

Some common nonmigrating slip additives include the following:

- Polytetrafluoroethylene (PTFE) in micropowder form imparts the lowest COF of any internal lubricant. Manufacturers and suppliers are many including DuPont™ Zonyl® and 3M Dyneon™.
- Molybdenum disulfide, commonly called "*moly*," is a solid lubricant often used in bearing applications.
- Graphite is a solid lubricant used like molybdenum disulfide.

2.10.3.5 Antiblock Additives

Blocking is a surface effect between adjacent film layers that stick to one another. Blocking is quantified by the force needed to separate two film layers under controlled conditions. Two situations where blocking is an issue are the opening of blown film tubes after extrusion and film layer separation after packing and storage. Antiblock additives are used to overcome these issues.

Antiblock additives can be divided into two classes: inorganic and organic. Chemically inert, inorganic antiblock additives migrate to the film surface and partially stick out of the surface to create a microroughness of the film surface. Figure 2.23 illustrates this principle.

The detailed mechanism of how organic antiblock additives work is not yet understood. It is thought that a barrier layer is formed on the plastic film surface, thus inhibiting the two adjacent plastic film layers' adhesion. Their usage is limited. Organic antiblock additives were partially discussed above and will not be further mentioned here.

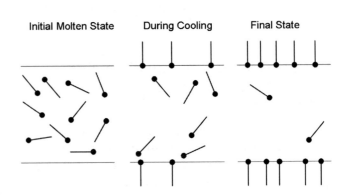

Figure 2.22 Mode of action of a typical migrating slip additive.

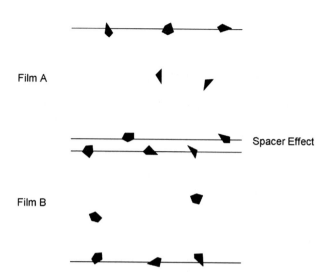

Figure 2.23 Antiblock additives maintain film separation.

2.10.3.6 Catalysts

Catalysts, substances that initiate or change the rate of a chemical reaction, do not undergo a permanent change in composition or become a part of the molecular structure of the final product. Occasionally used to describe a setting agent, hardener, curing agent, promoter, etc., catalysts are added in minute quantities, typically <1%.

2.10.3.7 Impact Modifiers and Tougheners

Many plastics do not have sufficient impact resistance for the use for which they are intended. Rather than change to a different type of plastic, they can be impact modified in order to fulfill the performance in use requirements. Addition of modifiers called impact modifiers or tougheners can significantly improve impact resistance. This is one of the most important additives. There are many suppliers and chemical types of these modifiers.

General purpose impact modification is a very low level of impact modification. It improves room temperature impact strength but does not take into account any requirements for low-temperature (<0 °C) impact strength. For most of these types of applications, only low levels of impact modifier will be required (<10%).

Low-temperature impact strength is required for applications that require a certain level of low-temperature flexibility and resistance to break. This is for example the case for many applications in the appliance area. For this purpose, modifier levels between 5% and 15% of mostly reactive modifiers will be necessary. Reactive modifiers can chemically bond to the base polymer.

Super tough impact strength may be required for applications that should not lead to a failure of the part even if hit at low temperatures (−30 to −40 °C) under high speed. This requirement can only be fulfilled with high levels (20−25%) of reactive impact modifier with low glass transition temperature.

2.10.3.8 UV Stabilizers

Another way plastics may degrade is by exposure to UV light. UV can initiate oxidation. Plastics that are used outdoors or exposed to lamps emitting UV radiation are subject to photooxidative degradation. UV stabilizers are used to prevent and retard photooxidation. Pigments and dyes may also be used in applications not requiring transparency. Photooxidative degradation starts at the exposed surface and propagates throughout the material.

2.10.3.9 Optical Brighteners

Many polymers have a slight yellowish color. They can be modified to appear whiter and brighter by increasing reflected bluish light (in the range of 400−600 nm). One way to accomplish this is by using an additive that absorbs in the UV range but reemits the energy at higher wavelength. This effect is called fluorescence, and these types of additives are called optical brighteners or fluorescent whitening agents.

2.10.3.10 Plasticizers

Plasticizers are added to help maintain flexibility in a plastic. Various phthalates are commonly used for this purpose. Since they are small molecules, they may extract or leach out of the plastic causing a loss of flexibility with time. Just as deliberately added, small molecules may leach out, small molecules from the environment may be absorbed by the plastic and act like a plasticizer. The absorption of water by nylons (polyamides) is an example.

Plasticizers increase the space between the polymers chains and so they are expected to increase the permeability as well. This is shown in Fig. 2.24, which shows the effect of plasticizer level on the permeability of cellulose acetate films.

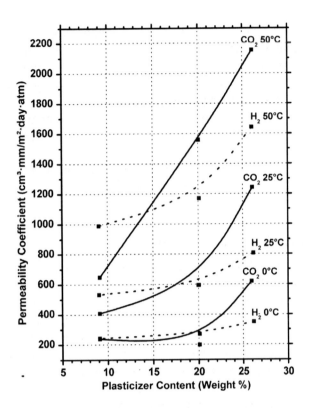

Figure 2.24 Effect of plasticizer content on the permeability of hydrogen and carbon dioxide through cellulose acetate at different temperatures.[12]

2.10.3.11 Pigments, Extenders, Dyes, and Mica

Pigments are added to give a color to the plastic, but they may also affect the physical properties. Extenders are usually cheap materials added to reduce the cost of plastic resins. Dyes are colorants that are chemically different from pigments. Mica is a special pigment added to impact sparkle or metallic appearance.

2.10.3.12 Coupling Agents

The purpose of adding fillers is either to lower the cost of the polymer, make it tougher or stiffer, or make it flame retardant, so that it does not burn when it is ignited. Often the addition of the filler will reduce the elongation at break, the flexibility, and in many cases the toughness of the polymer because the fillers are added at very high levels. One reason for the degradation of properties is that the fillers in most cases are not compatible with the polymers. The addition of coupling agents can improve the compatibility of the filler with the polymer. As a result, the polymer will like the filler more, the filler will adhere better to the polymer matrix, and the properties of the final mixture (e.g., elongation and flexibility) will be enhanced.

2.10.3.13 Thermal Stabilizers

One of the limiting factors in the use of plastics at high temperatures is their tendency to not only become softer but also to thermally degrade. Thermal degradation can present an upper limit to the service temperature of plastics. Thermal degradation can occur at temperatures much lower than those at which mechanical failure is likely to occur. Plastics can be protected from thermal degradation by incorporating stabilizers into them. Stabilizers can work in a variety of ways but discussion of these mechanisms is beyond the purpose of this book.

2.10.3.14 Antistats

Antistatic additives are capable of modifying properties of plastics in such a way that they become antistatic, conductive, and/or improve electromagnetic interference (EMI) shielding. Carbon fibers, conductive carbon powders, and other electrically conductive materials are used for this purpose.

When two (organic) substrates rub against each other, electrostatic charges can build up. This is known as tribocharging. Electrostatic charges can impact plastic parts in several ways; one of the most annoying being the attraction of dust particles. One way to counter this effect is to use antistats (or antistatic additives). This is principally a surface effect, although one potential counter measure (conductive fillers) converts it into a bulk effect.

Tools that decrease electrostatic charges and hence increase the conductivity of an organic substrate can be classified as follows:

- external antistat (surface effect)
- conductive filler (bulk and surface effect)
- internal antistat (surface effect)

An external antistat is applied via a carrier medium to the surface of the plastic part. The same considerations and limitations apply as with nonmigrating slip additives. Conductive filler is incorporated into the organic substrates and builds up a conductive network on a molecular level. Although both approaches are used in organic substrates, they are not the most common.

An internal antistat is compounded into the organic substrate and migrates to the plastic part surface. The same principle considerations apply as for migrating slip additives (see Fig. 2.22).

The need to protect sensitive electronic components and computer boards from electrostatic discharge during handling, shipping, and assembly provided the driving force for development of a different class of antistatic packaging materials. These are sophisticated laminates with very thin metalized films.

There are other additives used in plastics, but the ones discussed above are the most common.

2.11 Summary

The basis of all films is polymers. But as discussed in this chapter, the properties are complex and most have some effect on permeation, diffusion, and solubility. Most of this chapter did not go into the chemical structures of all the polymers used in films, containers, and membranes. The data of various polymers/plastics are presented beginning from Chapter 5 and the chemical structures of the polymers are discussed in the appropriate sections.

References

1. Available from: http://en.wikipedia.org/wiki/Dendronized_polymers
2. This is a file from the Wikimedia Commons which is a freely licensed media file repository.
3. Stepto RFT. Dispersity in polymer science (IUPAC Recommendations 2009). *Pure Appl Chem* 2009;**81**:351–3.
4. Matsumoto T, Horie S, Ochiumi T. Effect of molecular weight of ethylene–vinyl alcohol copolymer on membrane properties. *J Membr Sci* 1981;**9**:109–19.
5. Utracki LA. *Polymer blends handbook, Vols. 1–2*. Springer-Verlag; 2002. Available from: http://www.knovel.com/knovel2/Toc.jsp?BookID=1117&VerticalID=0.
6. Utracki LA. *Commercial polymer blends*. Springer-Verlag; 1998. Available from: http://www.knovel.com/knovel2/Toc.jsp?BookID=878&VerticalID=0.
7. Utracki LA. Encyclopaedic dictionary of commercial polymer blends. *ChemTec Publishing*; 1994. Available from: http://www.knovel.com/knovel2/Toc.jsp?BookID=285&VerticalID=0.
8. Flick EW. *Plastics additives—an industrial guide*. 2nd ed. William Andrew Publishing/Noyes; 1993. Available from: http://www.knovel.com/knovel2/Toc.jsp?BookID=353&VerticalID=0.
9. Pritchard G. *Plastics additives—an A-Z reference*. Springer-Verlag; 1998. Available from: http://www.knovel.com/knovel2/Toc.jsp?BookID=335&VerticalID=0.
10. Choudalakis G, Gotsis A. Permeability of polymer/clay nanocomposites: A review. *Eur Polym J* 2009;**45**:967–84.
11. Moggridge G. Barrier films using flakes and reactive additives. *Prog Org Coat* 2003;**46**:231–40.
12. Brubaker DW, Kammermeyer K. Flow of gases through plastic membranes. *Ind Eng Chem* 1953;**45**:1148–52.

3 Production of Films, Containers, and Membranes

Plastic resin is usually supplied in the form of beads as shown in Fig. 3.1. One of the ways these beads may be converted into film form (or many other shapes) is by processing them through an extruder equipped with a film die.

3.1 Extrusion

A schematic representation of a typical extruder is shown in Fig. 3.2. The plastic beads are loaded into a feed hopper and are gravity fed into a heated barrel that has a screw in it. The screw drives the material down the barrel, heated above the melt point of the polymer, where the plastic beads are melted. Usually, a heating profile is set for the barrel, which gradually increases the temperature of the barrel from the rear (where the plastic bead enters) to the front. This allows the plastic beads to melt gradually as they are pushed through the barrel. The gradual increase in temperature also lowers the risk of overheating, which may cause thermal degradation in the polymer. At the front end of the barrel, the molten plastic enters the die. The die is what gives the final product its profile and is designed such that the molten plastic evenly flows from a cylindrical profile to the product's profile shape, a film in this example. Figure 3.3 shows a schematic of a film die.

The extruded film is then cooled on a series of chilled drums or water bath to set the film and to reduce crystallinity. Once the film has cooled, it can be treated, trimmed to the desired width, and spooled (not shown in the figure). The extruded film thickness is often thicker at the edges.

Multilayer films are a common approach to minimizing permeation and effusion. Multiple extruders equipped with the special dies allow multilayer films to be produced. The extruder can put molten layer onto a solid layer, or the multiple molten layers may be laid down on each other. Mixing between molten layers usually does not occur. Figure 3.4 shows a picture of a stacked multilayer die. This process is called coextrusion.

The flow of molten polymer through dies is important to understand when planning film production using extrusion. Rheology is the science of deformation and flow. One common factor among solids, liquids, and all materials whose behavior is intermediate between solids and liquid is that if we apply a stress or load on any of them, they will deform or strain. Likewise when the stress is removed, they will react. The flow properties will depend on the properties of the polymers, including their molecular structure and molecular weight. They are also temperature dependent. Further information on this subject may be obtained from the Refs. 1 and 2.

3.2 Blown Film

Blown film is another extrusion process used to fabricate film products. In this case, the extrusion die is shaped as a circle, and air pressure is used to further expand the film. After it is expanded to the

Figure 3.1 Plastic resin is often supplied in bead form.[11]

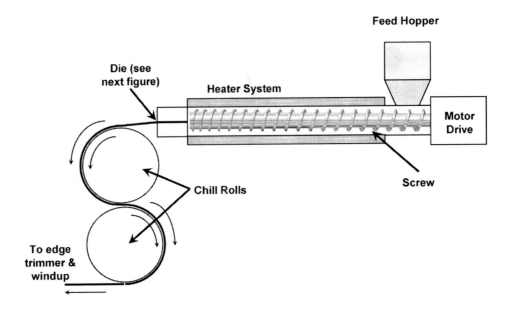

Figure 3.2 Schematic of an extruder.

desired dimensions, it is cooled to solidify the polymer. A schematic representation of the blown film process is shown in Fig. 3.5 and a photograph is shown in Fig. 3.6. Films are typically defined as less than 0.254 mm (10 mil) in thickness, although blown film can be produced as high as 0.5 mm (20 mil).

The blown film process is used to produce a wide variety of products, ranging from simple monolayer films for bags to very complex multilayer structures used in food packaging. Multilayer film structures may be made by blown film coextrusion and combines two or more molten polymer layers. In fact, equipment is commercially available to coextrude 7 to 11 layers.[3] Figure 3.7 is a schematic representation of a three-layer blown film die,[4] and Fig. 3.8 shows a photograph of a nine-layer blown film production machine.[12]

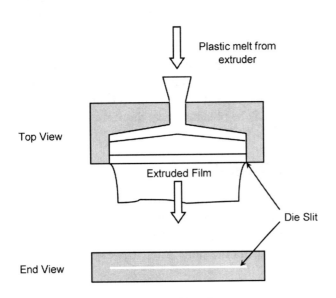

Figure 3.3 Schematic of a film die.

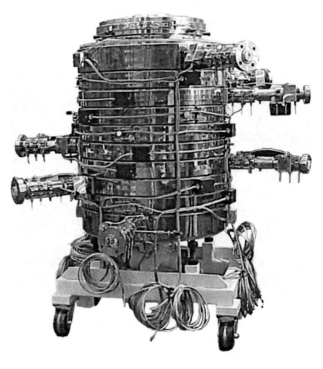

Figure 3.4 Photograph of Brampton Engineering 10-layer stacked die.[12]

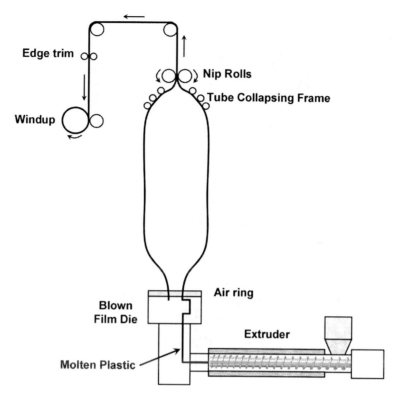

Figure 3.5 Schematic of a blown film process.

Figure 3.6 Photograph of film production by blowing process.[11]

3.3 Calendering

Calendering is a specialty process for the manufacture of sheet and film products, primarily polyvinyl chloride (PVC), although certainly other materials such as modified polyethylene (PE), polypropylene (PP), ABS, and other thermoplastics can also be calendered. Figure 3.9 is a schematic representation of a four-roller calender. An extruder, a kneeder, or the other mixer is used to compound the polymer mixture and additives. The compounded plastic material is transferred to a pair or heated calendering rolls where the film is formed by the pressure. The numbers of rolls used and their configuration may vary. In general, the film industry uses the "inverted L configuration," with four to seven rolls. The rolls are all individually driven and temperature controlled. The temperature and speed of the rolls influence the properties of the calendered film. Other roll configurations are also used such as those shown in Fig. 3.10.

After the film or sheet has been calendered to the proper thickness, other processes are frequently done to the film or sheet. Changing the surface texture is done by embossing rolls. The film passes between an

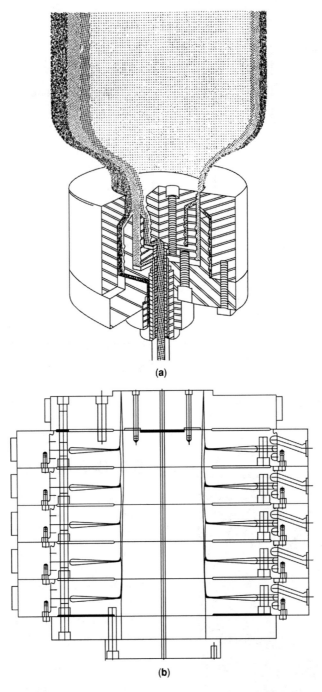

Figure 3.7 Drawing of a three-layer blown film die.[4]

embossing roll that has been machined into the desired surface pattern and a cooled rubber roll. When tempering rolls are heated or chilled, the physical properties of the final film/sheet products are affected.

During calendering, the film may also be stretched. Surface treatments may also be applied as a part of the calender process. At the end of the line is a winding station that includes cutting devices for edge trimming and in-line slitting of rolls.

3.4 Casting Film Lines

The oldest method of making films is the continuous solvent-casting process.[5] This was used to make

Figure 3.8 Nine-layer blown film die and bubble equipment (Image provided by Brampton Engineering).[12]

photographic film. A solution of the polymer is applied to a smooth continuous moving belt. As the cast solution is dried, the film is formed. When completely dry, the film is removed from the belt. A schematic representation of one such machine is shown in Fig. 3.11.

First, a solution of the polymer must be made. The solution is usually called "dope." These solutions are very carefully made. The solvent must dissolve enough polymer to make a solution that is reasonably concentrated. Table 3.1 shows some examples of polymers and solvents. The solvent usually needs to be very volatile, so that the film can be formed and dried quickly. The dope is usually made and kept in a tight temperature range and solids content, as these two factors can dramatically affect the viscosity. It must also be kept very clean to eliminate defects in the final film. It is often filtered several times and often contains additives.

The dope is applied to the moving belt as shown in Fig. 3.11. This part of the machine that lays down the dope is called the "caster." The caster is often a doctor blade die or slot die but can be of other configurations. The belt typically may be 1- to 2-m wide, 1- to 2-mm thick, and over 100-m long. Belts are commonly polished stainless steel or chrome plated steel, but may also be copper or even a polymer film such as polyester or fluoropolymer. The cast film must not stick to the belt when it has formed.

Drying air is introduced into the machine near the exit end, and it flows counter current to the moving film sweeping away the evaporating solvent. The solvent laden drying air leaves the machine and typically goes to a solvent recovery process. The solvent recovery minimizes air pollution and reduces cost by reusing the solvent.

Commercial casting operations often include a gauging system, which measures and controls the film thickness. Surface treatments, such as corona or flame treatment, are sometimes applied when the

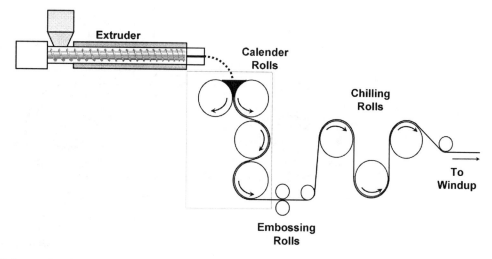

Figure 3.9 Schematic of an inverted L four-roll calender machine.

Figure 3.10 Schematic of three common calender roll configurations.

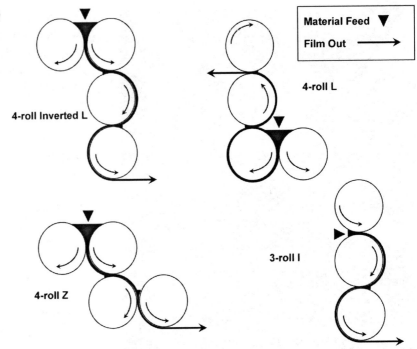

film exits the casting machine and prior to end trim and windup.

3.5 Post Film Formation Processing

The production of film is usually not the end of the production process. There are numerous other processes. The simplest are cut and trim processes and packaging-related processes that will not be discussed further. It is often desired to combine different films into multilayered structures that could not be produced directly by the film production methods just discussed.

Multilayered structures can be built by laminating two or more films together. When this is done sometimes, an adhesive must be applied between the

Figure 3.11 Schematic of a belt film casting machine.

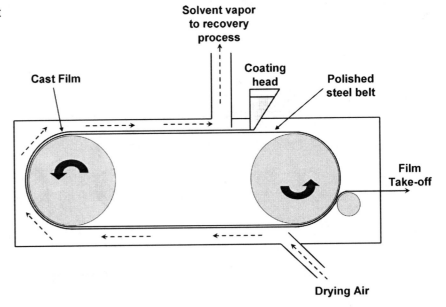

Table 3.1 Polymers and Solvents for Some Film-Casting Processes[5]

Polymer	Solvents
Cellulose nitrate	Ether and esters
Cellulose diacetate	Acetone and methanol
Cellulose triacetate	Methylene chloride and alcohols
PCs	Methylene chloride
Polyethersulfone	Methylene chloride
Polyetherimide	Methylene chloride
Polyvinylidene fluoride	Acetone
Polyvinyl chloride	Tetrahydrofuran and methyl ethyl ketone
Polyimides	Dimethylformamide
Polyvinyl alcohol	Water and methanol
Methyl cellulose	Water
Starch derivatives	Water
Gelatin	Water and methanol

layers to insure that they will be stuck together. The adhesive is often applied as a coating. Coating continuous film is called *web coating*. Other coatings are applied to change properties such as antistat, color, appearance, and permeability. The coatings are usually liquids that may require drying. Some coatings are deposited from vapor or from vacuum. Thin metal films may be deposited by a process called metallization.

Film physical properties may also be affected by stretching the film, in a process called orientation. All the processes need to be done at a high speed and minimal cost. These processes will be discussed in the following sections.

3.6 Web Coating

There are many web coating processes; most, but not all, of which are discussed briefly in this section. Web coating is used to apply thin layers to films for surface treatments or decoration. For more detail, see Refs. 6 and 7. The selection of which process to use depends on the properties of the film and the coating being applied.

3.6.1 Gravure Coating

The gravure process is similar to the process used for printing newspapers. It uses an engraved roller running in a coating bath, which fills the engraved dots or lines of the roller with the coating material, as shown in Fig. 3.12. The excess coating on the roller

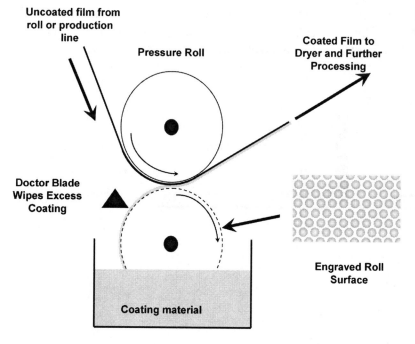

Figure 3.12 Schematic of gravure coating process.

Figure 3.13 Schematic of a reverse roller coating process.

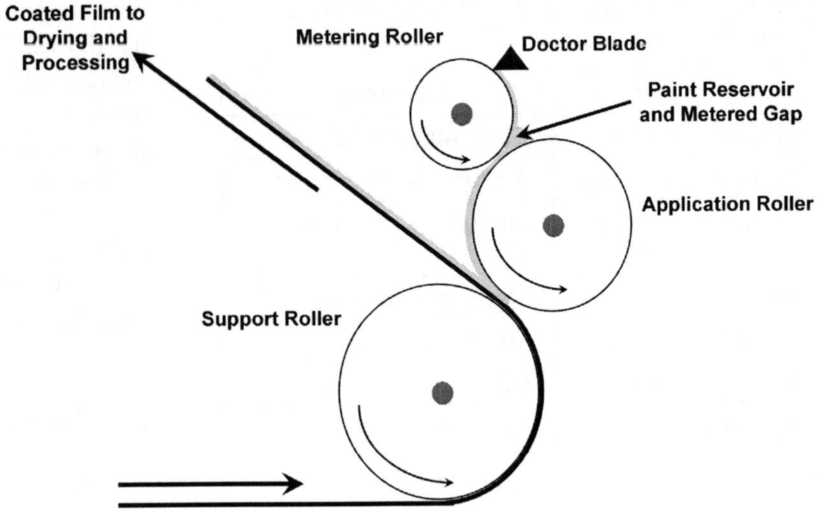

is wiped off by the Doctor Blade, and the coating in the engravings is then deposited onto the substrate as it passes between the engraved roller and a pressure roller. The coating may need to be dried if there is solvent in the coating formulation. This process is very common for applying adhesives to film surfaces.

3.6.2 Reverse Roll Coating

For the reverse roller coating process, the coating material is applied at a precise thickness onto the applicator roller by precision setting of the gap between the upper metering roller and the application roller below it. The coating is transferred off the application roller to the substrate as it passes around the support roller at the bottom as shown in Fig. 3.13. The diagram illustrates a three-roll reverse roller coating process, although four-roll versions are common.

3.6.3 Knife on Roll Coating

The knife coating process relies on a coating being applied to the substrate in a "puddle," which then passes through a "gap" between a "knife" and a support roller. As the coating and substrate pass through the gap, the excess is scraped off as shown in Fig. 3.14. This process can be used for high-viscosity coatings and very high coat weights, such as plastisols and rubber coatings. There are many variants of this relatively simple process.

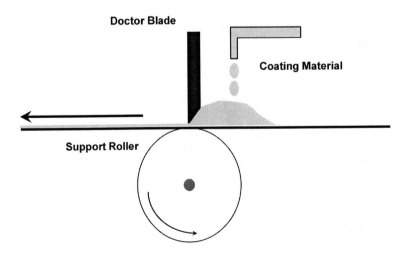

Figure 3.14 Schematic of a knife on roller coating process.

3.6.4 Metering Rod (Meyer Rod) Coating

For the Meyer rod coating process, an excess of the coating is deposited onto the substrate as it passes over the pick-up roller shown in Fig. 3.15. A wire-wound metering rod (or bar), known as a Meyer Rod, allows the desired quantity of the coating to remain on the substrate. The quantity of coating (film thickness) is determined by the diameter of the wire used on the rod.

The rods are identified by number. An increase in number means the rod is made from larger wire diameter. The wire and rod are commonly stainless steel. Table 3.2 shows the expected film thickness (or film weight that is related to thickness) that is applied for different rod sizes and for different paint dilutions.

3.6.5 Slot Die (Slot, Extrusion) Coating

In the slot die process, the coating is squeezed out by gravity or under pressure through a slot and onto the substrate film. If the coating is 100% solids, the process is termed "Extrusion," and in this case, the line speed is frequently much faster than the speed of the coating extrusion. This allows coatings to be considerably thinner than the width of the slot.

3.6.6 Immersion (Dip) Coating

Immersion or dip coating is a simple process in which the substrate is dipped into a bath of the coating. A schematic diagram is shown in Fig. 3.16. The coating is normally of a low viscosity to enable the coating to run back into the bath as the substrate emerges. This process is frequently used on porous substrates. It also coats both sides of the film. Variations of the dip coating process include passing the coated film through a slotted die or past an air knife to remove excess paint.

3.6.7 Vacuum Deposition

Plastic film with a metal appearance is very common in food bags such as those used to protect candy and potato chips. The metal not only looks good, but also one can imagine that it could dramatically reduce permeation of water vapor and oxygen through the film. That metal is typically aluminum, and it is deposited on the film by a process called *metallization*. This process occurs in high vacuum. Aluminum metal is vaporized usually by heating to its boiling point, in a ceramic pan called a *metallization boat*. The aluminum vapor is directed toward film that is moving by the aluminum vapor source as shown in Fig. 3.17. The schematic in this figure is one of a semicontinuous process in which film is fed off a pay-off roll, through a series of rolls to a chilled roll that passes by the aluminum vapor source. The metallization occurs on the chilled roll, and then the finish film is rewound up. Being semicontinuous means that after the roll is coated, the vacuum is released. The metalized roll is removed and a new roll is inserted. The vacuum must be reestablished before metallization can continue. True continuous processes have been designed that use sophisticated air locks to allow film to pass into and out of the vacuum chamber with very little introduction of air. The aluminum

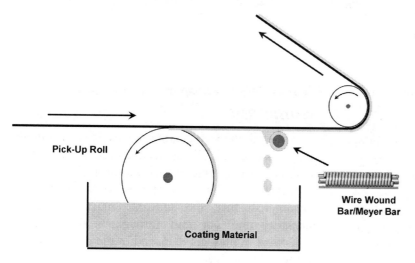

Figure 3.15 Schematic of a Meyer rod coating process.

Table 3.2 Meyer Rod Size and Expected Wet and Dry Film Weights (g/m² dry)

Rod Size	Wet Mils	Thickness (μm)	30% Solids	40% Solids	75% Solids	100% Solids
#3	0.3	7.6	2.18	2.91	5.47	7.29
#4	0.4	10.2	2.91	3.88	6.83	9.11
#5	0.5	12.7	3.64	4.85	9.11	12.14
#6	0.6	15.2	4.38	5.82	10.93	14.58
#7	0.7	17.8	5.09	6.80	12.74	16.99
#8	0.8	20.3	5.82	7.78	14.58	19.43
#9	0.9	22.9	6.56	8.34	16.40	21.87
#10	1.0	25.4	7.28	9.71	18.21	24.28
#11	1.1	27.9	8.00	10.69	20.04	26.72
#12	1.2	30.5	8.74	11.65	21.87	29.16
#13	1.3	33.0	9.47	12.63	23.71	31.60
#14	1.4	35.6	10.20	13.60	25.51	34.00
#15	1.5	38.1	10.93	14.58	27.32	36.41
#16	1.6	40.6	11.65	15.54	29.14	38.85
#17	1.7	43.2	12.38	16.51	30.99	41.29
#18	1.8	45.7	13.11	17.49	32.78	43.70
#19	1.9	48.3	13.85	18.45	34.61	46.14
#20	2.0	50.8	14.58	19.43	36.41	48.55
#22	2.2	55.9	16.03	21.36	40.07	53.43
#24	2.4	61.0	17.49	23.31	43.72	58.28
#26	2.6	66.0	18.94	25.25	47.38	63.16
#28	2.8	71.1	20.40	27.20	51.04	68.04
#30	3.0	76.2	21.85	29.14	54.65	72.86
#32	3.2	81.3	23.31	31.09	58.28	77.72
#34	3.4	86.4	24.76	33.04	61.62	82.55
#36	3.6	91.4	26.23	34.96	65.55	87.40

evaporation source is usually kept full of aluminum by feeding aluminum wire into the metallization boat.

Other materials can be applied besides aluminum may be vacuum deposited. A different evaporation source is used. For example, silicon oxide may be coated on plastic film if the evaporation source is replaced by an electron beam gun that targets a block of silicon oxide coating material, which is ionized/vaporized onto the film. Ceramis® technology developed by Amcor Limited is an example of silicon oxide technology.

3.6.8 Web Coating Process Summary

Although the web coating processes seem simple, they must be carefully controlled in order to produce quality results. One wants uniform thickness in machine direction and transverse direction. To accomplish this, the substrate must have good planarity and not curl throughout the process; otherwise, it may get damaged in the coating applicator or in the dryer.

3: Production of Films, Containers, and Membranes

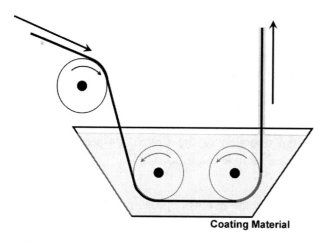

Figure 3.16 Schematic of an immersion or dip coating process.

Contamination must be avoided in the coating material and in the substrate. The coating must wet the surface to get adhered, so surface tension is important. Coating solution delivery must be precisely controlled. Vibration anywhere in the system can produce undesirable results. In the drying station, not only the temperature but also the relative humidity is important.

3.7 Lamination

In film lamination, a fabricated film is adhered to a moving film or substrate by application of heat and pressure. Film lamination methods include hot roll, belt, flame, calender lamination, and sheet extrusion; each type provide a different combination of heat and pressure.

3.7.1 Hot Roll/Belt Lamination

Hot roll and belt lamination use heat and pressure as their means of bonding. Usually one of the films has been coated with a heat-activated adhesive by one of the web coating processes discussed previously. As shown in Fig. 3.18, the two films are drawn onto heated rollers where the materials are heated and pressed together. The heat starts melting the adhesive activating it, creating a bond when pressed against the opposite material.

Heated roller temperature varies depending on product and adhesive, but it is typically from 200 to 300 °F. Circulating oil or electrical heaters are used to maintain heating of hot roll.

Heat-sensitive layer of laminating film never directly contacts heated roller, and we keep it on the other side.

Calender lamination is similar to hot roll lamination, except that a bank of hot rolls is used, and extrusion lamination produces one of the substrate films just prior to a second film being laminated to it by hot roll or calender process.

3.7.2 Flame Lamination

Flame lamination is often used to bond film and/or fabric to soft polyurethane foams. The process, shown in Fig. 3.19, involves the passing of the soft

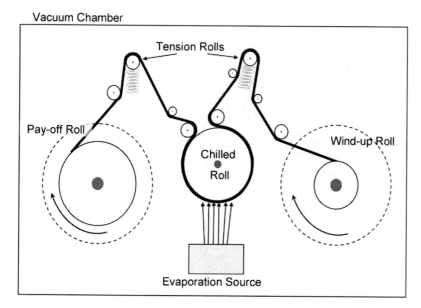

Figure 3.17 Schematic of a semicontinuous vapor deposition coating process.

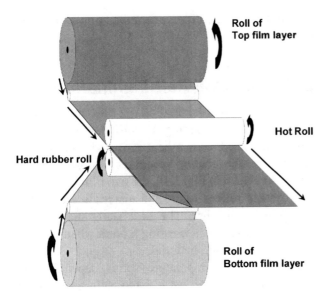

Figure 3.18 Schematic of a hot roll laminator.

foam over an open flame, which creates a thin layer of molten polymer. In this case, the molten polymer acts as the adhesive. The film and/or fabric are quickly pressed against the foam while it is still in the molten state. The strength of the bond depends on the film, fabric, and foam selected and the processing conditions.

3.8 Orientation

The process for making films usually results in the polymer chains arranging in a relatively random order. If one stretches a film, the polymer chains will tend to line up in the direction of the stretch. This process will affect the physical properties of the film in the direction of the stretch. Properties that are affected include the following:

- An increase in tensile properties, leading to better resistance break and tear
- Optical properties: improve clarity, reduced haze, and increased surface gloss; Fig. 3.20 shows the clarity difference of unoriented film compared to oriented film
- Controlled shrinkage
- Improved barrier properties, particularly to water and oxygen
- Film toughness and puncture resistance increases
- Stiffness increases.

Orientation in one direction is very common. This is often called machine direction orientation (MDO).

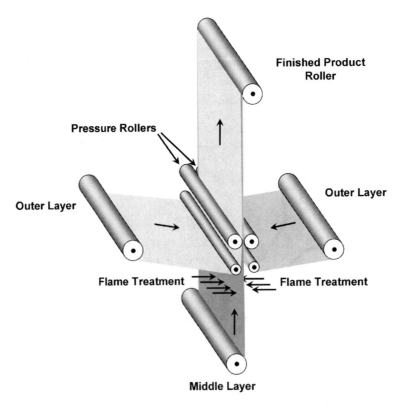

Figure 3.19 Schematic of flame lamination.

3: Production of Films, Containers, and Membranes

Figure 3.20 Orientation of film can improve clarity. Left: unorientated; right: orientated.

Orientation in two directions is also common and usually called biaxial orientation (BO). BO can occur sequentially or simultaneously.

3.8.1 Machine Direction Orientation

A schematic of a MDO process is shown in Fig. 3.21. There are four main steps in the MDO process.

1. Preheat—raises the film temperature uniformly without putting wrinkles in the film. This increases the polymer mobility and allows greater stretching in the next step without film failure.

2. Drawing—stretches the film (up to 10×) by making it longer, which forces alignment of polymer molecules in the film. In this section, one drawing roll runs faster than the other.

3. Annealing—locks in the properties and controls shrinkage.

4. Cooling.

Machine orientation processes are relatively compact and are often put in-line with film producing lines.

3.8.2 Biaxial Orientation

The next logical step in orientation is to stretch the film in the transverse direction, which is 90° to the MDO direction. This is done by a machine that grabs the edge of the film and pulls it along the width. This is done using what is called a tenter frame. The tenter frame, a schematic of which is shown in Fig. 3.22, is a series of tenter clips that form a chain on both sides of the film being oriented. The tenter clips close onto the film edge at the start. As the film moves in the machine direction, the tenter clips move further apart, stretching the film. Eventually, the clips let go of the film, which is now typically three to eight times wider and the clips wrap around a wheel and return to grab the film again. The tenter frame is typically located in an oven.

Figure 3.23 shows a schematic of DuPont™ LISIM® technology (linear motor simultaneous stretching), which stretches in two axes at once.

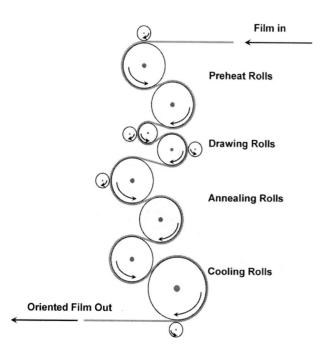

Figure 3.21 Schematic of a MDO process.

Figure 3.22 Schematic of a tenter frame for traverse direction orientation.

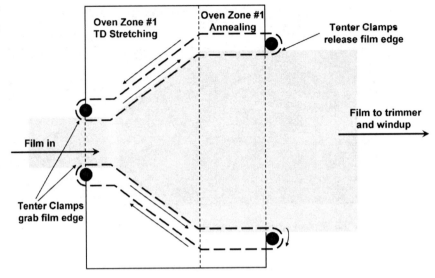

3.8.3 Blown Film Orientation

Blown film may be biaxially orientated using what is called the double bubble process, which is shown schematically in Fig. 3.24. The film is first formed in the die, and then it is inflated by air into a water quench. The films are reheated by radiant heater and then passed into the second bubble, which stretches the film in both directions. This process is commonly used to make heat shrink film and bags.

3.9 Membrane Production

Membranes are usually produced as hollow fibers, flat sheets, or are coated onto flat or tubular supports. Section 4.6.5 discusses how membranes are structured and built into devices. Hollow fiber membranes are common and are produced in a spinning process. The process starts with the polymer in solution as shown in Fig. 3.25. A spinneret casts the polymer solution over another inert sacrificial liquid (called a bore liquid), basically

Figure 3.23 Schematic of an LISIM® simultaneous stretching lines.[12]

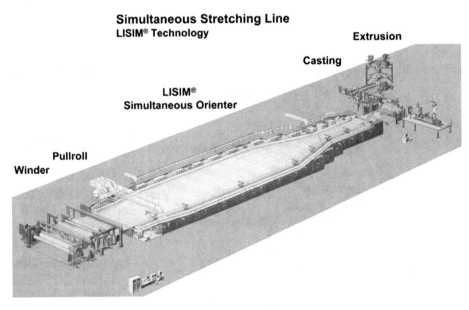

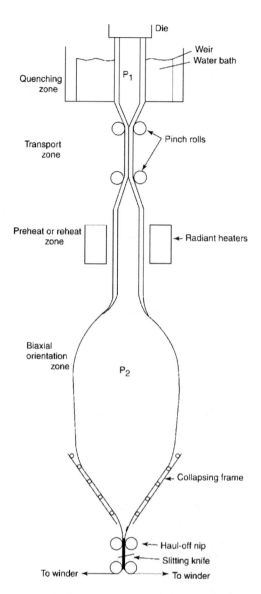

Figure 3.24 Schematic of BO during blown film process.[13]

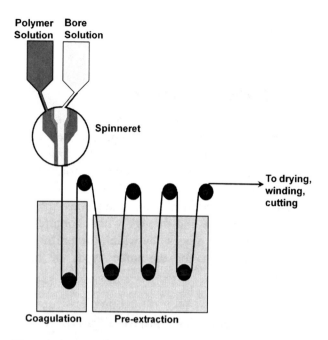

Figure 3.25 Schematic of the fiber spinning process used to make hollow fiber membranes.

Flat membranes are produced by any other of the film-forming processes including extrusion and solvent casting.

3.10 Molding of Containers

Containers are made from some kind of mold. The process by which the material is loaded into the mold

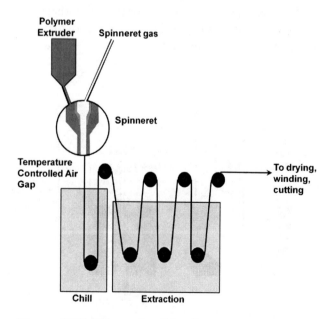

Figure 3.26 Schematic of the fiber extruder process used to make hollow fiber membranes.

forming a coated liquid stream. That stream is immediately put into a solution that coagulates the polymer solution that forms a firm film around the bore liquid. The fiber is processed through other liquids to remove the remainder of the solvent from the polymer solution and the bore liquid. The fibers are then dried, wound up, and cut for assembly later into modules. Polyethersulfone membranes are often made by this process.

For those polymers that cannot be dissolved such as PP, an extruder forms the hollow fiber around a pressurized gas stream as shown in Fig. 3.26. There are also other variations in this process where instead of spinning over gas, spinning can be done over a bar.

may vary considerably and these processes are summarized in the following sections.

3.10.1 Blow Molding

In general, there are three main types of blow molding: extrusion blow molding, injection blow molding (IBM), and stretch blow molding. The blow molding process begins with the forming of a preform (called a parison in the molding industry). The parison is a tube-like piece of plastic with a hole at one end through which compressed air can pass. The process by which the preform is made is different for the three types of blow molding.

3.10.2 Extrusion Blow Molding

The preform is made by extrusion of a tube directly into the open mold as shown in Fig. 3.27. Referring to the figure, the molten plastic is fed from the extruder at 1. The melted plastic enters the extruder head at 2 and 3 around an air tube at 4. The preform is shown at 5 and is a hot tubular shape. The mold 6 closes pinching at one end of the preform cylinder. Most extrusion blow molded parts will have an obvious point on its base where the pinch occurred. Air is blown into the preform at 7. The air pressure expands the parison against the walls of the mold and forces the plastic to conform to the shape of the mold. After the plastic part has cooled, it is removed from the mold at 8.

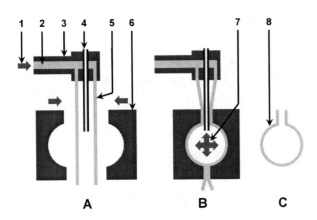

Figure 3.27 The principle of the extrusion blow molding process.

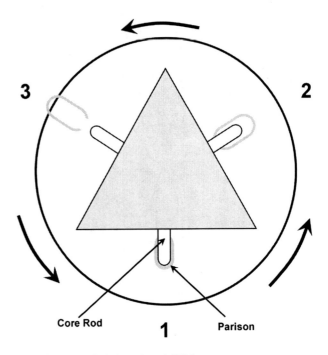

Figure 3.28 Schematic of IBM process.

3.10.3 Injection Blow Molding

IBM is similar to extrusion blow molding, except that the parison is formed by an injection molding process around a core rod or metal shank as shown in position 1 in Fig. 3.28. The preform consists of a fully formed bottle/jar neck with a thick tube of polymer attached, which will form the body of the bottle. The parison is transferred to the core rod into a blow mold, usually by simple rotation as shown at position 2 in Fig. 3.28. At the blow-molding station, the preform is inflated and cooled. When cooled enough, the mold opens at position 3 and the bottle is ejected. There are three sets of core rods, which allow concurrent preform injection, blow molding, and ejection. This is the least used of the three blow molding processes and is typically used to make small medical, cosmetic, and other single serve bottles.

3.10.4 Stretch Blow Molding

Stretch blow molding produces a plastic container from a preform that is stretched in all directions when the preform is blown into a container-shaped mold. The process is shown in Fig. 3.29.

1. The plastic is first molded into a "preform" by injection molding process. When making

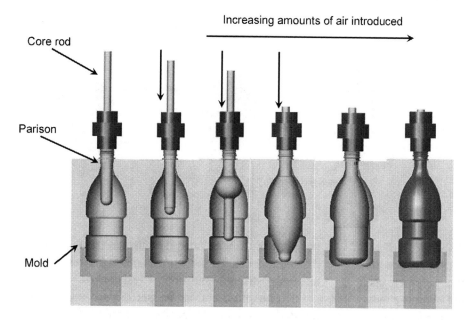

Figure 3.29 Schematic of the stretch blow molding process.[11]

bottle, the preforms are produced with the necks of the bottles, including threads on one end. The preform may also be produced by coextrusion, which will produce multilayer containers.

2. A perform is later fed into a reheat stretch blow molding machine.
3. The preforms are heated often by using infrared heaters above their glass transition temperature.
4. Usually, the preform is stretched with a core rod as shown in the figure.
5. Then high pressure air is blown into the preform, which expands the plastic into bottles using metal molds.

Common plastics for stretch blow molding include various thermoplastic materials such as acrylonitrile (AN), polystyrene (PS), PVC, polyamide (PA), polycarbonate (PC), polysulfone, acetal, polyarylate, PP, Surlyn®, and polyethylene terephthalate (PET).

Advantages of all blow-molding techniques include the following:

- Low tool and die cost
- Fast production rates
- Ability to mold complex part shapes
- Parts are usually recyclable.

Disadvantages of blow molding include the following:

- Limited to hollow parts
- Wall thickness is hard to control.

3.10.5 Rotational Molding/ Rotomolding

Rotational molding is a process that is used to make large and thick-walled containers, such as water cooler jugs, fuel tanks, and other plastic structures. The rotational molding process consists of distinct phases, which are shown in Fig. 3.30:

1. A measured quantity of polymer (usually in powder form) is loaded into the mold.
2. The mold is rotated on 2–3 axes such that the powder cascades onto all surfaces of the mold. While the mold is rotating, it is heated in an oven until all the polymer has melted and adhered to the mold wall. Rotation may be at different speeds, in order to avoid the accumulation of polymer powder in sections of the mold. The length of time the mold spends in the oven is critical. Time and temperature must be balanced to melt all the powder but to avoid thermal degradation.

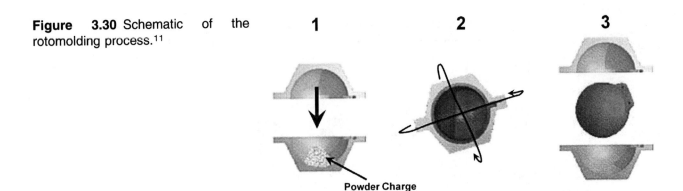

Figure 3.30 Schematic of the rotomolding process.[11]

3. The mold is cooled, usually while it is still rotating, at least until the temperature is well below the melt point of the polymer used. The part shrinks on cooling, coming away from the mold walls, and making easy removal of the part. The cooling rate must be kept within a certain range. Very rapid cooling may result in un-uniform shrinking producing a warped part.

3.11 Fluorination

High-density polyethylene (HDPE) and PP are used in rigid plastic containers. As compared to metal and glass, these plastics have advantages of light weight, low cost, high-stress crack resistance, high drop impact strength, tremendous flexibility in processing and design. They also offer some protection against moisture, solvents, and gases. The primary area where plastics have a drawback against metal or glass containers is in permeation and scalping. Materials that typically should not be packed in HDPE and PP would be chemicals such as fuels, brake fluids, solvents, solvent-based formulations, fuel additives, flavors, and fragrances (and perhaps others).

Fortunately, the permeation resistance of HDPE and PP containers can be dramatically improved by a process called fluorination. Fluorination of plastics is basically a surface modification process, which result in the substitution of some of the hydrogen atoms on the HDPE and PP polymers by fluorine atoms. The fluorination results in change of the surface properties minimizing the permeability of nonpolar solvents through a polymer surface. The bulk properties of fluorine-treated plastic container remain unchanged. Fluorination modifies only those polymer molecules near the surface, so there is no measurable change in the mechanical properties such as tensile strength and impact resistance.

Off-line fluorination is performed on already molded containers, by exposing them to fluorine gas at elevated temperature and pressure and controlled time.

3.11.1 Permeation Testing of Fluorinated Containers

The solvent permeation through containers is generally tested by an Accelerated Keeping Test (AKT) or as per IS: 2798 (Indian Standard, method of test for polyethylene containers), which involves measuring weight loss during high-temperature exposure of the filled container over a period of time. Normally, exposure at 50 °C for 28 days is considered equivalent to 1 year of normal exposure. After 28 days, a comparison of percent weight loss in the treated and untreated containers provides an indication of permeation barrier effectiveness. Table 3.3 lists the permeation results of this test performed for various common chemicals for nonfluorinated vs. fluorinated containers.

The improvement in permeation resistance is substantial.

Chemicals and products that presently use fluorinated bottles include the following:

- Acetone
- Auto additives
- Charcoal lighter
- Degreaser
- Electronics industry chemicals
- Health and beauty care products

Table 3.3 Permeation Data for Fluorinated and Untreated Containers When Filled with Various Organic Solvents[8]

Solvent	Untreated Container % Weight Loss	Fluorinated Container % Weight Loss
Carbon tetrachloride	28.26	0.05
Pentane	98.10	0.21
Hexane	61.29	0.19
Heptane	24.26	0.08
Xylene	42.52	0.21
Isooctane	4.54	0.03
Cyclohexane	22.34	0.15
Toluene	61.90	0.52
Paraxylene	59.20	0.54
1,3,5-Trimethylbenzene	15.85	0.18
Benzene	36.68	3.65
Chlorobenzene	32.05	5.41
1,2-Dichloroethane	11.55	2.89

- Insecticides
- Kerosene
- Lubricants
- Maintenance chemicals
- Paint thinners (turpentine, etc.)
- Plant growth products
- Waxes, cleaners, and polishes (various)
- Weed killers and herbicides
- Wood preservatives.

3.12 Coatings

There are many ways in which coatings may be applied, and discussion of coating is beyond the purpose of this book; there are many references that discuss coating processes thoroughly.[9,10]

3.13 Summary

The major production methods of films, membranes, and containers have been reviewed. The engineering of equipment for all these processes can be complex, and the small changes in plastic/polymer properties can have large effects on production. The next chapter will review how films, membranes, and containers are used.

References

1. Wagner JR. *Multilayer flexible packaging.* Elsevier; 2009. p. 57–72.
2. Cogswell FN. *Polymer melt rheology: A guide for industrial practice.* Woodhead Publishing Limited; 1981.
3. http://www.be-ca.com
4. Dooley J, Tung H. *Encyclopedia of polymer science and technology.* 4th ed. John Wiley & Sons; 2011.
5. Siemann U. Solvent cast technology—a versatile tool for thin film production, Progress. *Colloid Polymer Science* 2005;**130**:1–14.
6. Schweizer PM. *Liquid Film Coating - Scientific principles and their technological implications.* Springer; 1997.
7. Cohen ED. *Modern coating and drying technology.* Wiley-VCH; 1992.
8. Singh B. High Barrier Solutions for Plastic Containers Using Fluorination Process. *Popular Plastics and Packaging* 2007;**52**:84–9.

9. Goldschmidt A. *BASF handbook: basics of coating technology*. Vincentz; 2003.
10. McKeen LW. *Fluorinated coatings and finishes handbook - the definitive user's guide and databook*. William Andrew Publishing/Plastics Design Library; 2006.
11. This is a file from the Wikimedia Commons which is a freely licensed media file repository
12. Wagner JR. *Multilayer flexible packaging*. Elsevier; 2009.
13. Mount EM. *Encylopedia of polymer science and technology*. 4th ed. John Wiley & Sons; 2011.

4 Markets and Applications for Films, Containers, and Membranes

Generally, films are used as barriers, they keep liquid or gases on one side of the film. Membranes are made to allow only certain materials to pass through them but block others. Barrier film applications include packaging, coating, gloves, containers, hoses, and tubing. Membranes are discussed later.

Packaging may be classified into three categories of materials: flexible, semirigid, and sealants or adhesives.

- Flexible materials whose application may be wraps, lidding, pouches, or bags include films of a thickness ≤0.127 mm (5 mils).
- Semirigid materials are thicker than 0.127 mm. They are usually formed as sheets.
- Sealants or adhesives are used to adhere multiple layers together, typically requiring heat and/or pressure.

4.1 Barrier Films in Packaging

Polymeric packaging materials are used to surround a package completely, securing contents from gases and vapors, moisture, and biological effects of the outside environment, while providing a pleasing and often decorative appearance. Water vapor and atmospheric gases if allowed to permeate in or out of a package can alter the taste, color, and nutritional content of the packaged good. The effects of gas and vapors on food are complex and comprise a major branch of food science. Consequentially, the following is a brief overview for introductory purpose.

4.1.1 Water Vapor

Many products need to be protected against the gain or loss of moisture. Materials such as coated cellophane, polyethylene (PE), polypropylene (PP), polyvinylidene chloride, and polyester films are excellent barriers to water vapor and are used to block the transmission of water vapor through film. These materials are often used on the outside (and inside) layers of multilayer films. It should be noted, however, that even the most impermeable of these films has a measurable permeability.

Other products such as fresh vegetables need to breathe so as to avoid condensation of water or the growth of mold. Materials such as polyolefin plastomers and certain grades of cellophane are suited for these applications.

The rate of water vapor transmission will depend upon the vapor pressure gradient across the film. Dry contents in a humid environment would absorb moisture, wet contents in a dry environment would lose moisture, and if the relative humidity inside and outside the package is equal, there will be no transmission even through the most permeable of films.

4.1.2 Atmospheric Gases

Oxygen, carbon dioxide, and nitrogen within a package often must be controlled. If oxygen is allowed into a package, it will break down organic materials initiating or accelerating the decay process. Uncontrolled, this will promote staleness and loss of nutritive value.

In the case of fresh meat, a high rate of oxygen transmission is required to maintain the bright red color of meat. To meet this special requirement, special grades of cellophane, PEs, and nitriles have been developed to provide the low water vapor transmission needed to avoid drying of the meat, while providing high oxygen transmission to maintain the color.

This phenomenon of high transmission for oxygen combined with low transmission of water seems paradoxical but is very critical to these specialized needs. The reverse characteristics apply to nylon and other films that have a relatively high permeability to water vapor but a low permeability to oxygen, nitrogen, and carbon dioxide. Other films have high (or low) transmission rates for all gases, as well as water vapor.

4.1.3 Odors and Flavors

Packaging films are also used to control the permeation of many organic compounds that impart flavor and odor. This protects the package contents from either the absorption of unwanted odors or the loss of volatile flavoring ingredients. Two common flavoring ingredients used in breath fresheners and food flavors are d-limonene, a component in lemon and other citrus flavors, and methyl salicylate, a component in wintergreen, used in breath fresheners and food flavors. Aromas include allyl sulfide (garlic), acetic acid (vinegar), ethyl phenyl acetate (soaps and floral fragrances), P-pinene (household cleaners), ethyl acetate (food flavorings: citrus, berry, coconut, coffee, chocolate, and honey), and menthol (chewing gum and peppermint).

The permeation of flavors and odors is difficult to measure quantitatively because they contain many components. Many times, only a simple component of a flavor is measured if a quantitative value must be determined. Another important flavor consideration is commonly called "flavor scalping." Flavor scalping is the selective absorption of certain flavor constituents from the product. Polyolefins are known flavor scalpers.

Among good barriers to organic vapors are cellophane, saran, and vinyl. Cellulose acetate and PE are poor odor and flavor barriers unless coated with a good barrier material.

4.1.4 Markets and Applications of Barrier Films

Agricultural Chemicals: Fertilizers, insecticides, and herbicides are a few of the chemicals packaged in high-density polyethylene (HDPE) containers. Multilayer structures—up to 7—are used to provide reinforcement as well as additional protection against "pinholing," which often can occur in monolayer structures. Fluorination of the inner surface of the container provides a chemical resistant and tough inner coating. Fluorination is also applied to the outer skin of an HDPE container to provide additional shelf life. Nylon co-extrusions are recommended as a cost-effective alternative to fluorine gas treatment.

Agricultural Films: Advances in polyethylene technology for agricultural films have reduced the need for chemical fertilizers, pesticides, and herbicides. The most common types of agricultural films are greenhouse films, mulch films, and silage films. Greenhouse films are generally made by bubble extrusion of a blend of linear low-density polyethylene (LLDPE) and low-density polyethylene (LDPE), with additives. A low-melt-index LDPE (0.2) provides the best bubble stability, but the least favorable clarity. Adding a high-melt index LDPE (2.0) will give better clarity, but less bubble stability. The choice of LLDPE and LDPE also depends on equipment capabilities.

Mulch films are used to cover land in preparation for planting. They are used to reduce water consumption from evaporation, reduce weed growth, and improve herbicide retention. Silage films help to maintain the nutritional value of forage plants such as corn, vegetables, and grasses that continue to respire after cutting. They help to exclude the air, so that lactic acid fermentation can take place, leaving a feed rich in vitamins and carotene. When silage film is used, the feed can keep its nutrients for several months, depending on the amount of air left (the less air, the better). Thus, feed is available for use in periods when forage is not available in sufficient quantities.

Bag and Intermediate Bulk Containers Liner Film Polypropylene: Large PP-woven bags and intermediate bulk containers (IBC) are widely used to pack all types of materials, from powders to granules to liquids. Often, they are equipped with an inner liner to prevent leakage and/or to protect their contents (e.g., against moisture).

PP-based blown film is finding its way as inner liners for woven bags, wooden big boxes, carton octabins, and more. This is because a PP liner offers exceptional performance at reduced thickness as compared with conventional, polyethylene (PE) solutions. Typically, a PP liner can be up to 30% thinner than its PE counterpart while offering comparable mechanical properties.

Bakery, Convenience Food Items: Oriented polystyrene (OPS) is present in bakery and other food products, which require transparent, resistant, but flexible packaging. The food contact listing of polystyrene (PS) is also an important aspect of development of the material in this segment. A low amount of monomers is an absolute requirement, because of the special nature of the processing and its use in contact with food.

Chemical Products: Household cleaning supplies including liquid and solid laundry and dishwashing detergents and similar products for the industrial workplace are the primary chemical products packaged in HDPE containers. These containers usually do not require further barrier protection.

Compression Film Polypropylene: Film used for packaging of compressed products such as glass, wool, and baby diapers is called compression film. Its most important requirement is resistance to elongation under stress, otherwise known as creep resistance.

Condiments: Squeeze bottles containing condiments, ketchup, and mustard in particular have long been a major application for HDPE blow-molded bottles. The bottles have an inner layer of barrier material, primarily ethylene vinyl alcohol (EVOH), and also include nylon to protect the flavor in a shelf-stable, nonrefrigerated environment.

Consumer Bags and Wraps: Consumer bags and wraps protect products from contamination and damage during shipment. Overwraps must offer clarity, to reveal the visual quality of the product, and they must be printable.

Dairy Form Fill Seal: Extruded polystyrene sheets with a thickness ranging from 0.7 to 1.8 mm are used to package a variety of products on Form Fill Seal (FFS) machines. The FFS forms the container, fills the product, and seals the lids on the container in one processing step. Polystyrene is the choice material for the FFS sheet because it can be broken when twisted. Form fill and seal packaging are more common in Europe but are also used in North America.

Easy-Peel Film: Blending polybutene-1 (PB1) in PE results in an immiscible blend that forms the basis of a peelable seal formulation. An easy-open heat seal base allows for easy access to contents, and enhances the package appearance before and after opening. With this system, the consumer can peel the sealed package surfaces apart with a steady even force. The strength required to start peeling action (initiation peel strength) is similar to that required throughout the peeling process (propagation peel strength).

PB1 can be blended with either PE homopolymer or copolymers to form an easy-open system. HDPE, LDPE, and LLDPE can be components of the system. Additionally, the new metallocene polyethylenes (mPE) may also be used. These mPE systems offer enhanced performance (improved hot tack, better seal through contaminants, and low odor) and processing flexibility. The most commonly used PE copolymer is ethylene vinyl acetate (EVA); however, ethylene acrylic acid (EAA), ethylene methyl acrylate (EMA), and ethylene ethyl acrylate (EEA) may also be used. Minor amounts of one-third polyolefin may also be added to modify the performance properties. Processing aids such as slip and antiblock additives may be used as needed.

Extrusion Sheet for Consumer Packaging Polystyrene: The processing units are producing sheets with a width ranging from 200 to 850 mm and with a thickness ranging typically from 0.5 to 2 mm. This kind of production is made on large capacity extrusion lines (>1 T/h) in order to decrease variable costs. Such lines require a careful setup in order to adjust winding correctly, stress level, orientation, gloss, etc. If some barrier properties are required, a co-extrusion material must be adopted.

Extrusion Thermoforming for Disposables Polystyrene: The disposables market in polystyrene is mostly processed through the in-line thermoforming technology, because the latter is used in the case of huge production size, which is mostly the case of this market. This in-line process consists of extruding a sheet of polystyrene through a flat die and then some variations take place concerning technology:

- Either the sheet is pulled by a mini calender in order not to cool it down too much, and then directly fed to a forming station with off-mold cutting.

- Or the sheet is calendered, sometimes the edges are cut, and then this same sheet is fed to a standard thermoforming unit with heaters.

The first process is widely used to produce cups and the second is more widespread for lids and plates.

The in-line extrusion thermoforming process is the only solution to produce objects made out of pure crystal polystyrene, as this product cannot be wound up in a roll.

Fabric Film Laminates: The absorbent products sector, including disposable baby diapers, feminine hygiene products, and adult incontinence materials, along with the medical laminates segment are very important parts of the nonwovens industry. Films used for the diaper backsheet have evolved from monolayer PE film to blends of PE and PP, to multilayer films. In the 1990s, breathable films were adopted as backsheet materials, allowing higher water vapor to pass through the film.

Foam Extrusion Thermoforming for Consumer Packaging Polystyrene: Two main technologies are used today in order to produce thin foamed sheets. This sheet can be either immediately thermoformed (molten-phase thermoforming) or stored in huge rolls for some days and fed into a thermoforming unit specially designed to heat-foamed sheets. In the second case, there is a post expansion phenomenon, due to the fact that the sheet is reheated in the oven, causing gas trapped inside polystyrene to expand. The gases that are used are usually explosive like butane or pentane, which need special storing and handling solutions.

Food Wrap Film: A coextruded film with tacky skin layers and a PP core layer offers all the desired properties for this application. Food wrap film needs good puncture resistance as well as good elastic recovery. A coextruded film based on polyolefin resin has excellent puncture resistance and good elastic recovery.

A food wrap film also requires a good degree of oxygen and water vapor permeability. The coextruded solution combines a good oxygen and water vapor transmission rate, which most likely means a longer shelf life for products such as fresh meat.

Geomembranes: Geomembranes are sheet-like structures, which are commonly used in environmental and water protection applications. These membranes are used to prevent the release of gas or odors into buildings or into the environment and also help to protect groundwater against spoilage with contaminated water. Geomembranes are essential in waterproofing applications, helping to protect new construction against corrosion or water erosion. They are also used in containment, collection, and conveyance of drinking water, helping to prevent water loss.

Heat Resistant Film: Autoclavable Biohazard Waste Disposal Bags and Auto Paint Masking Films. Films manufactured into bags for autoclavable biohazard waste disposal are part of the overall industrial trash can liner market used to collect and dispose all types of waste. Autoclavable bags are used for infectious and regulated waste, also known as biohazard waste. Biohazard waste products are primarily generated by the medical industry (hospitals, nursing homes, clinics, doctors' offices, medical laboratories, etc.).

PP-based polymers are used to manufacture masking films and bags that are used to cover various sections of an automobile during an automotive body repainting process. These masking films act as a paint and temperature shield during a process that requires the painted part to pass through a series of baking ovens.

Heavy-Duty Shipping Sacks: Heavy-duty shipping sack producers manufacture a broad range of custom bags and films that provide moisture and barrier protection, reinforced strength properties, uniform gage control, and impact resistance. As a result, the resins used to produce the bags and films must be very robust, providing outstanding performance over a broad range of conditions.

Heat-Shrinkable Polypropylene Film: Thin, bio-riented PP shrink films produced using the Double-Bubble technology are widely used as display films for books, videos, toys, sweets, fruits, etc., as monolayer and coextruded film structures, and skin and core layers.

Heat Seal Resins: Heat seal resins are key to the production of heat sealable coextruded films. They are used extensively in the bi-axially oriented polypropylene (BOPP) market for sealable films, lacquered films, and metallized films. These resins have been designed to form a seal at a specific temperature, called the seal initiation temperature (SIT).

High-Clarity LLDPE Film: High-clarity PE films are often used to package products for retail sale, where the clarity and gloss of the film provide a better display and presentation for the enclosed product. These high-clarity PE films are predominantly produced from either specialty clarity LDPE grades, or metallocene LLDPEs.

High-Performance Food and Specialty Films: High-performance films are typically used in food and specialty packaging. Most are composed of a multilayer structure and include barrier materials such as EVOH copolymer, nylon, foil, HDPE, or oriented polypropylene (OPP). All PE structures may include combinations such as ULDPE/LLDPE/ULDPE used in liquid packaging. PE resins are typically incorporated into the sealant portion of the structure, or used as a toughness layer buried in the structure.

High-performance films are often used to package fresh produce, meat and cheese, liquids, dry foods, and frozen foods. Most of these are packaged utilizing form/fill/seal equipment. Other applications include pre-made pouches for bag-in-box applications, clarity films for bread bags, and lamination films. High-performance films are also used in medical packaging.

High-performance films are fabricated via a variety of processes, including blown and cast film co-extrusion, or adhesive and/or extrusion lamination of monolayer and coextruded films to barrier substrates.

Industrial Films: When used in industrial films, PE resins can provide desirable moisture barrier properties, tear strength, and puncture resistance. Industrial films made from PE are used in industrial sheet, wrap, and tubing, as well as fabricated industrial liners for shipping containers, steel drums, boxes, cans, tote bins, and truck beds.

The vast majority of industrial films are between 1 and 6 mils. Because industrial films are custom-made, each product has its own specifications.

LLDPE/LDPE blends dominate industrial films. LDPE resins with a fractional melt index are the most commonly used resins in industrial films, providing the bubble stability required to make thick films. A commodity-grade LLDPE resin is added to provide the proper bag toughness.

Injection Molding for Medical: Injection molding allows for production of high-tolerance parts in short cycle times. Small objects can be produced with very precise geometry to meet the needs of automatic diagnostic testing. Multicavity molds are used on high-speed injection molding machines to produce these Petri dishes and assay trays. Hot runners are used to ensure precise molding and eliminate scrap. Robots are used because of very short cycle time and the need for hygiene. White rooms are often needed to achieve the required cleanliness.

Gable-top paperboard cartons use EVOH almost exclusively for the barrier required for the packaging of fruit juices.

Modified Atmosphere Packaging: Modified atmosphere packaging (MAP) of fresh cut produce is one of the fastest growing food packaging segments. MAP film controls transmission of oxygen, carbon dioxide, and water vapor.

Medical Packaging: Most flexible packages for medical devices contain at least one part that is plastic film. This film provides a number of functions in a pouch: product visibility, puncture resistance, sealability, and peelability.

The most common material used in device packaging is a lamination of polyester and PE, typically 0.0127 mm oriented polyester film, adhesively laminated to low- to medium-density PE (0.038–0.051 mm) usually modified with EVA for better sealability. These films may be sealed to plain or coated DuPont Tyvek®, plain or coated papers, or other films.

A primary requirement of any sterile packaging material is that it provides a bacterial barrier. That is, any film or nonfibrous material must be pinhole free, and fibrous or porous material must have pores below a specified size to prevent passage of microorganisms through the material.

A barrier to gases or liquids is required for packages containing liquid or volatile substances, or materials that need protection from the environment. The degree of barrier required depends upon shelf life requirements and conditions to which the package will be subjected.

Porosity is required for packages that are used in sterilization or autoclaving to allow the sterilizing gases to enter and leave the package easily. Porosity is not essential for radiation sterilization.

Pharmaceutical blister packs are another plastics application in the medical marketplace. Two of these blister packs are shown in Fig. 4.1. The use of thermoformed blisters for the packaging of pharmaceutical products is a rapidly growing area, displacing traditional packaging media such as glass or plastic bottles. The ICH (International Conference on Harmonisation of Technical Requirements for Registration of Pharmaceuticals for Human Use) extended stability testing requirements for the pharmaceutical industry. Testing after a 6-month storage at ambient conditions, 40 °C and 75% RH, is required at a minimum, and for full validation,

Figure 4.1 Examples of pharmaceutical blister packs.

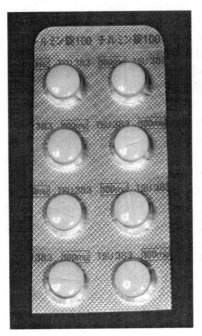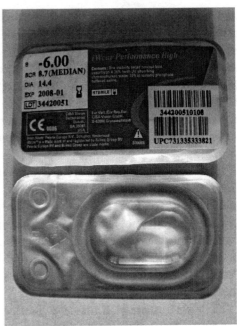

3 years of testing at ambient conditions is required. The chemical stability of the contents can be very sensitive to moisture, thus it is important that moisture penetration be as low as possible for several years.

PVC has traditionally been used in this market for products requiring little protection from moisture vapor, and PVC/polyvinylidiene chloride (PVDC) or PVC/polychlorotrifluoroethylene (PCTFE) for products requiring medium or high barrier. These products have been selected for their moisture vapor barrier, on the flat sheet prior to thermoforming. Cyclic olefin copolymer (COC), a new potential packaging material, has been developed for this market.

NonFusion Shrink Film: PE shrink film is often used for the covering of stacked pallets, particularly when pallet stability is crucial. Frequently, however, the heat used to shrink the film causes the hood to stick to the contents encased within. When the hood is removed, the packaging can be damaged and contents may spill. To prevent this problem, a thin nonfusion layer is often coextruded to the inside of the shrink film.

PP film is an alternative. PP sealing resins are modified to ensure that the film sticks to itself (at low SIT) but does not adhere to the PE film (bags, labels, etc.) due to the presence of the PP backbone.

To insure that no delamination occurs between the PP (nonfusion) and the PE (shrinkage) layer, one inserts a tie layer that contains a blend of the super-soft PP grade and a linear low-density PE.

OPS for Consumer Packaging: Bioriented polystyrene is used mainly in the consumer packaging area, bakery, and other food products that require transparent, resistant but flexible packaging. Sheets are produced with an orientation ratio ranging from 2×2 to 3×3. Most often, these sheets (colored or natural) are thermoformed by another processor. The sheets can be colored by way of masterbatch, and the formulation includes mainly a GPPS of high molecular weight, mixed with a small amount of elastomer, sometimes blended with an even smaller amount of HIPS in order to improve toughness and not to decrease clarity.

Shrink Bundling Film: The shrink bundling film market is dominated by LDPE films, typically blends of fractional melt index LDPE and LLDPEs. The most common blends run in the 25–50% LLDPE range. Adding more than 50% LLDPE typically results in a dramatic reduction in transverse direction (TD) shrinkage, to the point where shrink performance becomes unacceptable. EVA/LLDPE/EVA three-layered film is also used.

Stationery Films: Stationery films include applications such as photo albums, sheet protectors, book covers, and binders. In recent years, cast PP films have made significant inroads into this application, in particular, sheet protectors that had been dominated by PVC.

Stretch/Industrial Collation: Stretch film is an effective and inexpensive solution for protecting palletized products through storage and distribution. There are several constructions including three layers of PE (LDPE/LLDPE/LDPE) and EVA/LLDPE/EVA.

Trash bags: Trash bags are often three layers of PE, LDPE/LLDPE/LDPE.

4.1.5 Some Illustrated Applications of Multilayered Films

Illustrations in Fig. 4.2 show the multilayered film structures utilizing materials made by DuPont.

4.2 Containers

Beer Packaging: Beer packaging in polyethylene terephthalate (PET) has generated significant interest. The single most important limiting factor is the cost of the PET bottle with the appropriate barrier properties to prevent flavor scalping and to provide a 3- to 4-month shelf life for oxygen and carbon dioxide barrier.

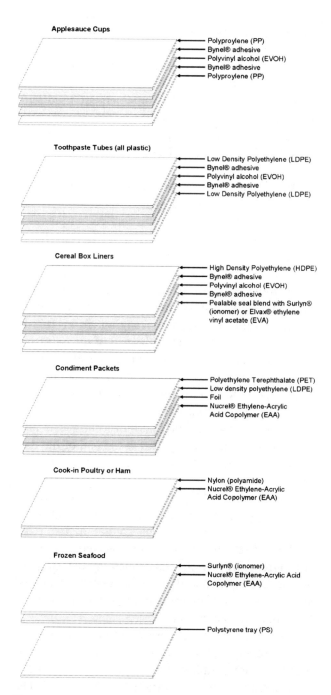

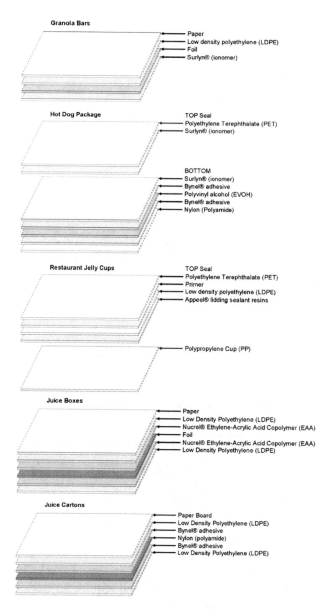

Figure 4.2 The multilayered film structures utilizing materials made by DuPont.

Figure 4.2 (continued)

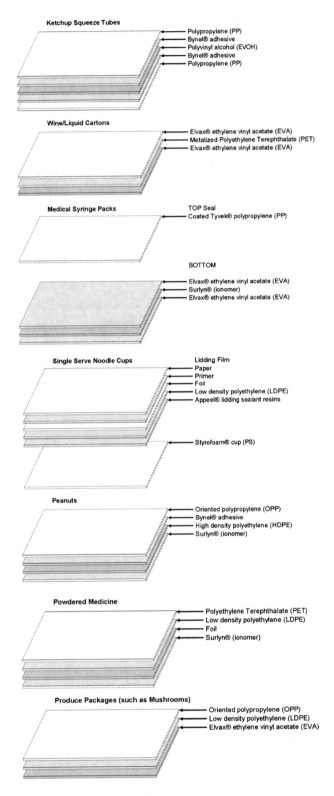

Figure 4.2 (continued)

Sugar Drinks: Single-serve sugar-based drinks in HDPE bottles are shelf stable and require no refrigeration. No barrier materials are used in the sugar-based drink market.

PET containers: For carbonated soft drinks, the major application of PET, as well as edible oils, peanut butter, juices, and isotonic sports drinks, no barrier materials are used. With some other food products, condiments, salad dressings, and dessert toppings, PET does not provide adequate oxygen barrier for a shelf-stable life. For these applications, multilayer structures with an EVOH barrier layer have been developed.

Milk Packaging: The largest food application for HDPE containers is milk packaging produced from homopolymer in a hazy color. Milk jugs are valued for their ability to be recycled. No barrier protection is required in this application.

Juice Packaging: HDPE containers are used with and without barrier layer. When nylon is used as the typical barrier layer, approximately 0.001 in. is coextruded with the HDPE outer wall of the container.

Dairy Containers, Food and Non-food Items, Drop Fill Seal: Thermoformed PS containers are present in many forms: yogurt and cream pots, ice cream containers, sour cream and cottage cheese containers, and coffee creamer portions. Thermoformed PS items are generally made by blending polystyrene to obtain flexible and ductile products. This two-step process involves the production of a sheet with a thickness ranging from 0.3 to 2.5 mm. This means that at the end of the extrusion line, a winding station will produce a roll. This roll will then be fed to a thermoforming station. In some cases, the sheet is made in one factory and the container is thermoformed in another. In other cases, the extrusion and thermoforming can take place in the same factory. But at the end of the day, preformed pots, ready to fill and seal, are delivered for filling.

Egg Cartons, Meat/Poultry/Veggie/Fruit Trays: Light but strong; the expanded foam PS trays are a mainstay of fresh food packaging in the retail market. Meat and poultry are presented on foam polystyrene trays either packaged in the store or prepackaged in a central location. Fresh products and vegetables are sold on trays and eggs are displayed in foam polystyrene cartons.

Foam sheet is made with crystal polystyrene on large specially equipped extrusion lines. The sheet is then thermoformed into the various trays and cartons to package all sizes and varieties of fresh food. The material choice depends on the required thickness, density, and rigidity.

Caps and Closures: A closure is an access-and-seal device, which attaches to glass, plastic, and metal containers. These include tubes, vials, bottles, cans, jars, tumblers, jugs, pails, and drums. The closure works in conjunction with the container fulfill two primary functions: to provide protection containment through a positive seal and to provide access and resealability according to varying requirements.

Molded plastic closures are divided into two groups, thermosets and thermoplastics. Thermoset materials cannot be recycled once they are molded. Thermoplastic materials can be softened or recycled by heat.

Drug Encapsulation: An application that is often overlooked is the use of degradable for drugs such as capsules or pill coatings. These are interesting because the capsule or pill coating must first protect the drug contents from the environment, particularly humidity and oxygen, but then dissolve or degrade quickly to release the drug once the pill or capsule is taken. Historically, these capsule materials were based on gelatin compositions. Gelatin is a protein produced by partial hydrolysis of collagen obtained from natural sources. However, better materials have been developed for many applications. Polyanhydrides are one of these polymers used.

4.3 Automotive Fuel Tanks and Hoses

Regulations involving automotive hydrocarbon emissions have become more stringent. The Sealed House for Evaporation (SHED) test sets as a target of 2 g of hydrocarbon emissions from the whole car during a 24-h period. Components of the fuel circuit are the greatest source of emissions, and the fuel circuit consists primarily of polymeric components. The use of oxygen-containing fuels and blends of fuels make the situation more complex since many blends can be more aggressive than unleaded fuel alone on polymers. Methanol content of the fuel influences permeability of the polymeric components since methanol contributes to swelling of many polymers.

Nylon 12, polyamide (PA12), has been used for most fuel-line systems but nylon 12 will not meet the SHED requirements. Thus, multilayer tubes are often the solution to mechanical and barrier needs. Multilayer tubes generally consist of a primary inner barrier layer to decrease diffusion of the fuel with an outer layer of nylon 12 to provide mechanical properties such as toughness, flexibility, and impact strength. These layers are often joined with a "tie" layer adhesive to prevent delamination.

Plastics are also used as containers for fuels, primarily blow-molded containers. The development of new materials and the improved permeability of existing materials are the focuses of today's automotive industry and those manufacturers and researchers who support the industry. The following data are an overview of the permeability of several materials to automotive fuels designed to show trends between polymeric families. This chapter is not designed to be a comprehensive resource for polymers in automotive applications, rather a supporting chapter to present general permeability trends of polymers used in automotive fuel applications. Society of Automotive Engineers (SAE) and the manufacturers of the materials continue to remain the best source for specific needs.

HDPE Fuel Tanks: HDPE with a high molecular mass has been widely accepted as a material for fuel tanks. It permits substantial rationalization on automotive production lines because of the great scope it allows in styling, the savings in weight that it achieves over its steel counterparts, the ease with which it can be produced by extrusion blow molding, and assembled in the vehicle. As compared to steel, PE is not completely impermeable to gasoline, but it does not rust.

The permeability to gasoline can be reduced by more than 90% by fluorinating or sulfonating the fuel tanks. Since the thickness of the impermeable fluorinated or sulfonated layer is of the order of only a few micrometers, the fuel tanks retain their high level of mechanical properties. Permeability may also be reduced by using multilayer co-extrusion with EVOH copolymer (trade name EVAL™). The barrier properties of EVAL™ are 4400 times better than those of HDPE. Adding one EVAL™ layer in a multilayer fuel tank creates a very significant reduction in polluting hydrocarbon emissions.

Another way to reduce PE fuel tank emissions is through the use of amorphous nylon (Selar® RB, DuPont™). The patented "Laminar Barrier Technology" works by adding Selar® RB to the PE in-line at the blow-molding machine. With controlled mixing of the PE and Selar® RB blend, the fuel tank is produced with many large discontinuous and overlapping barrier platelets within the PE structure. This technology can reduce up to 98% of fuel permeation that normally occurs with PE.

There are number of reference fuels used in fuel permeation testing. Reference fuels are used to establish the quality or performance characteristics of various fuels used in commercial and industrial applications. Reference fuels are not generally used as fuels in transportation vehicles. The compositions of these fuels are summarized in Appendix B.

4.4 Coatings

Coatings are used for both decorative and functional purposes, and frequently both at the same time. The most common functional purpose of a coating is to provide corrosion protection. The chemistry of corrosion is discussed in Section 1.4, which also points out that oxygen and water permeation of the coating are critical. When coatings are used for other purposes, permeation to other materials might also be important. However, this book is focused mostly on films and membranes, so that there will be no more detail on coatings beyond what has already been discussed.

4.5 Gloves

Gloves are an important application for elastomeric/plastic materials. There are many kinds of chemically resistant gloves in the market. However, these gloves may often provide a false sense of security. None of the glove can provide protection against all chemicals, so selecting the appropriate glove and knowing its limitations are critically important.

The usefulness of a glove to protect against chemical exposure is based on the following three factors:

1. *Breakthrough time*: The time it takes for the chemical of interest to travel through the glove material. This is only recorded at the detectable level on the inner surface of the glove and is typically reported in minutes.
2. *Degradation*: This is the physical change that happens to the glove as it is affected by chemical exposure, including, but not limited to swelling, shrinking, hardening, and cracking of the glove material.
3. *Permeation rate*: The rate at which the chemical material passes through the glove, typically reported in micrograms per square centimeter per minute.

Glove selection becomes a complex problem because each of these properties must be evaluated for each chemical. Glove materials are also rated whether light, intermediate, or heavy direct contact with chemicals is anticipated. Glove thickness plays a role in determining breakthrough time. Gloves may also be disposable or reusable. Reusable gloves are generally thicker, offering more protection, than disposable. However, dexterity often lessens with increased protection level, and reusable gloves are more expensive than disposable. Manufacturers typically provide degradation/permeation/breakthrough time charts that list the performance characteristics of a glove material to a given chemical. When comparing manufacturers, the permeation rate and breakthrough time correlate reasonably well as they are usually determined by the test standard, ASTM F739-07 Standard Test Method for Permeation of Liquids and Gases through Protective Clothing Materials under Conditions of Continuous Contact. The glove test cell for permeation is shown in Fig. 4.3. Degradation tests vary with each manufacturer and it is very important to consult each manufacturer's glove chart for their rating system criteria. Appendix C contains a *Permeation/Degradation Resistance Guide for Ansell Gloves*, 8th

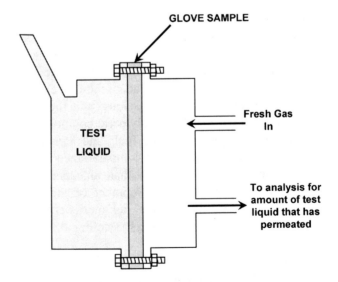

Figure 4.3 Schematic of the permeation cell used for testing gloves.

Edition. Ansell also has a useful online tool called Specware® found at http://ansellpro.com/specware/index.asp with which up to five chemicals may be selected and the best overall glove is suggested.

4.6 Membranes

Following are four commercial membrane processes in which diffusion, permeation, and solubility are important.

- Dialysis
- Reverse osmosis
- Pervaporation
- Gas separation

Each of these is summarized in the following sections. Other industrial membranes processes are discussed in the literature.[1]

4.6.1 Dialysis

Dialysis is a medical process that is primarily used to provide an artificial replacement for lost kidney function in people with renal failure. Dialysis basically cleans the blood of accumulated materials and excess water, which healthy kidneys would usually remove. Dialysis relies on the diffusion of solutes and ultrafiltration of fluid across semipermeable membranes. Blood is pumped on one side of a semipermeable membrane and a dialysate (special dialysis fluid) flows by the opposite side. A semipermeable membrane is a critical component of the dialysis equipment. The process and equipment are shown in Fig. 4.4. This is called *hemodialysis*. The dialyzer removes wastes and water by circulating blood outside the body. The blood flows in one direction and the dialysate flows in the opposite direction, the counter-current flow of the blood and dialysate maximizes the concentration gradient of solutes between the blood and dialysate. The increased gradient helps to remove more potassium, phosphorus, urea, and creatinine from the blood. The dialysis solution has minerals like potassium and calcium whose levels are similar to their natural concentration in healthy blood, so that these materials are not completely removed from the blood stream.

Polymers used to produce membranes for dialysis include

- Cellulose/modified cellulose
- Cellulose acetate
- Polycarbonate

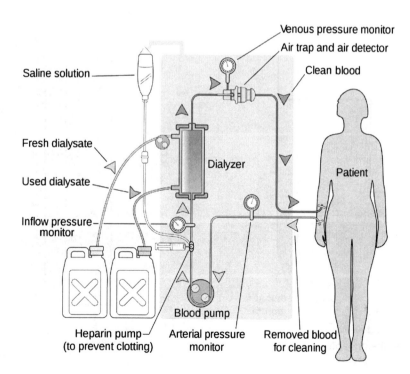

Figure 4.4 Schematic of the dialysis equipment and process.[3]

- Polyacrylonitrile
- Polyamide
- Polysulfone
- Polyethersulfone

4.6.2 Reverse Osmosis

Consider a solvent in a U-tube separated by a membrane as shown in Fig. 4.5. On one side of the membrane is the pure solvent, and on the opposite side is the same solvent, but with a solute dissolved in it. The levels are equal at the beginning. An example is water as the solvent and sugar being the solute. The solvent naturally moves from the area of low solute concentration through the membrane to the area of high solute concentration. The membrane allows the solvent to pass but not the solute. The movement of a pure solvent to equalize solute concentrations on each side of a membrane generates a pressure, h, which is called the "osmotic pressure."

If the experiment is changed such that the U-tube is filled only on one side, right side, with solvent and solute, such as water and salt and a pressure greater than the osmotic pressure is applied to the solvent/solute mixture, pure solvent will pass through the membrane and gets collected on the left side of the U-tube in the diagram. This is called *reverse osmosis* and is a way to make pure solvent. Reverse osmosis is a filtration method that is most commonly known for its use in drinking water purification from seawater. But it is also used in water and wastewater purification as an economical operation for concentrating food liquids (such as fruit juices), concentration of milk, production of wine, commercial car washes for water recycling, maple syrup production, and aquariums. Membranes are often multiple layers with polyamide being common.

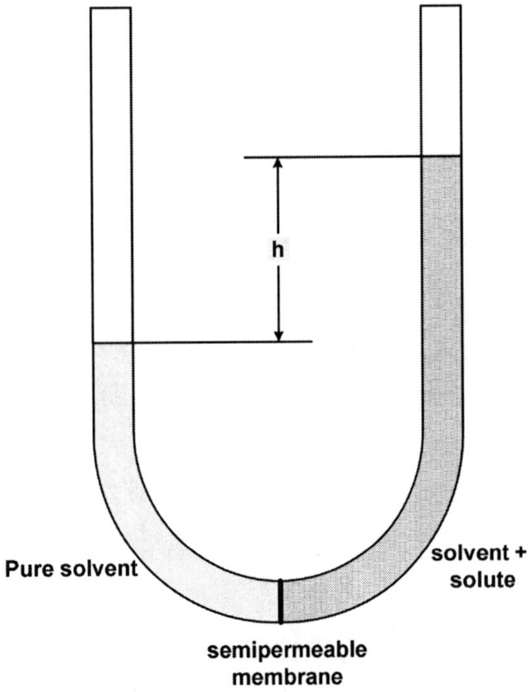

4.6.3 Pervaporation

Pervaporation is a membrane-based process that involves a liquid mixture from which a separation of two or more components takes place across a membrane into a vacuum. The name of this process is derived from the two basic steps of the process, first the permeation through the membrane by the permeate and then its evaporation into the vapor phase. A concentration and vapor pressure gradient is used to allow one component to preferentially permeate across the membrane. The process is schematically shown in Fig. 4.6.

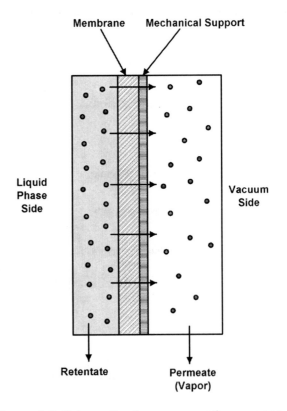

Figure 4.5 Demonstration of osmotic pressure.

Figure 4.6 Schematic of a pervaporation process.

Usually, liquid phase side of the membrane is at ambient pressure and the downstream side is under vacuum to allow the evaporation of the selective component after permeation through the membrane. The retentate is the remainder of the feed leaving the membrane feed chamber, which is not permeated through the membrane.

Pervaporation is typically suited to separate a minor component of a liquid mixture, thus high selectivity through the membrane is essential. This process is used by many industries for several different processes, including purification and analysis, due to its simplicity and in-line nature.

Pervaporation takes place under very mild conditions and is quite effective for separation of mixtures that cannot survive the harsh conditions of distillation.

- Dehydration of ethanol/water and isopropanol/water azeotropes
- Continuous ethanol removal from yeast fermentors
- Continuous water removal from condensation reactions such as esterifications that allow enhanced conversion and rate of the reaction
- Membrane introduction mass spectrometry
- Removing organic solvents from industrial waste waters
- Concentration of hydrophobic flavor compounds in aqueous solutions

4.6.4 Gas Separation

Because permeation through membranes is dependent on the chemical identity of the gas, mixtures of gases have components that permeate at different rates. This means that gas mixtures may be separated to some degree by taking advantage of the different permeation rates through membranes.

A critical parameter in determining the feasibility of using a membrane for a gas separation is the *permselectivity* of the membrane to the gases of interest. The permselectivity is defined as the ratio of the permeation coefficients of each gas pair as shown in Eqn (1).

$$\alpha_{A/B} = \frac{P_A}{P_B} \quad (1)$$

It is important to keep in mind that the permeability coefficients and permselectivity are functions of temperature, pressure, and other factors.

Membranes for gas separations are very important applications and are commercially used for the following:

- Separation of hydrogen from gases like nitrogen and methane
- Recovery of hydrogen from product streams of ammonia plants
- Recovery of hydrogen in oil refinery processes
- Separation of methane from biogas
- Enrichment of air by oxygen for medical or metallurgical purposes
- Enrichment of ullage (the unfilled space or headspace in a container of liquid) by nitrogen in inerting systems designed to prevent fuel tank explosions
- Removal of water vapor from natural gas
- Removal of carbon dioxide from natural gas
- Removal of hydrogen sulfide from natural gas
- Removal of volatile organic liquids (VOL) from air of exhaust streams
- Desiccation

Membrane separations have advantages over other separation technologies, which include

- Low energy use (no phase change such as in distillations/condensations)
- Reliability, there are no moving parts
- Small footprint

Membrane separations have drawbacks that include

- Incomplete separation (membranes need higher permselectivity)
- Membranes may have chemical and thermal stability limits (need more resistant materials)

Industrial scale membrane separations often require hundreds to thousands of square meters of membranes to be practical. The configurations of the membranes used in separation devices vary considerably.

4.6.5 Membrane Structures

Several commercial membrane structures are used and they are summarized in the following sections. Membranes are often very thin and are almost always supported and shown in Fig. 4.7. The support can be rigid or flexible. The membrane structures can be configured in many ways and some of these are discussed in the following sections.

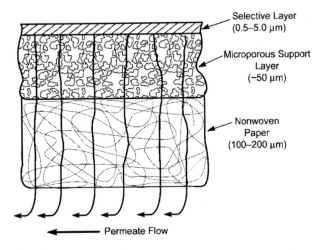

Figure 4.7 Drawing of a supported membrane structure.[4]

4.6.5.1 Plate and Frame Modules

One of the earliest and simplest modules is the plate and frame configuration. An early example is the separation from helium from natural gas using a fluorinated ethylene propylene (FEP) membrane as shown in Fig. 4.8.[2] The membrane is on both sides on a plate in the form of a heavy mesh screen. A large number of plates are stacked, but spaced apart from each other, so that they do not touch except at the permeate outlets. The stack is put inside a chamber that contains flowing high pressure (1000 psi) natural gas. The helium is the permeate that is removed from the natural gas and it is collected. Plate and frame modules are now generally used only in electrodialysis and pervaporation systems and in a limited number of reverse osmosis and ultrafiltration applications with high fouling conditions.

4.6.5.2 Hollow Fiber Modules

Hollow fiber is one of the most popular membranes used in industry. The hollow fiber usually has a membrane coated on the outside of a porous fiber support as shown in Fig. 4.9. Hollow fiber modules are characteristically 4–8 in. (10–20 cm) in diameter and 3–5 ft (1.0–1.6 m) long. The membrane is often applied to the outside fiber by a dip process in

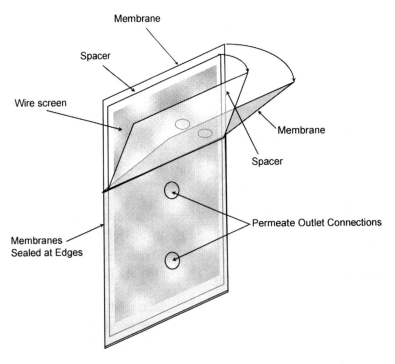

Figure 4.8 An example of a plate and frame membrane gas separation configuration.

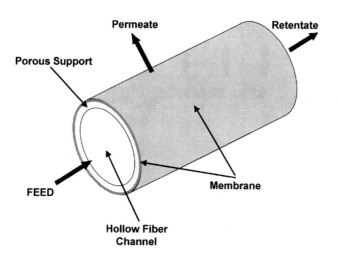

Figure 4.9 Schematic of a hollow fiber used in gas separation.

which the polymer membrane is dissolved in a solvent. Although Fig. 4.6 shows the feed-stream on the inside of the fiber, it is more commonly on the outside of the fiber, so that the membrane is pressurized against the porous support. These fibers are bundled together by cementing the fibers together at the ends as shown in Fig. 4.10.

Hollow fibers gas separators have several extra features, beyond those inherent to all membrane systems, which make them especially interesting to industry. Most important is that they offer a large membrane surface per module volume. It is also a flexible system in that it can carry out the filtration by two ways, either "inside-out" or "outside-in".

It has a few disadvantages. Membrane fouling of hollow fiber is more frequent than other membrane configurations. Contaminated feed will increase the rate of membrane fouling, especially for hollow fiber. The hollow fiber system is more expensive than other membrane systems available in market because of its fabrication method. Both polymer hollow fiber and the porous support may be affected by high temperature conditions and corrosive gases during use.

Hollow fiber modules are commonly used in gas separations and pervaporation processes.

4.6.5.3 Tubular Membrane Modules

There are several ways to construct membranes in a tubular shape. If the membrane is on the inner surface of a cylinder, it is called a tubular module and it is schematically shown in Fig. 4.11. This configuration is primarily used for ultrafiltration. Besides their rugged construction, tubular membranes have the advantage of being able to process high suspended solids without plugging.

Tubular membranes are suited to:

- Metalworking oily waste
- Wastewater minimization and recovery from industrial processes
- Juice clarification
- Degreaser recovery
- Pulp and paper water treatment and recovery
- Paint, pigment, dye recovery, and purification
- Dialysis procedures
- Mining operations wastewater treatment
- Food processing
- Cosmetics manufacturing wastewater treatment
- Pharmaceuticals manufacturing

4.6.5.4 Spiral Wound Modules

Industrial scale spiral wound modules contain several membrane envelopes as shown in Fig. 4.12, each with an area of 1 to 2 m^2, wrapped around the central collection pipe. Multienvelope design minimizes the pressure drop encountered by the permeate traveling toward the central pipe. The standard industrial spiral wound module is 8 in. in diameter and 40 in. long.

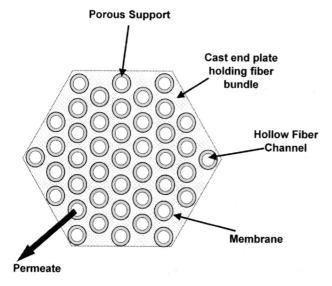

Figure 4.10 Schematic of a hollow fiber bundle used in gas separation.

Figure 4.11 Tubular membrane configuration.

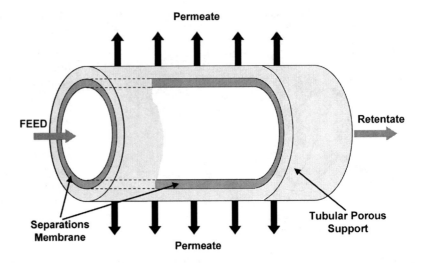

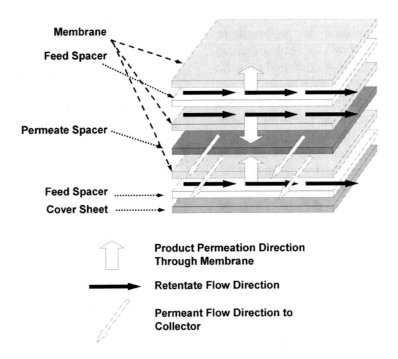

Figure 4.12 Material flow in a spiral wound membrane configuration.

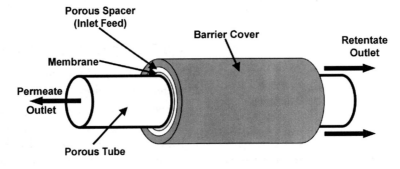

Figure 4.13 Spiral wound membrane configuration.

The module is placed inside a tubular pressure vessel. The feed solution passes across the membrane surface and a portion of the feed permeates into the membrane envelope where it spirals toward the center and exits through the collection tube as shown in Fig. 4.13.

Four to six spiral wound membrane modules are normally connected in series inside a single pressure vessel. A typical 8-in.-diameter tube containing six modules has 100–200 m^2 of membrane area.

Membrane applications are sure to find more applications in the future.

References

1. Keith Scott. *Handbook of industrial membranes.* Elsevier; 1998.
2. Stern S, Sinclair T, Gareis P, Vahldieck NP, Mohr PH. Helium recovery by permeation. *Ind Eng Chem* 1965;**57**:49–60.
3. Yassine Mrabet. This is a freely licensed media file from the Wikimedia Commons.
4. Baker RW. *Advanced membrane technology and applications: vapor and gas separation by membranes.* John Wiley & Sons; 2008.

5 Styrenic Plastics

This chapter on styrenic plastics covers a broad category of polymeric materials of which styrene is an important part. Styrene, also known as vinyl benzene, is an organic compound with the chemical formula $C_6H_5CH=CH_2$. Its structure is shown in Fig. 5.1.

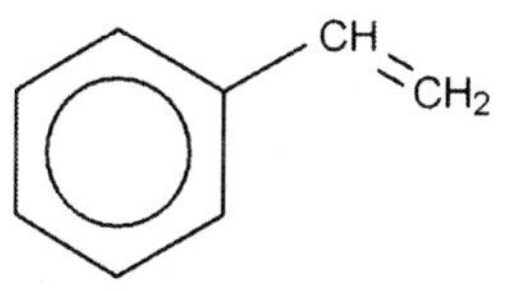

Figure 5.1 Chemical structure of styrene.

It is used as a monomer to make plastics such as polystyrene (PS), acrylonitrile–butadiene–styrene (ABS) copolymer, styrene–acrylonitrile (SAN) copolymer, and the other polymers mentioned in this chapter.

5.1 Acrylonitrile–Butadiene–Styrene Copolymer

Acrylonitrile–butadiene–styrene, or ABS, is a common thermoplastic used to make light, rigid, molded products such as pipe, automotive body parts, wheel covers, enclosures, protective head gear.

Styrene–acrylonitrile (SAN) copolymers have been available since the 1940s and while its increased toughness over PS made it suitable for many applications, its limitations led to the introduction of a rubber, butadiene, as a third monomer producing the range of materials popularly referred to as ABS plastics. These became available in the 1950s and the availability of these plastics and ease of processing led ABS to become one of the most popular engineering polymers.

The chemical structures of the monomers are shown in Fig. 5.2. The proportions of the monomers typically range from 15 to 35% acrylonitrile, 5 to 30% butadiene, and 40 to 60% styrene. It can be found as a graft copolymer, in which SAN polymer is formed in a polymerization system in the presence of polybutadiene rubber latex; the final product is a complex mixture consisting of SAN copolymer, a graft polymer of SAN and polybutadiene, and some free polybutadiene rubber. The CAS number is 9003-56-9.

Manufacturers and trade names: SABIC Innovative Polymers Cycolac®; INEOS Lustran® and Novodur®; Perrite Ronfalin®; and BASF AG Teluran®.

Applications: Medical devices, cosmetics, housewares, automobiles, business equipment, cabinets and casings, baths, shower trays, pipes, boat hulls, and vehicle components (Tables 5.1–5.7 and Figs. 5.3 and 5.4).

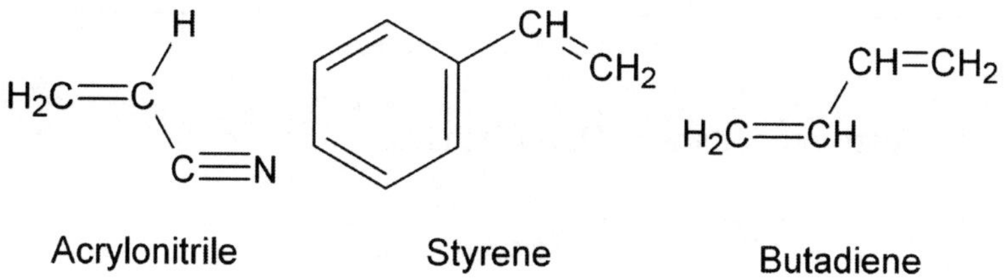

Figure 5.2 Chemical structures of ABS raw materials.

Acrylonitrile   Styrene   Butadiene

Table 5.1 Permeation of Oxygen and Water Vapor through Sabic Innovative Plastics Cycolac® ABS[1]

Temperature (°C)	25	24
Relative humidity (%)		90
Penetrant	Oxygen	Water vapor
Source document units		
Permeability coefficient (cc mil/24 h 100 in.^{2} atm)	100	
Vapor transmission rate (g mil/24 h 100 in.^{2})		12
Normalized units		
Permeability coefficient (cm^3 mm/m^2 day atm)	39.3	
Vapor transmission rate (g mm/m^2 day)		5.88

Table 5.2 Permeation of Oxygen, Nitrogen, and Carbon Dioxide at 24 °C through Dow Chemical Low Acrylonitrile ABS Film[2]

	Permeability Coefficient	
Penetrant Gas	Source Document Units (cm^3 mil/100 in.^{2} day)	Normalized Units (cm^3 mm/m^2 day atm)
Oxygen	200–260	79–102
Nitrogen	25–35	9.8–13.8
Carbon dioxide	900–1200	354–472

Table 5.3 Permeation of Oxygen, Nitrogen, and Carbon Dioxide at 24 °C through Dow Chemical Medium Acrylonitrile Content ABS Film[2]

	Permeability Coefficient	
Penetrant Gas	Source Document Units (cm^3 mil/100 in.^{2} day)	Normalized Units (cm^3 mm/m^2 day atm)
Oxygen	120–140	47–55
Nitrogen	10–15	3.9–5.9
Carbon dioxide	400–600	157–236

Table 5.4 Permeation of Water Vapor through Dow Chemical ABS Film[2]

Temperature (°C)	24–38
Source document units	5–16
Vapor transmission rate (g mil/100 in.^{2} day)	
Normalized units	2.0–6.3
Vapor transmission rate (g mm/m^2 day)	

5: Styrenic Plastics

Table 5.5 Permeation of Water Vapor at 23 °C through BASF AG Terluran® ABS Films[3]

Terluran® Product Code	997 VE	967 K	887 M
Source document units	27	27	31
Vapor transmission rate (g/m² day)			
Normalized units	2.7	2.7	3.1
Vapor transmission rate (g mm/m² day)			

Sample thickness: 0.1 mm; test method: DIN 53122. Test note: Values for permeability depend on the conditions under which the film was produced and may differ by as much as 50% from those given.

Table 5.6 Permeation of Oxygen, Nitrogen, Carbon Dioxide and Water Vapor at 23 °C through BASF AG Terluran® ABS Films[3]

Penetrant Gas	Permeability Coefficient					
	Source Document Units (cm³ 100 µm/m² day bar)			Normalized Units (cm³ mm/m² day atm)		
Terluran® Product Code	997 VE	967 K	887 M	997 VE	967 K	887 M
Oxygen	800	500	450	81	50.7	45.6
Nitrogen	200	100	100	20.3	10.1	10.1
Carbon dioxide	3000	2000	2000	304	203	203

Sample thickness: 0.1 mm; test method: DIN 53380. Test note: Values for permeability depend on the conditions under which the film was produced and may differ by as much as 50% from those given.

Table 5.7 Permeation of Oxygen, Nitrogen, Carbon Dioxide, and Methane for INEOS ABS Lustran®246 ABS Containing 60% Styrene, 27% Acrylonitrile, and 13% Butadiene[4] (see also Figs 5.3 and 5.4)

Source Document Units Permeation Coefficient				
	Carbon Dioxide	Methane	Oxygen	Nitrogen
T (K)	$P \times 10^{10}$ (cm³ (STP)/cm² s cm Hg)			
293	2.97	0.126	0.697	0.103
303	3.57	0.175	0.950	0.162
313	3.97	0.248	1.240	0.210
323	4.95	0.355	1.550	0.264
Normalized Units Permeation Coefficient				
	Carbon Dioxide	Methane	Oxygen	Nitrogen
T (°C)	(cm³ mm/m² day atm)			
20	195.0	8.3	45.8	6.8
30	234.4	11.5	62.4	10.6
40	260.7	16.3	81.4	13.8
50	325.0	23.3	101.8	17.3

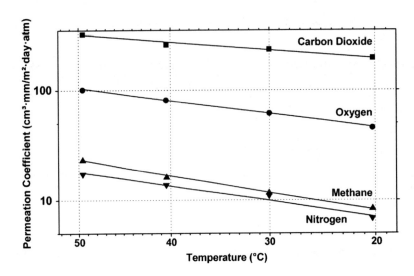

Figure 5.3 Temperature dependence of gas permeability coefficients for INEOS ABS Lustran®246 ABS containing 60% styrene, 27% acrylonitrile, and 13% butadiene.[4]

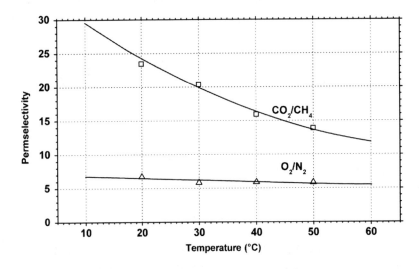

Figure 5.4 Permselectivity vs. temperature for INEOS ABS Lustran®246 ABS containing 60% styrene, 27% acrylonitrile, and 13% butadiene.[4]

5.2 Acrylonitrile–Styrene–Acrylate

Acrylonitrile–styrene–acrylate (ASA) is a terpolymer that can be produced by either a reaction process of all three monomers or by a graft process. The CAS number is 26299-47-8. ASA is usually made by introducing a grafted acrylic ester elastomer during the copolymerization of styrene and acrylonitrile, known as SAN. SAN is described in the next section of this chapter. The finely divided elastomer powder is uniformly distributed and grafted to the SAN molecular chains. The outstanding weatherability of ASA is due to the acrylic ester elastomer. ASA polymers are amorphous plastics, which have mechanical properties similar to those of the ABS resins described in Section 5.1. However, the ASA properties are far less affected by outdoor weathering.

ASA resins are available in natural, off-white, and a broad range of standard and custom-matched

Table 5.8 Permeation of Various Gases at 23 °C through BASF Luran® S 776 S Injection Molding Grade with Enhanced Toughness and Lower Flowability ASA Film[5–7]

Permeant Gas	Film Type	Standard Test Method	Permeability Coefficient	
			Source Document Units (cm³/m² day bar)	Normalized Units (cm³ mm/m² day atm)
Hydrogen	Blown film	DIN 53380	5000	507
Methane	Blown film	DIN 53380	110	11.1
Nitrogen	Film[a]	DIN 53380, Part 2 Method N	100	10.1
Nitrogen	Blown film	DIN 53380	70	7.1
Oxygen	Film[a]	DIN 53380, Part 2 Method N	550	55.7
Oxygen	Blown film	DIN 53380	180	18.2
Carbon dioxide	Film[a]	DIN 53380, Part 2 Method N	2300	233
Carbon dioxide	Blown film	DIN 53380	1400	142

Film thickness: 0.1 mm.
[a]Values depend on conditions under which film was produced. Figures may differ by as much as 50%.

colors. ASA resins can be compounded with other polymers to make alloys and compounds that benefit from ASA's weather resistance. ASA is used in many products including lawn and garden equipment, sporting goods, automotive exterior parts, safety helmets, and building materials.

Manufacturers and trade names: BASF Luran® S
Applications and uses: Automotive components, electrical equipment subjected to high temperatures, parabolic reflectors, solar energy systems, movement sensors, surfboards, and golf cars (Tables 5.8–5.11).

Table 5.9 Permeation of Nitrogen, Hydrogen, and Methane at 23 °C through BASF Luran® S 757 R High Stiffness and Medium Flowability ASA Film[5,6]

Permeant	Film Type	Standard Test Method	Permeability Coefficient	
			Source Document Units (cm³/m² day bar)	Normalized Units (cm³ mm/m² day atm)
Hydrogen	Blown film	DIN 53380	5000	507
Methane	Blown film	DIN 53380	100	10.1
Nitrogen	Blown film	DIN 53380	60	6.1
Oxygen	Blown film	DIN 53380	150	15.2
Carbon dioxide	Blown film	DIN 53380	1000	101

Film thickness: 0.1 mm.

Table 5.10 Permeation of Nitrogen, Hydrogen, and Methane at 23 °C through BASF Luran® S 797 S Injection Molding Grade with Very High-Impact Strength ASA Film[5,6]

Permeant	Film Type	Standard Test Method	Permeability Coefficient	
			Source Document Units (cm^3/ m^2 day bar)	Normalized Units (cm^3 mm/ m^2 day atm)
Nitrogen	Film[a]	DIN 53380, Part 2 Method N	75	7.6
Oxygen	Film[a]	DIN 53380, Part 2 Method N	500	50.7
Carbon dioxide	Film[a]	DIN 53380, Part 2 Method N	2000	203

Film thickness: 0.1 mm.
[a]Values depend on conditions under which film was produced. Figures may differ by as much as 50%.

Table 5.11 Permeation of Water Vapor at 23 °C through BASF Luran® S ASA Films[5,6]

Material	Film Type	Pressure Gradient (Mbar)	Vapor Transmission Rate	
			Source Document Units (g/m^2 day)	Normalized Units (g mm/m^2 day)
Luran® S 757 R	Blown film	19.86	30	3
Luran® S 776 S	Film[a]	23.87	35	3.5
Luran® S 797 S	Film[a]	23.87	30	3

Film thickness: 0.1 mm; relative humidity gradient: 85–0%; test method: DIN 53122.
[a]Values depend on conditions under which film was produced. Figures may differ by as much as 50%.

5.3 Polystyrene

Polystyrene (PS) is the simplest plastic based on styrene. Its structure is shown in Fig. 5.5. Its CAS number is 9003-53-6.

Pure solid PS is a colorless, hard plastic with limited flexibility. Polystyrene can be transparent or can be made in various colors. It is economical and is used for producing plastic model assembly kits, plastic cutlery, CD "jewel" cases, and many other objects where a fairly rigid, economical plastic is desired.

Polystyrene's most common use, however, is as expanded polystyrene (EPS). EPS is produced from a mixture of about 5–10% gaseous blowing agent (most commonly pentane or carbon dioxide) and 90–95% PS by weight. The solid plastic beads are expanded into foam through the use of heat (usually steam). The heating is carried out in a large vessel holding 200–2000 L. An agitator is used to keep the

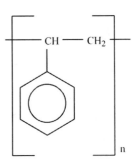

Figure 5.5 Chemical structure of PS.

beads from fusing together. The expanded beads are lighter than unexpanded beads so they are forced to the top of the vessel and removed. This expansion process lowers the density of the beads to 3% of their original value and yields a smooth-skinned, closed cell structure. Next, the pre-expanded beads are usually "aged" for at least 24 h in mesh storage silos. This allows air to diffuse into the beads, cooling them, and making them harder. These expanded beads are excellent for detailed molding. Extruded polystyrene (XPS), which is different from EPS, is commonly known by the trade name Styrofoam™. All these foams are not of interest in this book.

Three general forms are:

- General purpose polystyrene (PS or GPPS),
- Oriented polystyrene (OPS),
- High-impact polystyrene (HIPS).

One of the most important plastics is HIPS. This is a PS matrix that is imbedded with an impact modifier, which is basically a rubber-like polymer such as polybutadiene. This is shown in Fig. 5.6.

Manufacturers and trade names: BASF Polystyrene and Polystyrol, Dow Chemical Trycite™, Styron Styron™

Applications and uses:

General purpose: Yogurt, cream, butter, meat trays, egg cartons, fruit and vegetable trays, as well as cakes, croissants, and cookies. Medical and packaging/disposables, bakery packaging, and large and small appliances, particularly where clarity is required.

Oriented: OPS films can be printed and laminated to foams for food service plates and trays offering improved esthetics. The films can also be used as a laminate to PS sheet for a high gloss shine and to prepare bakery, convenience food items.

High impact: Refrigeration accessories, small appliances, electric lawn and garden equipment, toys, and remote controls (Tables 5.12—5.15, Figs. 5.7—5.9).

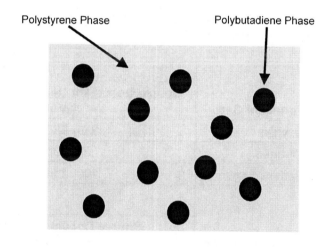

Figure 5.6 The structure of HIPS.

Table 5.12 Permeation of Gases at 24 °C through Styron Styron™ PS[8]

Permeant Gas	Permeability Coefficient	
	Source Document Units (cm³ mil/100 in.² day atm)	Normalized Units (cm³ mm/m² day atm)
Oxygen	300–400	118–157
Nitrogen	40–50	16–20
Carbon dioxide	1000–1500	394–590
Permeant Vapor	Vapor Transmission Rate	
	Source Document Units (g mil/m² day)	Normalized Units (g mm/m² day)
Water	2–10	0.8–3.9

Table 5.13 Permeation of Oxygen, Nitrogen, Carbon Dioxide, and Water Vapor at 23 °C through BASF AG Polystyrol 168 N GPPS Film[9]

Permeant Gas	Permeability Coefficient	
	Source Document Units (cm^3/100 $in.^2$ day bar)	Normalized Units (cm^3 mm/m^2 day atm)
Oxygen	1000	101
Nitrogen	250	25.3
Carbon dioxide	5200	527
Permeant Vapor	Vapor Transmission Rate	
	Source Document Units (g/m^2 day)	Normalized Units (g mm/m^2 day)
Water	12	1.2

Thickness: 0.1 mm; test methods: DIN 53380 and DIN 53122.

Table 5.14 Permeation of Oxygen, Nitrogen, Carbon Dioxide, and Water Vapor at 24 °C through Dow Chemical Trycite™ Oriented PS Film[8]

Permeant Gas	Permeability Coefficient	
	Source Document Units (cm^3 mil/100 $in.^2$ day bar)	Normalized Units (cm^3 mm/m^2 day atm)
Oxygen	250–350	98–138
Nitrogen	50–60	19.7–23.6
Carbon dioxide	700–1100	276–433
Permeant Vapor	Vapor Transmission Rate	
	Source Document Units (g mil/m^2 day)	Normalized Units (g mm/m^2 day)
Water	9	3.5

Table 5.15 Permeation of Oxygen, Nitrogen, Carbon Dioxide, and Water Vapor at 23 °C through BASF AG Polystyrol 476L HIPS Film[9]

Permeant Gas	Permeability Coefficient	
	Source Document Units (cm^3/m^2 day bar)	Normalized Units (cm^3 mm/m^2 day atm)
Oxygen	1600	162
Nitrogen	400	40.5
Carbon dioxide	10,000	1013
Permeant Vapor	Vapor Transmission Rate	
	Source Document Units (g/m^2 day)	Normalized Units (g mm/m^2 day)
Water	13	1.3

Thickness: 0.1 mm; test methods: DIN 53380 and DIN 53122; see also Figs 5.7–5.9.

5: Styrenic Plastics

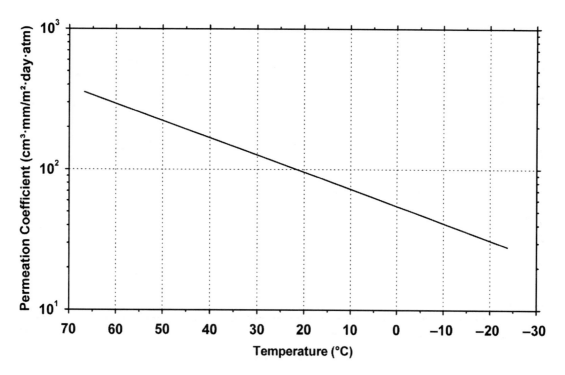

Figure 5.7 Permeation of oxygen vs. temperature through Dow Chemical Styron™ PS film.[8]

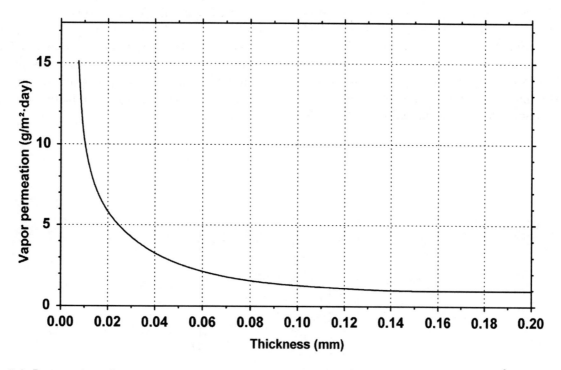

Figure 5.8 Permeation of water vapor vs. thickness through Dow Chemical Styron™ PS film.[8]

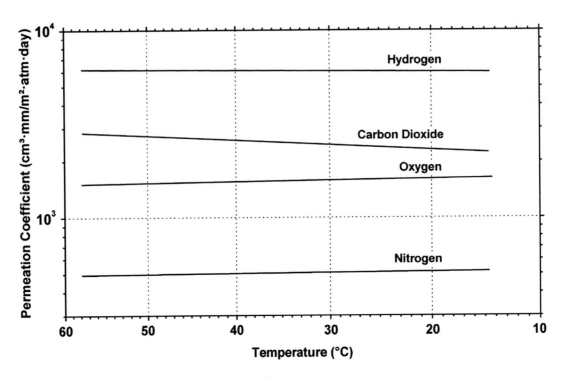

Figure 5.9 Gas permeation vs. temperature of PS.[12]

5.4 Styrene–Acrylonitrile Copolymer

Styrene and acrylonitrile monomers can be copolymerized to form a random, amorphous copolymer that has good weatherability, stress crack resistance, and barrier properties. The copolymer is called styrene–acrylonitrile or SAN copolymer. The SAN copolymer generally contains 70–80% styrene and 20–30% acrylonitrile. It is a simple random copolymer. This monomer combination provides higher strength, rigidity, and chemical resistance than PS, but it is not quite as clear as crystal PS and its appearance tends to discolor more quickly. The general structure is shown in Fig. 5.10. Its CAS number is 9003-54-7.

Manufacturers and trade names: BASF Luran®, Dow Chemical TYRIL Resins.

Applications and uses: Household: Mixing bowls, electric mixers, refrigerator inserts, tableware, vacuum flask casings, food storage containers, toiletries, cosmetics packaging, writing implements, and industrial batteries (Tables 5.16–5.18).

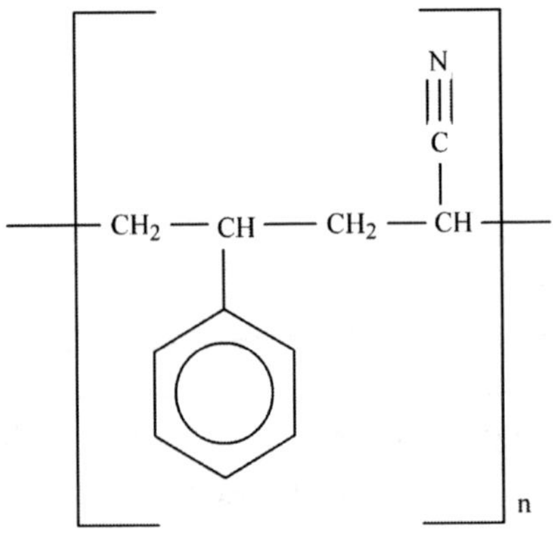

Figure 5.10 Chemical structure of SAN.

Table 5.16 Permeation of Oxygen at 23 °C of BASF Luran® SAN[10]

Luran® Grade	Permeability Coefficient	
	Source Document Units (cm³/m² day bar)	Normalized Units (cm³ mm/m² day atm)
358N	200–500	20.2–50.6
378P	200–300	20.2–30.4

Thickness: 0.100 mm; test method: DIN 53380.

Table 5.17 Permeation of Water Vapor at 23 °C of BASF Luran® SAN[11]

Luran® Grade	Permeability Coefficient	
	Source Document Units (g²/m² day)	Normalized Units (g mm/m² day)
358N	20–25	2.0–2.5
378P	20–25	2.0–2.5

Thickness: 0.100 mm; test method: DIN 53122; relative humidity gradient: 85–0%.

Table 5.18 Permeation of Gases at 24 °C through Dow Chemical Tyril® Low Acrylonitrile Content SAN[11]

Permeant Gas	Permeability Coefficient	
	Source Document Units (cm³ mil/100 in.² day atm)	Normalized Units (cm³ mm/m² day atm)
Oxygen	80–100	31.5–39.4
Nitrogen	10	3.9
Carbon dioxide	400	157

References

1. Oxygen and water permeability, Cycolac Resins. *GE Plastics*; 1997–2002.
2. *Permeability of polymers to gases and vapors. Supplier technical report (P302-335-79, D306-115-79)*. Dow Chemical Company; 1979.
3. *Terluran product line, properties, processing. Supplier design guide [B 567e/(8109) 9.90]*. BASF Aktiengesellschaft; 1990.
4. Marchese J. Gas sorption, permeation and separation of ABS copolymer membrane. *J Memb Sci* 2003;**221**(1–2):185–97. Available from, http://linkinghub.elsevier.com/retrieve/pii/S0376738803002588.
5. *Luran® S Acrylonitrile Styrene Acrylate product line, properties, processing. Supplier design guide (B 566e/1 1.90)*. BASF Aktiengesellschaft; 1990.
6. *Luran® S Acrylonitrile Styrene Acrylate product line, properties, processing. Supplier design guide (B 566e/10.83)*. BASF Aktiengesellschaft; 1983.
7. *Luran® S Plastic Plus acrylonitrile–styrene–acrylate copolymer (ASA and ASA + PC)*. 2007.
8. *Permeability of polymers to gases and vapors. Supplier technical report (P302-335-79, D306-115-79)*. Dow Chemical Company; 1979.

9. *Polystyrol product line, properties, processing. Supplier design guide (B 564e/2.93)*. BASF Aktiengesellschaft; 1993.
10. *Luran® brochure, LSIL 0701 BE*. BASF; 2010.
11. *Permeability of polymers to gases and vapors. Supplier technical report (P302-335-79, D306-115-79)*. Dow Chemical Company; 1979.
12. Brubaker DW, Kammermeyer K. Separation of gases by means of permeable membranes. Permeability of plastic membranes to gases. *Ind Eng Chem* 1952;**44**(6):1465–74. Available from, http://pubs.acs.org/cgi-bin/doilookup/?10.1021/ie50510a071.

6 Polyesters

Polyesters are formed by a condensation reaction that is very similar to the reaction used to make polyamide or nylons. A diacid and dialcohol are reacted to form the polyester with the elimination of water as shown in Fig. 6.1.

While the actual commercial route for making the polyesters may be more involved, the end result is the same polymeric structure. The diacid is usually aromatic. Polyester resins can be formulated to be brittle and hard, tough and resilient, or soft and flexible. In combination with reinforcements such as glass fibers, they offer outstanding strength, a high strength-to-weight ratio, chemical resistance, and other excellent mechanical properties. The three dominant materials in this plastics family are PC, polyethylene terephthalate (PET), and polybutylene terephthalate (PBT). Thermoplastic polyesters are similar in properties to Nylon 6 and Nylon 66 but have lower water absorption and higher dimensional stability than the nylons.

6.1 Liquid Crystalline Polymers

Liquid crystalline polymers (LCP) are a relatively unique class of partially crystalline aromatic polyesters based on 4-hydroxybenzoic acid and related monomers shown in Fig. 6.2. Liquid crystal polymers are capable of forming regions of highly ordered structure while in the liquid phase. However, the degree of order is somewhat less than that of a regular solid crystal. Typically, LCPs have outstanding mechanical properties at high temperatures, excellent chemical resistance, inherent flame retardancy, and good weatherability. Liquid crystal polymers come in a variety of forms from sinterable high temperature to injection moldable compounds.

LCPs are exceptionally inert. They resist stress cracking in the presence of most chemicals at elevated temperatures, including aromatic or halogenated hydrocarbons, strong acids, bases, ketones, and other aggressive industrial substances. Hydrolytic stability in boiling water is excellent. Environments that deteriorate these polymers are high-temperature steam, concentrated sulfuric acid, and boiling caustic materials.

As an example, the structure of Ticona Vectra® A950 LCP is shown in Fig. 6.3.

Manufacturers and trade names: Eastman Thermx®, DuPont Engineering Polymers Zenite®, Ticona Vectran™ and Vectra®, Solvay Advanced Polymers Xydar®, Sumitomo Sumikasuper®, and Toray Siveras®.

Applications and uses: Electrical/electronic connectors: fiber optic cables, chip carriers, printed circuit boards, and surface mount parts. Health care: sterilizable trays, dental tools, and surgical instruments. Industrial-consumer: printers, copiers, fax machine components, and business machine housings. Chemical process industry: pumps, meters, and valve liners. High barrier retort: pouches, closures, trays, and lids (Tables 6.1–6.7).

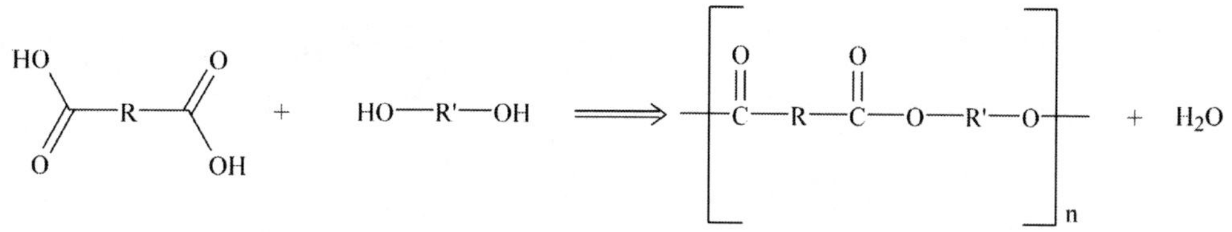

Figure 6.1 Chemical structure of PC polyester.

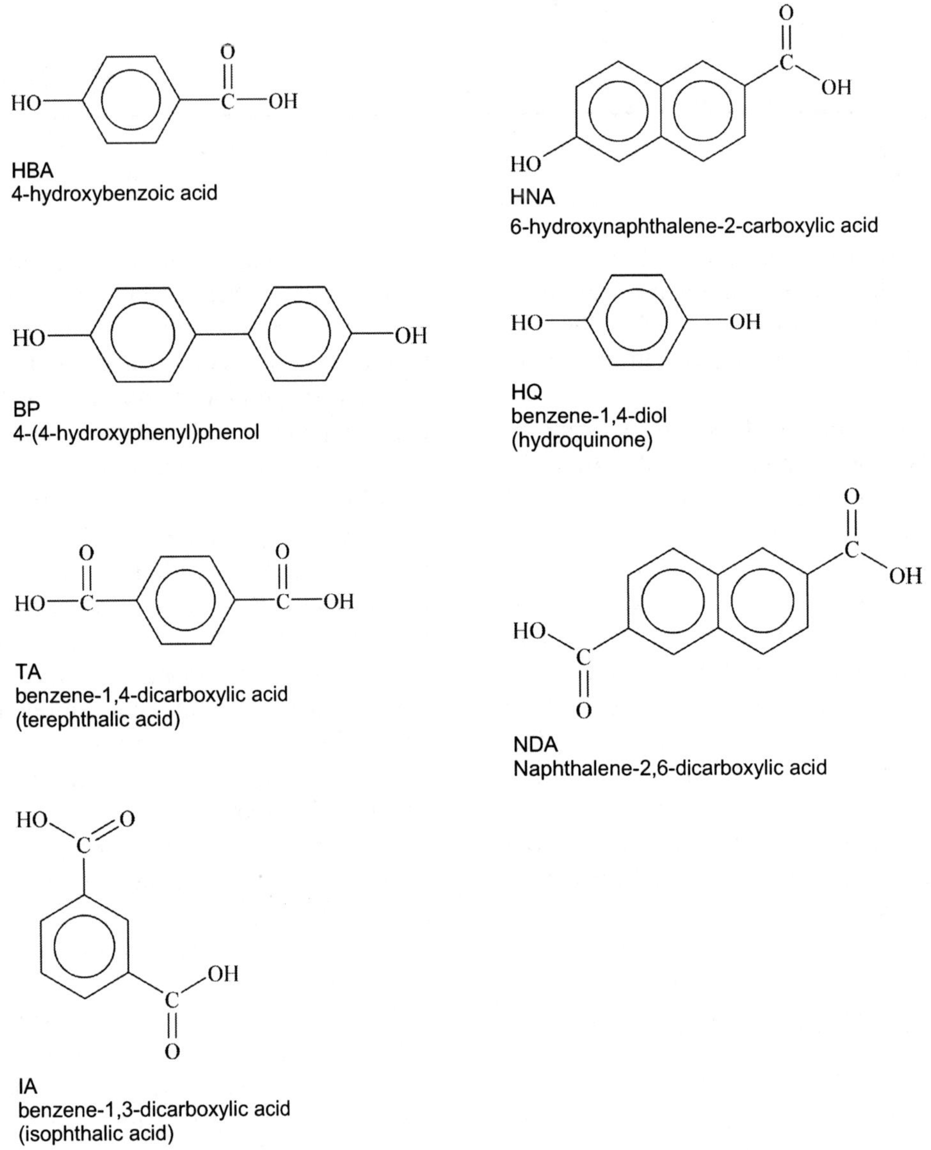

HBA
4-hydroxybenzoic acid

HNA
6-hydroxynaphthalene-2-carboxylic acid

BP
4-(4-hydroxyphenyl)phenol

HQ
benzene-1,4-diol
(hydroquinone)

TA
benzene-1,4-dicarboxylic acid
(terephthalic acid)

NDA
Naphthalene-2,6-dicarboxylic acid

IA
benzene-1,3-dicarboxylic acid
(isophthalic acid)

Figure 6.2 Chemical structures of monomers used to make liquid crystalline polymer polyesters.

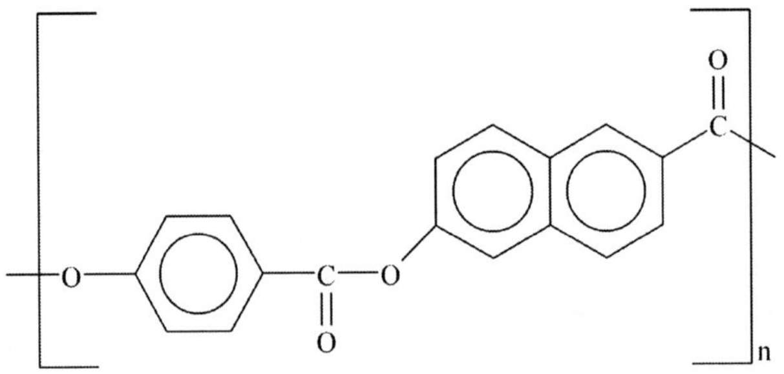

Figure 6.3 Chemical structure of Ticona Vectra® A950 LCP.

6: Polyesters

Table 6.1 Permeation of Hydrogen through Ticona Vectra® LCP[1]

Grade	Test Conditions	Film Thickness (mm)	Permeability Coefficient	
			Source document Units (cm³/ m² day bar)	Normalized Units (cm³ mm/ m² day atm)
Vectra A950	40 °C, 0% RH	0.50	78	40
Vectra A950	150 °C, 0% RH	2.5	98	248
Vectra E130i	150 °C, 0% RH	2.5	104	263

Table 6.2 Permeation of Oxygen through Ticona Vectra® A950 LCP[2]

Temperature (°C)	Relative Humidity (%)	Permeability Coefficient	
		Source Document Units (cm³ mil/100 in.² day atm)	Normalized Units (cm³ mm/m² day atm)
23	0	0.08	0.031
23	100	0.045	0.018
38	0	0.35	0.138
38	100	0.145	0.057

Table 6.3 Permeation of Oxygen at 23 °C through Ticona Vectran™ LCP Films[3]

Grade	Permeability Coefficient			
	Source Document Units (cm³ mil/100 in.² day atm)		Normalized Units (cm³ mm/m² day atm)	
	0% RH	100% RH	0% RH	100% RH
V100P	0.07	0.06	0.03	0.02
V200P	0.04	0.04	0.02	0.02
V300P	0.12	0.10	0.05	0.04
V400P	0.09	0.08	0.04	0.03

Test method: DIN 53380 Part 3.

Table 6.4 Permeation of Oxygen at 23 °C and 0% Relative Humidity Through Ticona Vectra® LCP[4]

Grade	Permeation Coefficient (cm³ mm/m² day atm)
Vectra A950	0.014
Vectra B950	0.006
Vectra L950	0.041
Vectra LKX1107	0.016
Vectra RD802	0.019

Table 6.5 Permeation of Carbon Dioxide at 25 °C and 100% Relative Humidity through Ticona Vectran™ LCP Films[3]

Grade	Permeability Coefficient	
	Source Document Units (cm³ mil/100 in.² day atm)	Normalized Units (cm³ mm/m² day atm)
V100P	0.13	0.05
V300P	0.24	0.09

Test method: ISO/CD 15105 Part 2 Annex C.

Table 6.6 Water Vapor Transmission at 23 °C and 100% Relative Humidity Through Ticona Vectra® LCP

Grade	Vapor Transmission Rate (g mm/m² day)
Vectra A950	0.0054
Vectra B950	0.0034
Vectra L950	0.014
Vectra LKX1107	0.010
Vectra RD802	0.0054

Table 6.7 Water Vapor Transmission at 38 °C Through Ticona Vectran™ LCP

Grade	Vapor Transmission Rate	
	Source Document Units (g mil/100 in.² day)	Normalized Units (g mm/m² day)
V100P	0.02	0.008
V200P	0.015	0.006
V300P	0.04	0.016
V400P	0.03	0.012

Test method: DIN 53122 Part 2.

6.2 Polybutylene Terephthalate (PBT)

Polybutylene terephthalate (PBT) is semicrystalline, white or off-white polyester similar in both composition and properties to PET. It has somewhat lower strength and stiffness than PET, is a little softer but has higher impact strength and similar chemical resistance. As it crystallizes more rapidly than PET, it tends to be preferred for industrial scale molding. Its structure is shown in Fig. 6.4.

PBT performance properties include:

- High mechanical properties
- High thermal properties
- Good electrical properties
- Dimensional stability

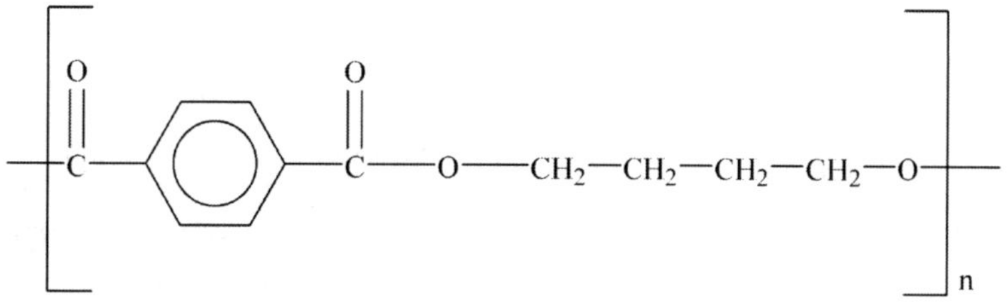

6: Polyesters

Figure 6.4 Chemical structure of PBT polyester.

Table 6.8 Water Vapor, Nitrogen, Oxygen, and Carbon dioxide at 23 °C through BASF AG Ultradur® PBT[5]

Permeant	Test Method	Relative Humidity (%)	Permeation Coefficient	
			Source Document Units (cm^3/m^2 day bar)	Normalized Units (cm^3 mm/m^2 day atm)
Nitrogen	DIN 53380	50	12	3.04
Oxygen	DIN 53380	50	60	15.2
Carbon dioxide	DIN 53380	50	550	139
			Vapor Transmission Rate	
			Source Document Units (g/m^2 day)	Normalized Units (g mm/m^2 day)
Water vapor	DIN 53122	85–0% Gradient	10	2.5

Thickness: 0.25 mm.

- Excellent chemical resistance
- Flame retardancy

Manufacturers and trade names: BASF Ultradur®, DuPont Crastin®, PolyOne Burgadur™, SABIC Innovative Plastics Enduran, Ticona Celanex®.

Applications and uses: Packaging, automotive, electrical, and consumer markets (Table 6.8).

6.3 Polycarbonate (PC)

Theoretically, PC is formed from the reaction of bis-phenol A and carbonic acid. The structures of these two monomers are given in Fig. 6.5.

Commercially, different routes are used, but the PC polymer of the structure shown in Fig. 6.6 is the result.

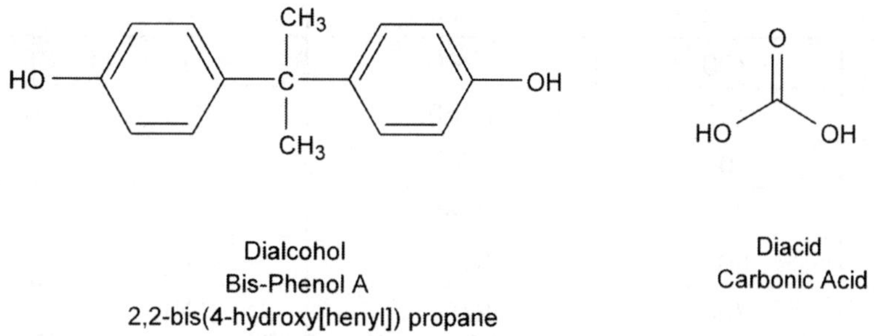

Dialcohol
Bis-Phenol A
2,2-bis(4-hydroxy[henyl]) propane

Diacid
Carbonic Acid

Figure 6.5 Chemical structures of monomers used to make PC polyester.

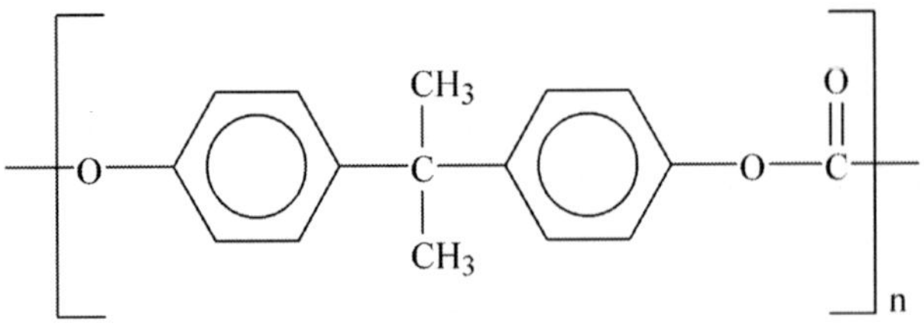

Figure 6.6 Chemical structure of PC polyester.

PC performance properties include:

- Very impact resistant and is virtually unbreakable and remains tough at low temperatures
- "Clear as glass" clarity
- High heat resistance
- Dimensional stability
- Resistant to ultraviolet light, allowing exterior use
- Flame retardant properties

Manufacturers and trade names: Bayer Material Science Makrolon®, Dow Calibre™.

Applications and uses: Glazing, safety shields, lenses, casings and housings, light fittings, kitchenware (microwaveable), medical apparatus (sterilizable), and CD's (the disks). Packaging: milk bottles, baby bottles, food containers. Medical: dialyzers and artery cannulas. Electrical: distribution box lids, fuses, sockets, lamp holders, and covers (Tables 6.9 and 6.10).

All product grade listed in Table 6.11 are reported to have a water permeation of 15 g/m^2·day for a 0.1 mm film, which normalized is 1.5 g mm/m^2 day (Tables 6.12–6.14, Figs. 6.7–6.9).

Table 6.9 Oxygen Permeation at 23 °C through Bayer MaterialScience Makrolon® PC[6]

Makrolon® Grade	Source Document Units, Permeability (cm^3/m^2 day bar)		Normalized Units, Permeability Coefficient (cm^3 mm/m^2 day atm)	
Film Thickness (mm)	0.1	0.0254	0.1	0.0254
AL2247	800	3150	81	81
AL2647	700	2760	71	71
LQ2687	700	2760	71	71
LQ2687	700	2760	71	71
LQ3147	700	2760	71	71
LQ3187	700	2760	71	71
2205	800	3150	81	81
2207	800	3150	81	81
6717	700	2760	71	71
AG2677	700	2760	71	71
2405	800	3150	81	81
2407	800	3150	81	81
2605	700	2760	71	71
2607	700	2760	71	71
2805	700	2760	71	71
2807	700	2760	71	71
3105	700	2760	71	71
3107	700	2760	71	71
2456	800	3150	81	81
2656	700	2760	71	71
2806	700	2760	71	71

Table 6.9 (*Continued*)

Makrolon® Grade	Source Document Units, Permeability (cm³/m² day bar)		Normalized Units, Permeability Coefficient (cm³ mm/m² day atm)	
2856	700	2760	71	71
3106	700	2760	71	71
3156	700	2760	71	71
3206	700	2760	71	71
2665	700	2760	71	71
2667	700	2760	71	71
2865	700	2760	71	71
2867	700	2760	71	71
6555	700	2760	71	71
6557	700	2760	71	71

Test method: ISO 2556.

Table 6.10 Nitrogen Permeation at 23 °C through Bayer MaterialScience Makrolon® PC[6]

Makrolon® Grade	Source Document Units, Permeability (cm³/m² day bar)		Normalized Units, Permeability Coefficient (cm³ mm/m² day atm)	
Film Thickness (mm)	0.1	0.0254	0.1	0.0254
AL2247	160	630	16	16
AL2647	130	510	13	13
LQ2687	130	510	13	13
LQ2687	130	510	13	13
LQ3147	130	510	13	13
LQ3187	130	510	13	13
2205	160	630	16	16
2207	160	630	16	16
6717	130	510	13	13
AG2677	130	510	13	13
2405	160	630	16	16
2407	160	630	16	16
2605	130	510	13	13
2607	130	510	13	13
2805	130	510	13	13
2807	130	510	13	13
3105	130	510	13	13
3107	130	510	13	13
2456	160	630	16	16
2656	130	510	13	13
2806	130	510	13	13

(*Continued*)

Table 6.10 (*Continued*)

Makrolon® Grade	Source Document Units, Permeability (cm³/m² day bar)		Normalized Units, Permeability Coefficient (cm³ mm/m² day atm)	
2856	130	510	13	13
3106	130	510	13	13
3156	130	510	13	13
3206	130	510	13	13
2665	130	510	13	13
2667	130	510	13	13
2865	130	510	13	13
2867	130	510	13	13
6555	130	510	13	13
6557	130	510	13	13

Test method: ISO 2556.

Table 6.11 Carbon dioxide permeation at 23 °C through Bayer MaterialScience Makrolon® PC[6]

Makrolon® Grade	Source Document Units, Permeability (cm³/m² day bar)		Normalized Units, Permeability Coefficient (cm³ mm/m² day atm)	
Film Thickness (mm)	0.1	0.0254	0.1	0.0254
AL2247	4800	18,900	486	486
AL2647	4300	16,900	436	435
LQ2687	4300	16,900	436	435
LQ2687	4300	16,900	436	435
LQ3147	4300	16,900	436	435
LQ3187	4300	16,900	436	435
2205	4800	18,900	486	486
2207	4800	18,900	486	486
6717	4300	16,900	436	435
AG2677	4300	16,900	436	435
2405	4800	18,900	486	486
2407	4800	18,900	486	486
2605	4300	16,900	436	435
2607	4300	16,900	436	435
2805	4300	16,900	436	435
2807	4300	16,900	436	435
3105	4300	16,900	436	435
3107	4300	16,900	436	435
2456	4800	18,900	486	486
2656	4300	16,900	436	435
2806	4300	16,900	436	435

Table 6.11 (Continued)

Makrolon® Grade	Source Document Units, Permeability (cm^3/m^2 day bar)		Normalized Units, Permeability Coefficient (cm^3 mm/m^2 day atm)	
2856	4300	16,900	436	435
3106	4300	16,900	436	435
3156	4300	16,900	436	435
3206	4300	16,900	436	435
2665	4300	16,900	436	435
2667	4300	16,900	436	435
2865	4300	16,900	436	435
2867	4300	16,900	436	435
6555	4300	16,900	436	435
6557	4300	16,900	436	435

Test method: ISO 2556.

Table 6.12 Nitrogen Permeation at 23 °C through Dow Calibre™ PC[7]

Calibre™ Grade	Permeability Coefficient	
	Source Document Units (cm^3 mil/100 in.^{2} day)	Normalized Units (cm^3 mm/m^2 day atm)
300-4	31	12.2
300-15	27	10.6
800-6	57	22.4

Test method: ASTM 2752 (see also Fig. 6.7).

Table 6.13 Oxygen Permeation at 23 °C through Dow Calibre™ PC[7]

Calibre™ Grade	Permeability Coefficient	
	Source Document Units (cm^3 mil/100 in.^{2} day)	Normalized Units (cm^3 mm/m^2 day atm)
300-4	260	102
300-15	230	91
800-6	314	124

Test method: ASTM 2752 (see also Fig. 6.8).

Table 6.14 Carbon dioxide Permeation at 23 °C through Dow Calibre™ PC[7]

Calibre™ Grade	Permeability Coefficient	
	Source Document Units (cm^3 mil/100 in.^{2} day)	Normalized Units (cm^3 mm/m^2 day atm)
300-4	1950	768
300-15	1720	677
800-6	2100	827

Test method: ASTM 2752 (see also Fig. 6.9).

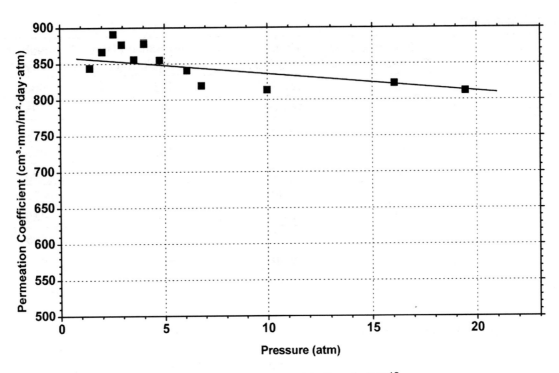

Figure 6.7 Effect of driving pressure on the permeability of helium in PC.[12]

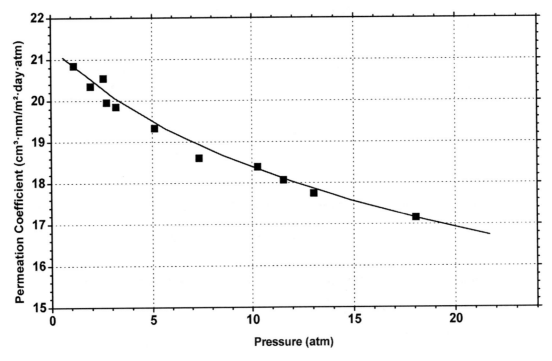

Figure 6.8 Effect of driving pressure on the permeability of methane in PC.[8]

6: Polyesters

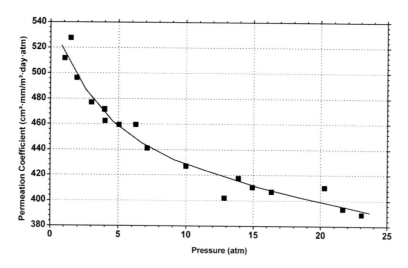

Figure 6.9 Effect of driving pressure on the permeability of carbon dioxide in PC.[8]

6.4 Polycyclohexylene-dimethylene Terephthalate

Polycyclohexylene-dimethylene terephthalate (PCT) is high-temperature polyester that possesses the chemical resistance, processability, and dimensional stability of polyesters PET and PBT. However, the aliphatic cyclic ring shown in Fig. 6.10 imparts added heat resistance. This puts it between the common polyesters and the LCP polyesters described in the previous section. At this time only DuPont makes this plastic under the trade name Thermx®.

Manufacturers and trade names: Eastman Eastar, DuPont Thermx®.

Applications and uses: Bags, credit cards, gaming cards, identification cards, plastic cards, rigid medical, blister packaging, debit cards, gift cards, phone cards, rapid deployment flood walls, and smart cards (Table 6.15).

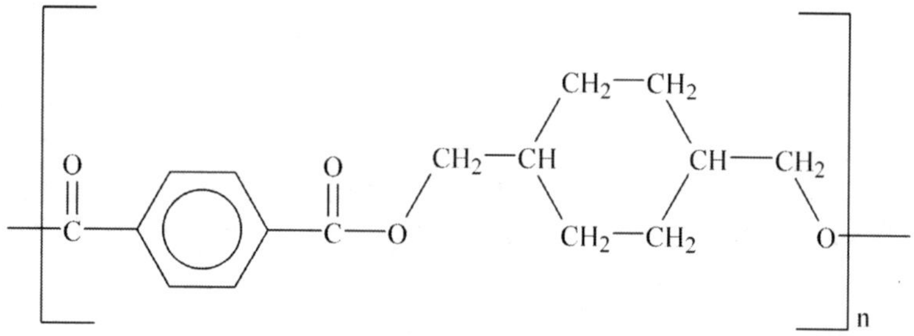

Figure 6.10 Chemical structure of PCT polyester.

Table 6.15 Permeation through Eastman Eastar 5445 PCTG Copolyester Film

Permeant	Permeation Coefficient Units	Permeation Coefficient	Test Method
Water vapor	g/m^2 day	7	ASTM F372
Carbon dioxide	cm^3 mm/m^2 day atm	49	ASTM D1434
Oxygen	cm^3 mm/m^2 day atm	11	ASTM D3985

Thickness: 0.250 mm.

6.5 Polyethylene Naphthalate

Polyethylene naphthalate (PEN) is similar to PET but has better temperature resistance. The CAS number for PEN is 25853-85-4. The structure of this polyester is shown in Fig. 6.11.

Manufacturers and trade names: DuPont™ Teijin Films TEONEX® and Eastman Eastar®.

Applications and uses: Electrical, industrial, general purpose, high-value-added applications in labels, laminates, circuitry, and release (Table 6.16).

6.6 Polyethylene Terephthalate (PET)

PET polyester is the most common thermoplastic polyester and is often called just "polyester". This often causes confusion with the other polyesters in this chapter. PET exists both as an amorphous (transparent) and as a semicrystalline (opaque and white) thermoplastic material. The semicrystalline PET has good strength, ductility, stiffness, and hardness. The amorphous PET has better ductility

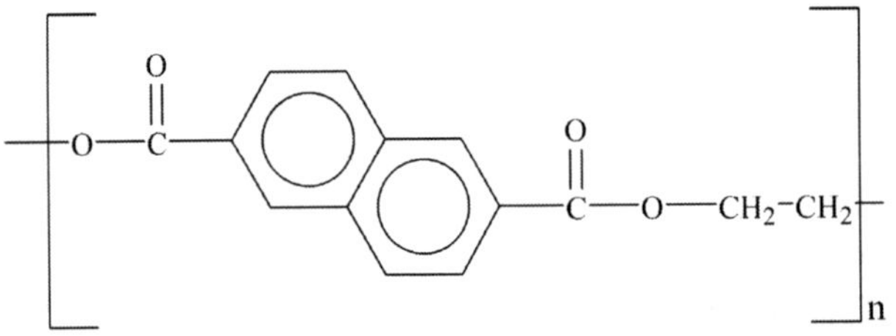

Figure 6.11 Structure of PEN.

Table 6.16 Gas and Water Vapor Permeability of DuPont™ Teijin Films TEONEX® Q51 PEN Biaxially Oriented Film[8]

Permeant Gas	Permeability Coefficient	
	Source Document Units, $\times 10^{-12}$ (cm^3 cm/cm^2 s cm Hg)	Normalized Units (cm^3 mm/cm^2 day atm)
Carbon dioxide	3.7	2.4
Oxygen	0.8	0.5
	Permeability Coefficient	
	Source Document Units (g/m^2 day)	Normalized Units (g mm/m^2 day)
Water vapor	6.7	4.2

Test method: ASTM D1434, JIS Z-0206. Film thickness: 0.025 mm.

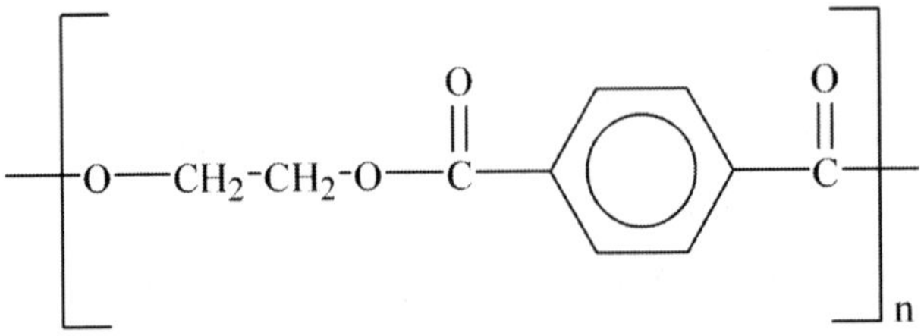

Figure 6.12 Chemical structure of PET polyester.

6: Polyesters

but less stiffness and hardness. It absorbs very little water. Its structure is shown in Figure 6.12.

Manufacturers and trade names: DuPont Teijin Films™ Mylar® and Melinex®, Mitsubishi Polyester Film Hostaphan®.

Applications and uses: Bottles for soft drinks and water, food trays for oven use, roasting bags, audio/video tapes, and mechanical components (Tables 6.17–6.24, Figs. 6.13 and 6.14).

Table 6.17 Permeation of Carbon dioxide at 23 °C and 75% Relative Humidity through DuPont™ Teijin Films™ Mylar® PET Films[9]

Grade	Thickness (Gauge)[a]	Thickness (mm)	Source Document Units, Permeability (cc/100 in.^{2})	Normalized Units, Permeability Coefficient (cm^3 mm/m^2 day atm)
Mylar® 800	48	0.0122	31	5.86
Mylar® 813	48	0.0122	31	5.86
Mylar® 840	48	0.0122	31	5.86

Test method: ASTM D1434.
[a]1 gauge = 0.01 mil for plastic film.

Table 6.18 Permeation of Nitrogen at 23 °C and 75% Relative Humidity through DuPont™ Teijin Films™ Mylar® PET Films[10]

Grade	Thickness (Gauge)[a]	Thickness (mm)	Source Document Units, Permeability (cc/100 in.^{2})	Normalized Units, Permeability Coefficient (cm^3 mm/m^2 day atm)
Mylar® 800	48	0.0122	1.6	0.30
Mylar® 813	48	0.0122	1.6	0.30
Mylar® 840	48	0.0122	1.6	0.30

Test method: ASTM D1434.
[a]1 gauge = 0.01 mil for plastic film.

Table 6.19 Permeation of Oxygen at 23 °C and 75% Relative Humidity through DuPont™ Teijin Films™ Mylar® PET Films[10]

Grade	Film Treatment	Thickness (Gauge)[a]	Thickness (mm)	Source Document Units, Permeability (cc/100 in.^{2})	Normalized Units, Permeability Coefficient (cm^3 mm/m^2 day atm)
Mylar® 800	None	48	0.0122	6	1.13
	Metalized	48	0.0122	0.08	0.02
Mylar® 800C	None	48	0.0122	6	1.13
	Metalized	75	0.0190	4	1.18
Mylar® 813	None	48	0.0122	6	1.13
	Metalized	48	0.0122	0.08	0.02

(Continued)

Table 6.19 (Continued)

Grade	Film Treatment	Thickness (Gauge)[a]	Thickness (mm)	Source Document Units, Permeability (cc/100 in.^{2})	Normalized Units, Permeability Coefficient (cm^3 mm/ m^2 day atm)
Mylar® 822	None	48	0.0122	6	1.13
Mylar® 823	None	48	0.0122	6	1.13
Mylar® 850	None	48	0.0122	6	1.13
Mylar® 850	None	80	0.0203	3.7	1.17
Mylar® 851H	None	60	0.0150	5	1.18
Mylar® 854	None	48	0.0122	6	1.13
Mylar® 854	None	60	0.0150	5	1.18
Mylar® 864	None	48	0.0122	6	1.13
Mylar® 814	None	48	0.0122	6	1.13

Test method: ASTM D1434.
[a] 1 gauge = 0.01 mil for plastic film.

Table 6.20 Water Vapor Permeation at 38 °C and 90% Relative Humidity through DuPont™ Teijin Films™ Mylar® PET Films[10]

Grade	Film Treatment	Thickness (Gauge)[a]	Thickness (mm)	Source Document Units, Vapor Transmission (g/100 in.^{2}/ day)	Normalized Units, Vapor Transmission Rate (g mm/ m^2 day)
Mylar® 800	None	48	0.0122	2.8	0.53
	Metalized	48	0.0122	0.05	0.01
Mylar® 800C	None	48	0.0122	2.8	0.53
Mylar® 800C	None	75	0.0190	1.8	0.53
Mylar® 813	None	48	0.0122	2	0.38
	Metalized	48	0.0122	0.05	0.01
Mylar® 822	None	48	0.0122	2.8	0.53
Mylar® 850	None	48	0.0122	2.8	0.53
Mylar® 850	None	80	0.0203	1.8	0.57
Mylar® 854	None	48	0.0122	2.8	0.53
Mylar® 854	None	60	0.0150	2.3	0.54
Mylar® 864	None	48	0.0122	2.8	0.53

Test method: ASTM F1249.
[a] 1 gauge = 0.01 mil for plastic film.

6: Polyesters

Table 6.21 Vapor Permeation Through DuPont™ Teijin Films™ Mylar® PET Films[10]

		Vapor Transmission Rate	
Permeant Vapor	**Temperature (°C)**	**Source Document Units (g mil/100 in.² day)**	**Normalized Units (g mm/m² day)**
Acetone	40	2.22	0.87
Benzene	25	0.36	0.14
Carbon tetrachloride	40	0.08	0.03
Ethyl acetate	40	0.08	0.03
Benzene	40	0.12	0.05

Test method: ASTM E96.

Table 6.22 Permeation of Gases at 23 °C Through Mitsubishi Polyester Film Hostaphan® RN 25 Biaxially Oriented PET Release Film[11]

Permeant Gas	**Test Method**	**Source Document Units, Permeation (cm³/m² day bar)**	**Normalized Units, Permeation Coefficient (cm³ mm/m² day bar)**
Air	DIN 53380	30	0.4
Ammonia, dry	In-house method	4000	48.6
Argon	DIN 53380	25	0.3
Carbon dioxide	DIN 53380	240	2.9
Chlorine	DIN 53380	60	0.7
Ethylene oxide	In-house method	650	7.9
Freon® 11	DIN 53380ª 24.5 °C	<4	<0.1
Freon® 12	DIN 53380	12	0.1
Freon® 13	DIN 53380	14	0.2
Freon® 21	DIN 53380	7	0.1
Freon® 22	DIN 53380	7	0.1
Freon® 114	DIN 53380	6	0.1
Freon® 502	DIN 53380	<6	<0.1
Helium	DIN 53380	2000	24.3
Hydrogen	DIN 53380	1100	13.4
Hydrogen sulfide	In-house method	500	6.1
Methyl bromide	DIN 53380	50	0.6
Nitrogen	DIN 53380	20	0.2
Oxygen	DIN 53380	70	0.9
Phosgene	DIN 53380	50	0.6
Prussic acid	DIN 53380	8000	97.3
Sulfur dioxide	In-house method	1000	12.2

Film thickness: 0.012 mm.

Table 6.23 Vapor Permeation at 23 °C Through Mitsubishi Polyester Film Hostaphan® RN 25 Biaxially Oriented PET Release Film[11]

Permeant Vapor	Source Document Units, Vapor Permeation (g/m² day)	Normalized Units, Vapor Permeation Rate (g mm/m² day)
Acetone	<0.1	<0.001
Benzene	<0.1	<0.001
Carbon disulfide	3	0.036
Carbon tetrachloride	0.2	0.002
Ethyl acetate	<0.1	<0.001
Ethyl alcohol	0.005	<0.001
Formaldehyde (30% solution)	0.003	<0.001
Hexane	<0.1	<0.001
Methyl alcohol	0.7	0.008
Water	8	0.096

Film thickness: 0.012 mm. Test method: In-house method except for water: DIN 53122.

Table 6.24 Permeation of "Aromas" at 20 °C Through Mitsubishi Polyester Film Hostaphan® RN 25 Biaxially Oriented PET Release Film[11]

Aromas	Source Document Units, Vapor Permeation (g/m² day)	Normalized Units, Vapor Permeation Rate (g mm/m² day)
Camphor	3.0×10^{-6}	3.6×10^{-8}
Cinnamaldehyde	5.0×10^{-2}	6.0×10^{-4}
Diphenylmethane	4.0×10^{-3}	4.8×10^{-5}
Eucalyptol	8.0×10^{-3}	9.6×10^{-5}
Eugenol	1.6×10^{-4}	1.9×10^{-6}
Geraniol	1.3×10^{-4}	1.6×10^{-6}
Menthol	7.0×10^{-4}	8.4×10^{-6}
Vanillin	1.0×10^{-5}	1.2×10^{-7}

Film thickness: 0.012 mm. Test method: In-house method.

6: Polyesters

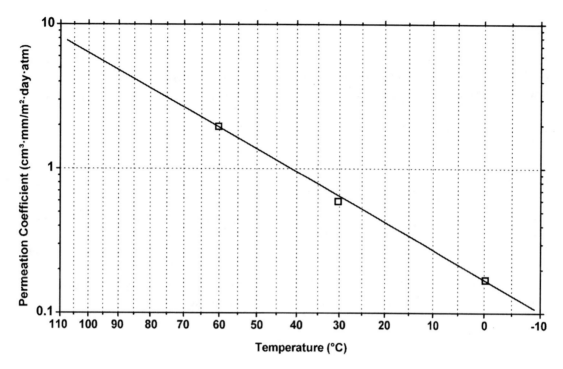

Figure 6.13 Permeation of hydrogen sulfide vs. temperature through DuPont™ Mylar® PET.[13]

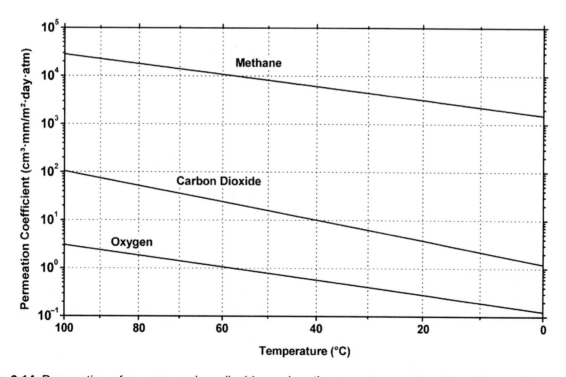

Figure 6.14 Permeation of oxygen, carbon dioxide, and methane vs. temperature through amorphous PET.[14]

References

1. Vectra. Liquid crystal polymer (LCP); 2006.
2. Linstid HC, et al. Liquid crystal polymers: an overview of technology and typical applications, NPE 2000. *Ticona*; 2000.
3. Vectran™. LCP packaging. Ticona; 2000.
4. Vectra. Liquid crystal polymer (LCP); 2003.
5. *Ultradur® polybutylene terephthalate (PBT) product line, properties, processing, supplier design guide [B 575/1e-(819) 4.91]*. BASF Aktiengesellschaft; 1991.
6. *Makrolon® technical information sheets*. Bayer Material Science; 2008.
7. *Calibre engineering thermoplastics basic design manual, supplier design guide (301-1040-1288)*. Dow Chemical Company; 1988.
8. Teonex® product information. DuPont™ Teijin Films; 2004.
9. Mylar® specification sheets. DuPont Teijin Films™; 2010.
10. *Mylar Polyester Film, supplier technical report (E-99499)*. DuPont Company; 1988.
11. Mitsubishi Polyester Film Americas.
12. Chiou JS, et al. Gas permeation in polyethersulfone. *J Appl Polym Sci* 1987;**33**:1823–8.
13. Heilman W, Tammela V, Meyer J, Stannett V, Szwarc M. Permeability of polymer films to hydrogen sulfide gas. *Ind Eng Chem* 1956;**48**: 821–4.
14. Wood-Adams P. Permeation: essential factors for permeation. In: *Polymer. Course Notes, Physical Chemistry of Polymers*. Concordia University; 2006. p. 1–16.

7 Polyimides

This chapter covers a series of plastics of which the imide group is an important part of the molecule. The imide group is formed by a condensation reaction of an aromatic anhydride group with an aromatic amine as shown in Fig. 7.1.

This group is very thermally stable. Aliphatic imides are possible, but the thermal stability is reduced, and thermal stability is one of the main reasons to use an imide-type polymer.

7.1 Polyamide—Imide

Polyamide—imides (PAIs) are thermoplastic amorphous polymers that have useful properties such as:

- Exceptional chemical resistance
- Outstanding mechanical strength
- Excellent thermal stability
- Performs from cryogenic up to 260 °C
- Excellent electrical properties.

The monomers used to make PAI resin are usually a diisocyanate and an acid anhydride such as those shown in Fig. 7.2.

When these two types of monomers are reacted, carbon dioxide is generated along with a PAI polymer. The closer the monomer ratio is to 1:1, the higher the molecular weight of the polymer as shown in Fig. 7.3 (Tables 7.1 and 7.2).

Manufacturers and trade names: Solvay Advanced Polymers Torlon®.

Applications and uses: Electrical connectors, switches, and relays; thrust washers, spline liners, valve seats, bushings, bearings, wear rings, cams; and other applications requiring strength at high temperature and resistance to wear (Tables 7.3 and 7.4, Figs. 7.4 and 7.8).

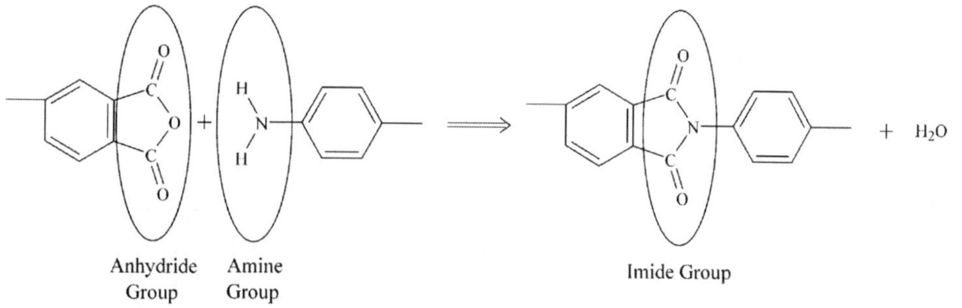

Figure 7.1 Reaction of amine with anhydride to form an imide.

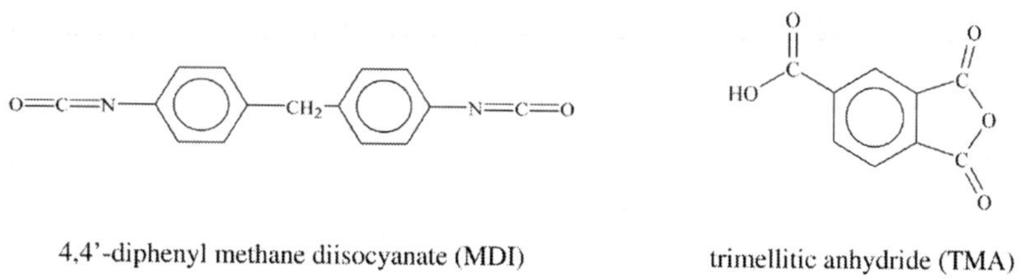

Figure 7.2 Chemical structures of monomer used to make polyamide–imides.

4,4'-diphenyl methane diisocyanate (MDI)

trimellitic anhydride (TMA)

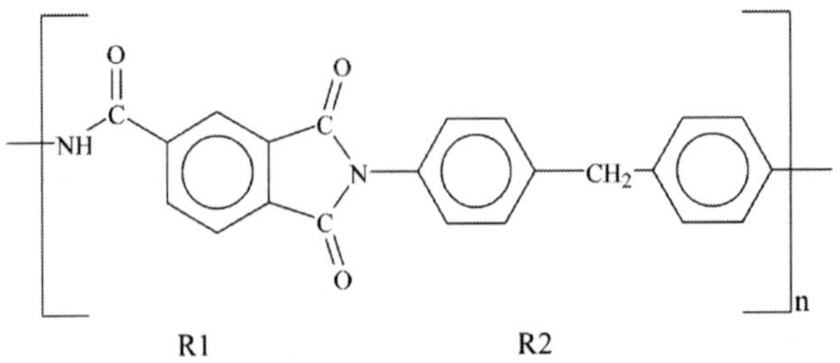

Figure 7.3 Chemical structure of a typical polyamide–imide.

Table 7.1 The Polymer Units of Various Amide–Imide Polymers (Refer Fig. 7.3 for Polymer Structure)[1]

PAI code	R1 from Acid Anhydride	R2 from Diisocyanate
PAI(TMI/DPA)	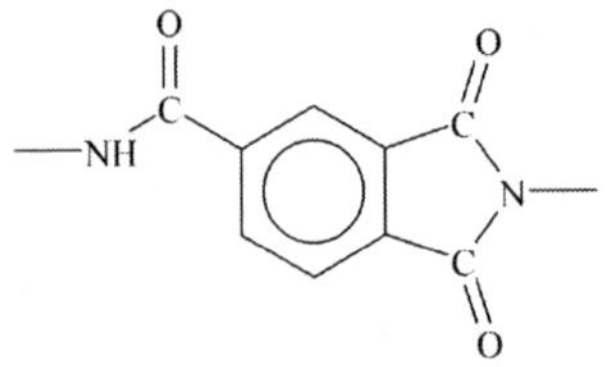	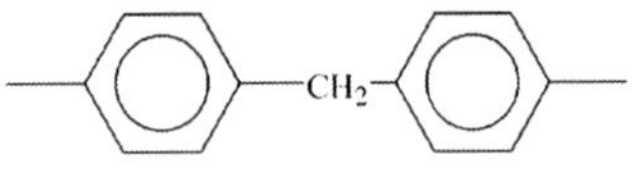
PAI(TMI/HEA)	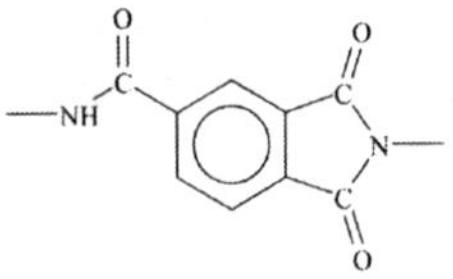	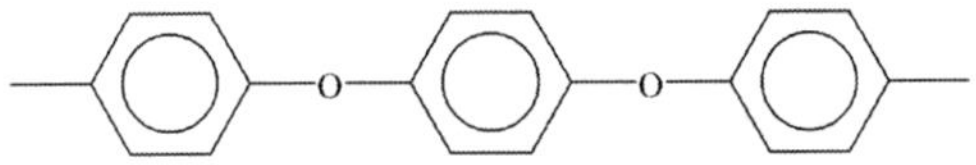
PAI(TMI/TFA)	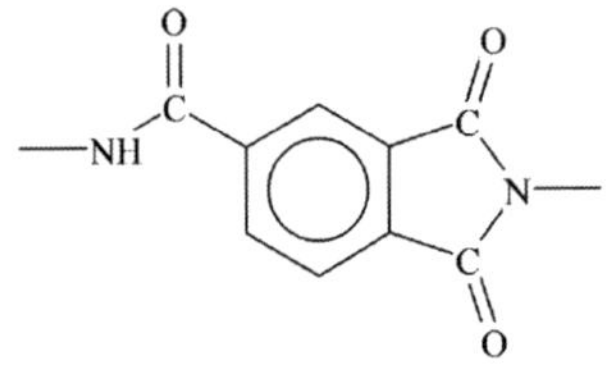	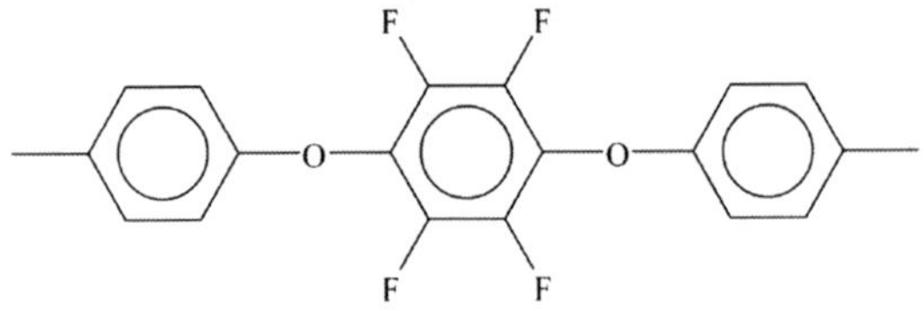
PAI(TMI/CDA)	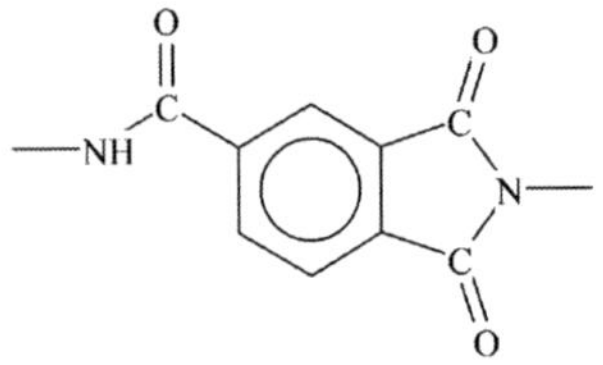	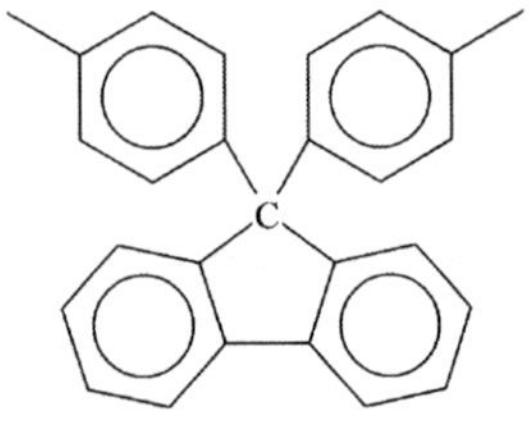
PAI(PMI/CDA)	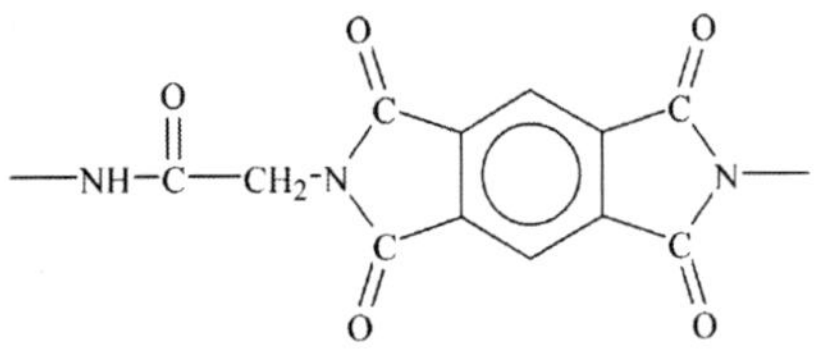	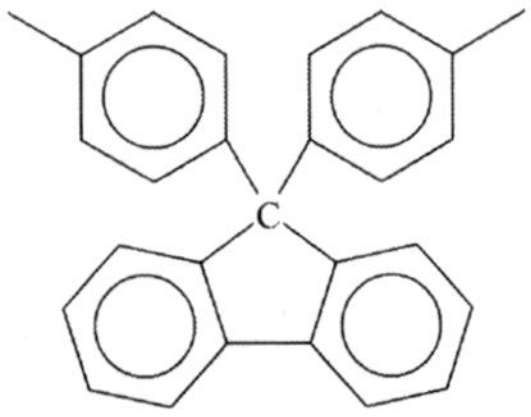

Table 7.2 The Polymer Units of Various Amide—Imide Polymers (Refer Fig. 7.3 for Polymer Structure)[1]

PAI code	R1 from Acid Anhydride	R2 from Diisocyanate
PAP	—NH—C(=O)—[phthalimide ring]	—[C₆H₄]— (para-phenylene)
PAO	—NH—C(=O)—[phthalimide ring]	—[C₆H₄]—O—[C₆H₄]—
PAM	—NH—C(=O)—[phthalimide ring]	—[C₆H₄]—CH₂—[C₆H₄]—
PAD	—NH—C(=O)—[phthalimide ring]	3,3'-dimethyl biphenylene (H₃C substituents)
PAT	—NH—C(=O)—[phthalimide ring]	3,3',5,5'-tetramethyl biphenylene (four H₃C substituents)

Table 7.3 Permeability Coefficients of Gases at 35 °C and of Various Polyamide–Imide Resins[2]

Polymer Sample	Permeability Coefficient							
	Barrer [1 × 10^{-10} cm^3 (STP) cm/cm^2 s cm Hg]				Normalized Units (cm^3 mm/m^2 day atm)			
	Carbon Dioxide	Methane	Oxygen	Nitrogen	Carbon Dioxide	Methane	Oxygen	Nitrogen
PAI(TMI/DPA)	0.264	0.01	0.065	0.017	17.3	0.7	4.3	1.1
PAI(TMI/HEA)	0.749	0.015	0.185	0.02	49.2	1.0	12.1	1.3
PAI(TMI/TFA)	1.273	0.013	0.284	0.071	83.6	0.9	18.6	4.7
PAI(TMI/CDA)	1.539	0.017	0.347	0.064	101.1	1.1	22.8	4.2
PAI(PMI/CDA)	1.698	0.076	0.272	0.065	111.5	5.0	17.9	4.3

Refer Table 7.1 for polymer definitions; pressure = 2 bar.

Table 7.4 Permeabilities of Gases of Various PAI Polymers at 30 °C and 1 atm[1]

Polymer	Permeability Coefficient						
	Barrer [1 × 10^{-10} cm^3 (STP) cm/cm^2 s cm Hg]			Normalized Units (cm^3 mm/m^2 day atm)			
	Hydrogen	Oxygen	Nitrogen	Hydrogen	Oxygen	Nitrogen	
PAP	0.0412	0.00205	0.000331	2.7	0.1	0.0	
PAO	0.38	0.0173	0.00302	25.0	1.1	0.2	
PAM	1.97	0.103	0.0194	129.4	6.8	1.3	
PAD	3.12	0.171	0.0328	204.9	11.2	2.2	
PAT	15.9	0.894	0.175	1044.1	58.7	11.5	

See also Figs. 7.4–7.8; Refer Table 7.2 for polymer definitions.

7: Polyimides

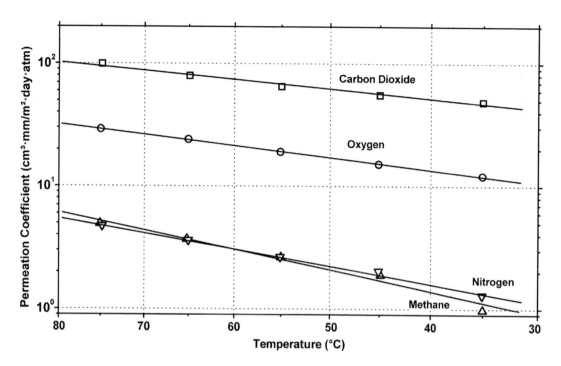

Figure 7.4 Dependence of gas permeability on temperature; pressure = 2 bar, PAI(TMI/HEA).[2]

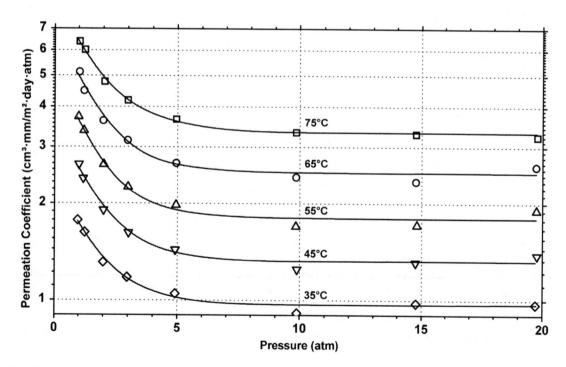

Figure 7.5 Dependence of nitrogen permeability on pressure and temperature through PAI(TMI/HEA).[2]

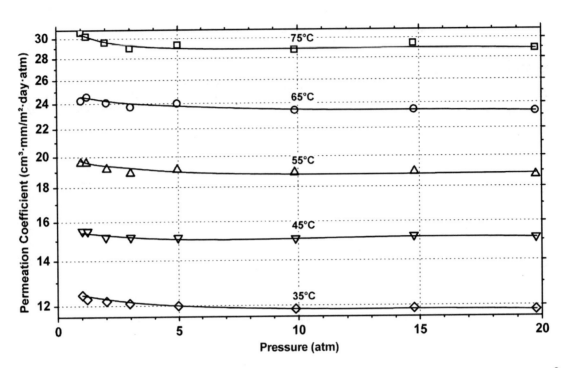

Figure 7.6 Dependence of oxygen permeability on pressure and temperature through PAI(TMI/HEA).[2]

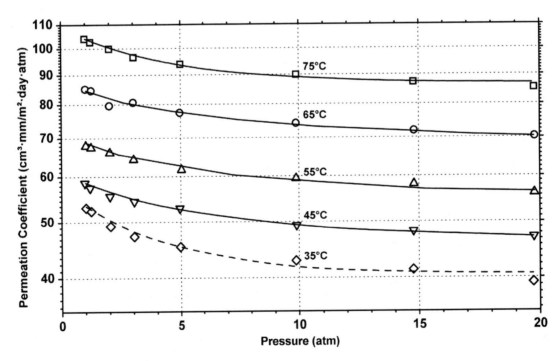

Figure 7.7 Dependence of carbon dioxide permeability on pressure and temperature through PAI(TMI/HEA).[2]

7: Polyimides

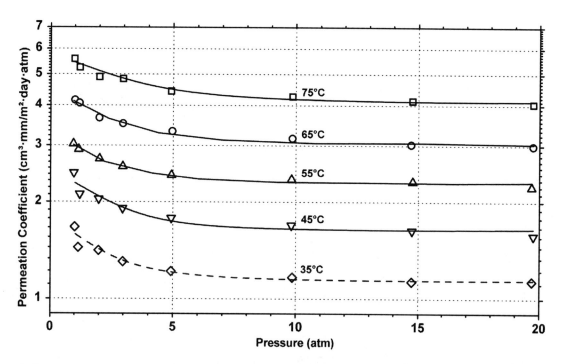

Figure 7.8 Dependence of methane permeability on pressure and temperature through PAI(TMI/HEA).[2]

7.2 Polyetherimide

Polyetherimide (PEI) is an amorphous engineering thermoplastic. Thermoplastic PEIs provide the strength, heat resistance, and flame retardancy of traditional polyimides (PIs) with the ease of simple melt processing seen in standard injection-molding resins like polycarbonate and ABS.

The key performance features of PEI resins include

- Excellent dimensional stability at high temperatures under load
- Smooth as-molded surfaces
- Transparency, though slightly yellow
- Good optical properties
- Very high strength and modulus
- High continuous-use temperature
- Inherent ignition resistance without the use of additives
- Good electrical properties with low-ion content.

There are several different polymers that are offered in various PEI plastics. The structures of these are shown in Figs 7.9–7.13 with references to one of the product lines that utilize that molecule. The CAS number is 61128-46-9.

The acid dianhydride used to make most of the PEIs is 4,4′-bisphenol A dianhydride (BPADA), the structure of which is shown in Fig. 7.14.

Some of the other monomers used in these PEIs are shown in Fig. 7.15.

Many products are called thermoplastic polyimide (TPI) by their manufacturer. These can usually be classified as PEIs.

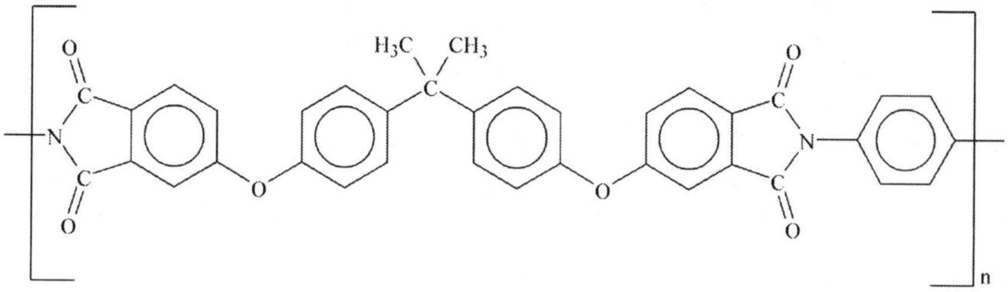

Figure 7.9 Chemical structure of BPADA–PPD PEI (Ultem® 5000 Series).

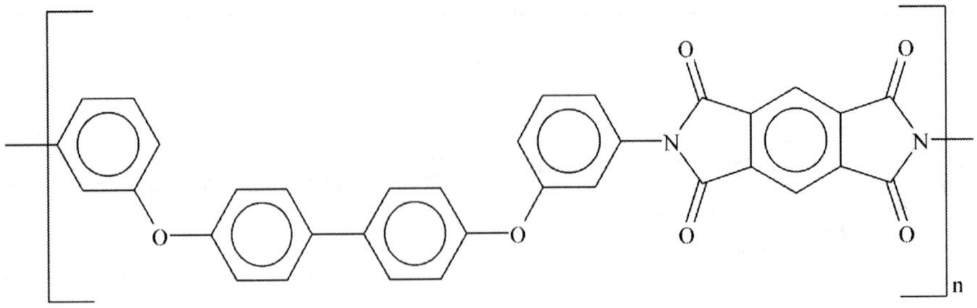

Figure 7.10 Chemical structure of bisphenol diamine PMDA PEI (Aurum®, Vespel® TP-8000 series).

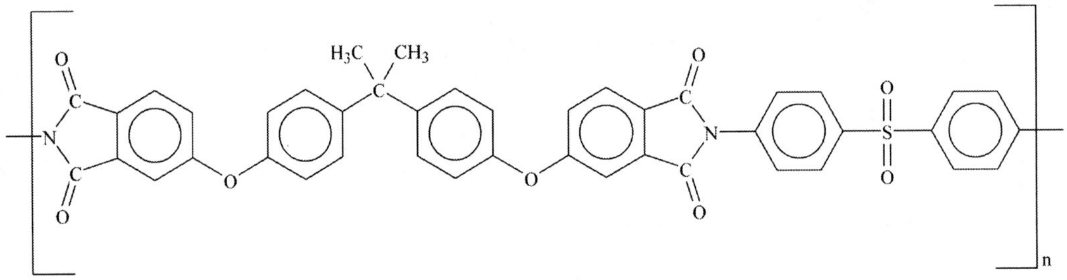

Figure 7.11 Chemical structure of BPADA–DDS PEI sulfone (Ultem® XH6050).

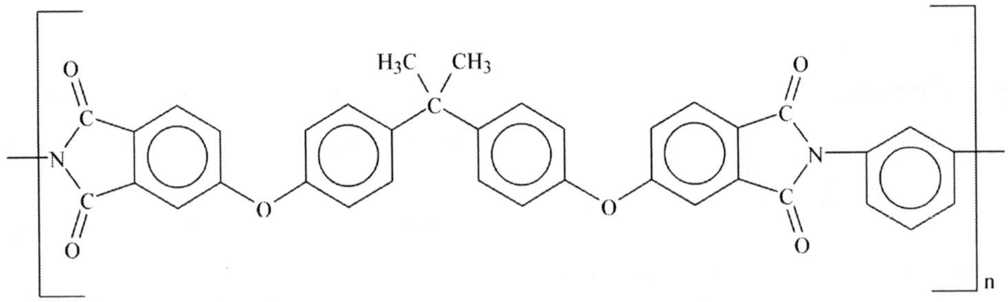

Figure 7.12 Chemical structure of BPADA–MPD PEI (Ultem® 1000 Series).

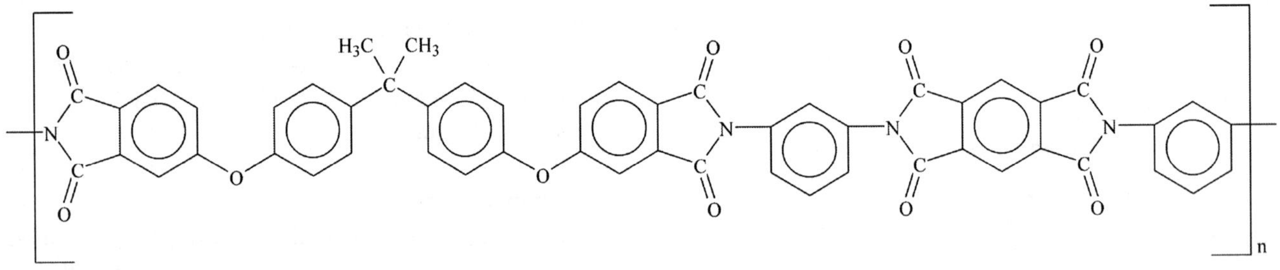

Figure 7.13 Chemical structure of BPADA–PMDA–MPD coPEI (Ultem® 6000 Series).

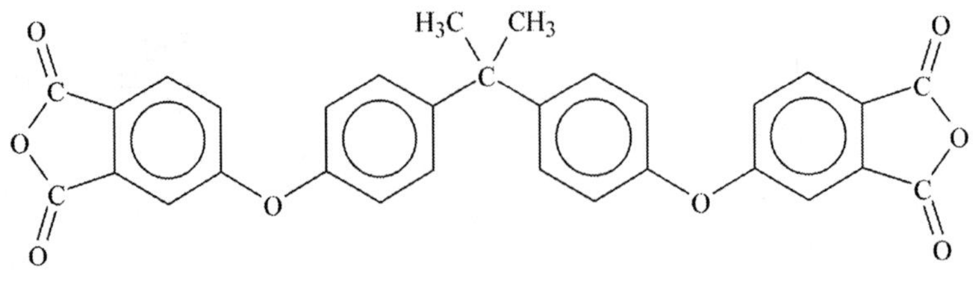

Figure 7.14 Chemical structure of BPADA monomer.

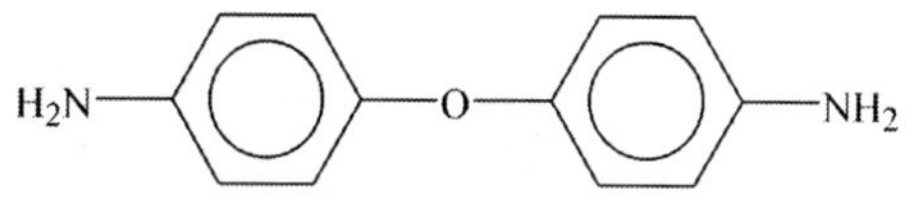

Oxydianiline
(ODA)

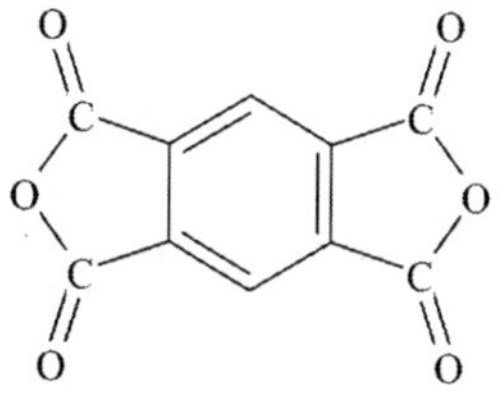

Pyromellitic Dianhydride (PMDA)

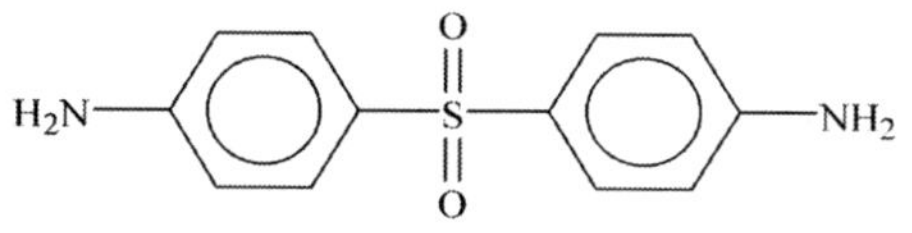

Diamino Diphenyl Sulfone
(DDS)

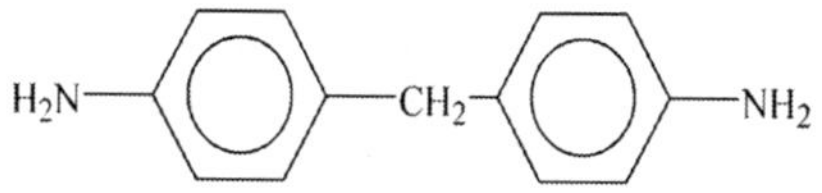

Methylene Dianiline
(MDA)

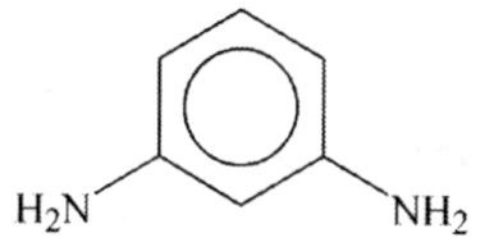

-Phenylene Diamine
(MPD)

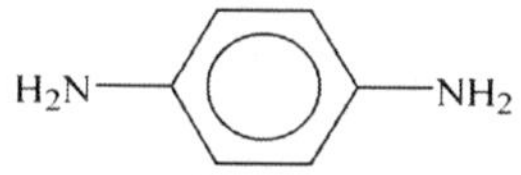

p-Phenylene Diamine
(PDA)

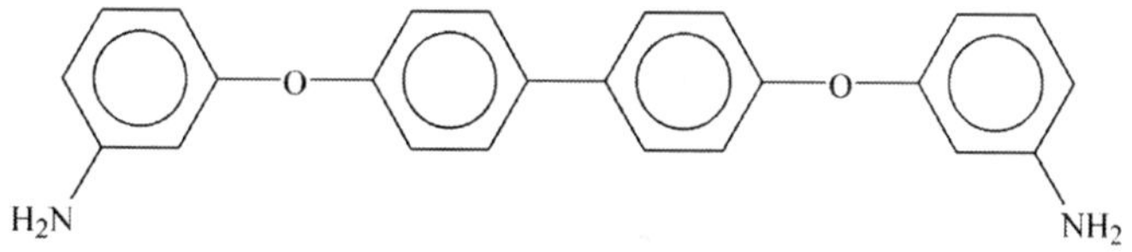

Biphenol Diamine
(BP Diamine)

Figure 7.15 Chemical structures of other monomers used to make polyimides.

Manufacturers and trade names: SABIC Innovative Polymers Ultem®, DuPont™ Vespel®, and AURUM®.

Applications and uses: Surgical probes, pharmaceutical process equipment manifolds, high-frequency insulators used in microwave communications equipment, clamps used to connect printed circuit boards to video display units used in airplanes, tanks, and ships (Tables 7.5–7.8, Figs. 7.15–7.18).

Table 7.5 Permeation of Gases and Water Vapor through Westlake Plastics Tempalux® PEI Film[3]

Permeant Gas	Source Document Units; Permeability Coefficient (cm^3 mil/100 in.^{2} day atm)	Normalized Units; Permeability Coefficient (cm^3 mm/m^2 day atm)
Carbon dioxide	171	67
Oxygen	25	10
Permeant Vapor	**Source Document Units; Vapor Transmission Rate (g mil/100 in.^{2} day)**	**Normalized Units; Vapor Transmission Rate (g mm/m^2 day)**
Water	5.8	2.3

Table 7.6 Permeation of Oxygen and Water Vapor at 25 °C through Sabic Innovative Plastics Ultem® 1000 PEI[4]

Permeant Gas	Source Document Units; Permeability Coefficient (cm^3 mil/100 $in.^2$ day atm)	Normalized Units; Permeability Coefficient (cm^3 mm/m^2 day atm)
Oxygen	37	14
Permeant Vapor	**Source Document Units; Vapor Transmission Rate (g mil/100 $in.^2$ day)**	**Normalized Units; Vapor Transmission Rate (g mm/m^2 day)**
Water	7.9	3.0

Table 7.7 Permeation of Gases through Sabic Innovative Plastics Ultem® and Extem® XH 1015 PEI Membranes[5]

	Permeability Coefficient			
	Source Document Units (Barrer)		Normalized Units (cm^3 mm/m^2 day atm)	
Permeant Gas	Extem®	Ultem®	Extem®	Ultem®
Helium	11.1	9.4	729	617
Carbon dioxide	3.28	1.33	215	87
Oxygen	0.81	0.41	53	27
Nitrogen	0.13	0.051	8.5	3.3
Methane	0.13	0.036	8.5	2.4

Table 7.8 Selectivity of Gas Pair through Sabic Innovative Plastics Ultem® and Extem® XH 1015 PEI Membranes[3]

Resin	Helium/Methane	Carbon Dioxide/Methane	Oxygen/Nitrogen
Extem®	85	25.2	6.2
Ultem®	261	36.9	8.0

See Figs 7.16–7.18.

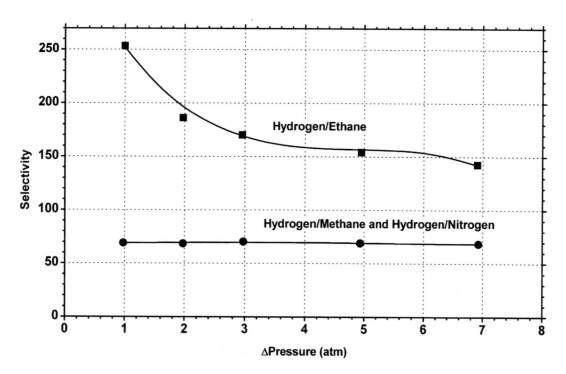

Figure 7.16 Permeation selectivity of hydrogen/ethane, hydrogen methane and hydrogen/nitrogen vs. pressure differential through PEI membranes.[8]

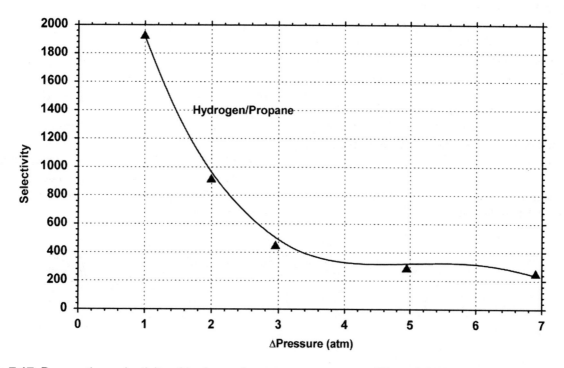

Figure 7.17 Permeation selectivity of hydrogen/propane vs. pressure differential through PEI membranes.

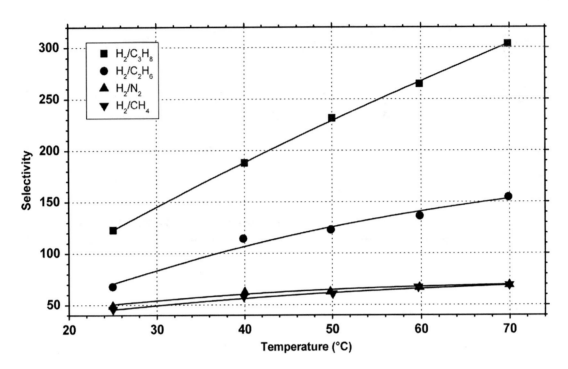

Figure 7.18 Permeation selectivity of various gas pairs vs. temperature through PEI membranes.

7.3 Polyimide

PIs are high-temperature engineering polymers originally developed by the DuPont™ company. PIs exhibit an exceptional combination of thermal stability (>500 °C), mechanical toughness, and chemical resistance. They have excellent dielectric properties and inherently low coefficient of thermal expansion. They are formed from diamines and dianhydrides such as those shown in Fig. 7.19.

Many other diamines and several other dianhydrides may be chosen to tailor the final properties of a polymer whose structure is like that shown in Fig. 7.20.

Manufacturers and trade names: DuPont™ Kapton®, UBE Industries Upilex®-S.

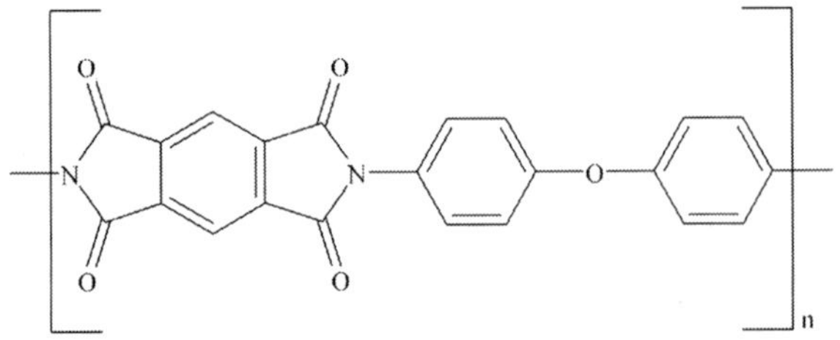

Figure 7.20 Chemical structure of a typical polyimide.

Applications and uses: Aerospace, flexible printed circuits, automotives, heaters, bar code labels, pressure sensitive tape, electrical insulation, and safety (Tables 7.9–7.13).

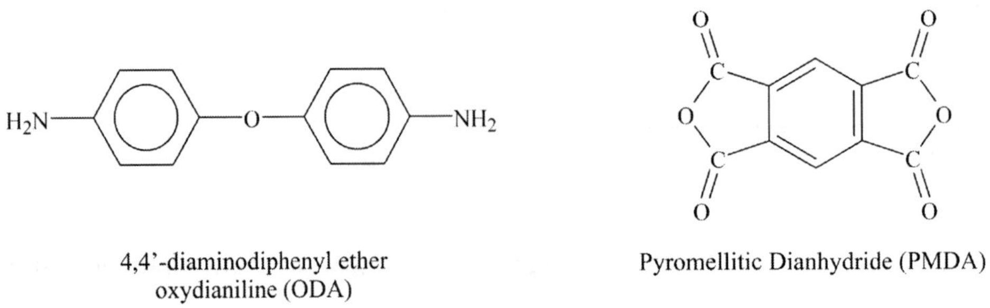

4,4'-diaminodiphenyl ether
oxydianiline (ODA)

Pyromellitic Dianhydride (PMDA)

Figure 7.19 Chemical structures of monomer used to make polyimides.

7: Polyimides

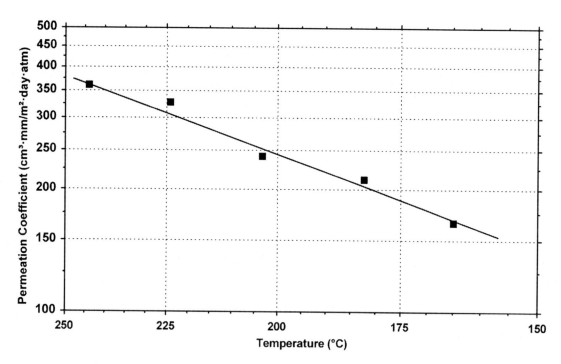

Figure 7.21 Permeation of oxygen vs. temperature for DuPont™ Kapton® film.[9]

Table 7.9 Permeability of Various Gases at 23 °C (73 °F) and 50% RH through DuPont™ Kapton®[6]

Gas	Permeability Coefficient	
	Source Documents Units (ml/m²/24 h/MPa)	Normalized Units (cm³ mm/m² day atm)
Carbon dioxide	6840	17
Oxygen	3800	10
Hydrogen	38,000	100
Nitrogen	910	2
Helium	63,080	163

Test standard: ASTM D1434.

Table 7.10 Water Vapor Permeation of DuPont™ Kapton®[7]

Vapor transmission rate	
Source Document Units (g/m²/24 h)	Normalized Units (g mm/m² day)
54	1.35

Test standard: ASTM E-96-92.

Table 7.11 Water Vapor Permeability of FEP Coated DuPont™ Kapton® Films[7]

Property	120FN616	150FN019	400FN022	200FWR919
FEP/PI/FEP thickness (μm)	2.5/25.4/2.5	0/25.4/12.7	0/50.8/50.8	12.7/25.4/12.7
Source document units (g/m²/24 h)	17.5	9.6	2.4	8.4
Normalized units (g mm/m² day)	0.53	0.37	0.24	0.43

Test standard: ASTM E96.

Table 7.12 Oxygen, Nitrogen, Carbon dioxide, and Helium at 30 °C through UBE Industries Upilex® Films[7]

Product	UBE UPILEX R			UBE UPILEX S and VT		
Penetrant	Oxygen	Nitrogen	Carbon Dioxide	Helium	Oxygen	Carbon Dioxide
Source document units (cm^3/mil/m^2/day/atm)	100	30	115	2200	0.8	1.2
Normalized permeability coefficient (cm^3 mm/m^2 day atm)	2.54	0.76	2.92	55.9	0.02	0.03

Film thickness: 0.025 mm; test standard: ASTM D1434.

Table 7.13 Water Vapor Permeation Rates at 38 °C and 90% Relative Humidity through UBE Industries Upilex Films[8]

Product	UBE UPILEX R	UBE UPILEX S and VT
Source document units (g mil/m^2 day atm)	22	1.7
Normalized units; permeability coefficient (g mm/m^2 day)	0.56	0.04

See Fig. 7.21; film thickness: 0.025 mm; test standard: ASTM D1434.

References

1. Cao X, Lu F. Structure/permeability relationships of polyamide–imides. *J Appl Polym Sci* 1994;**54**:1965–70.
2. Kresse I, et al. Gas transport properties of soluble poly(amide–imide)s. *J Polym Sci B Polym Phys* 1999;**37**:2183–93.
3. Tempalux® Film Product Bulletin, Westlake Plastic Products; 2001.
4. Oxygen and Water Permeability Data, Sabic Innovative Plastics; 2008.
5. Xia J, Liu S, Pallathadka PK, Chng ML, Chung T. Structural determination of Extem XH 1015 and its gas permeability comparison with polysulfone and Ultem via molecular simulation. *Ind Eng Chem Res* 2010:100722143436002. Available at: http://pubs.acs.org/doi/abs/10.1021/ie901906p; 2010.
6. Kapton summary of Properties, Dupont, H-34892–2; 2006.
7. Ube Ultra-High Heat-Resistant Polyimide Film Upilex, supplier marketing literature. Ube Industries, Ltd, 2010.
8. Wang D, Teo WK, Li K. Permeation of H_2, N_2, CH_4, C_2H_6, and C_3H_8 through asymmetric polyetherimide hollow-fiber membranes. *J Appl Polym Sci* 2002;**86**(3):698–702. Available at: http://doi.wiley.com/10.1002/app.10966.
9. Koros WJ, Wang J, Felder RM. Oxygen permeation through FEP Teflon and Kapton polyimide. *J Appl Polym Sci* 1981;**26**(8):2805–9. Available at: http://doi.wiley.com/10.1002/app.1981.070260832.

8 Polyamides (Nylons)

High-molecular-weight polyamides are commonly known as nylon. Polyamides are crystalline polymers typically produced by the condensation of a diacid and a diamine. There are several types and each type is often described by a number, such as nylon 66 or polyamide 66 (PA66). The numeric suffixes refer to the number of carbon atoms present in the molecular structures of the amine and acid, respectively (or a single suffix if the amine and acid groups are part of the same molecule).

The polyamide plastic materials discussed in this book and the monomers used to make them are given in Table 8.1.

The general reaction is shown in Fig. 8.1.

The −COOH acid group reacts with the $-NH_2$ amine group to form the amide. A molecule of water is given off as the nylon polymer is formed. The properties of the polymer are determined by the R and R′ groups in the monomers. In nylon 6,6, R′ = 6C, and R = 4C alkanes, but one also has to include the two carboxyl carbons in the diacid to get the number it designates to the chain.

The structures of these diamine monomers are shown in Fig. 8.2, the diacid monomers are shown in Fig. 8.3. Figure 8.4 shows the amino acid monomers. These structures only show the functional groups, the CH_2 connecting groups are implied at the bond intersections.

All polyamides tend to absorb moisture that can affect their properties. Properties are often reported as dry as molded (DAM) or conditioned [usually at equilibrium in 50% relative humidity (RH) at 23 °C]. The absorbed water tends to act like a plasticizer and can have a significant effect on the plastics properties.

Table 8.1 Monomers Used to Make Specific Polyamides/Nylons

Polyamide/Nylon Type	Monomers Used to Make
Nylon 6	Caprolactam
Nylon 11	Aminoundecanoic acid
Nylon 12	Aminolauric acid
Nylon 66	1,6-Hexamethylene diamine and adipic acid
Nylon 610	1,6-Hexamethylene diamine and sebacic acid
Nylon 612	1,6-Hexamethylene diamine and 1,12-dodecanedioic acid
Nylon 666	Copolymer based on nylon 6 and nylon 66
Nylon 46	1,4-Diaminobutane and adipic acid
Nylon amorphous	Trimethyl hexamethylene diamine and TPA
PPA	Any diamine and IPA and/or TPA

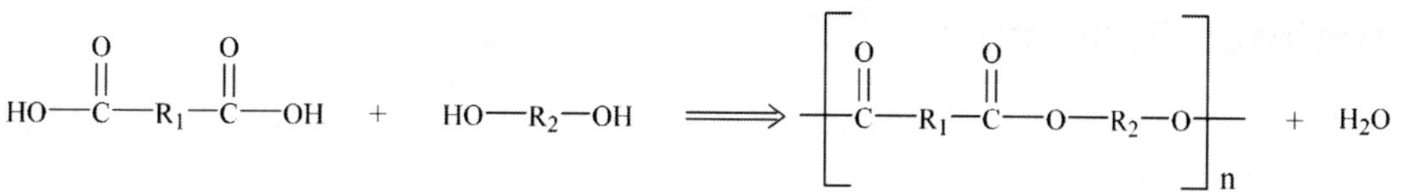

Figure 8.1 Generalized polyamide reaction.

Permeability Properties of Plastics and Elastomers. DOI: 10.1016/B978-1-4377-3469-0.10008-6

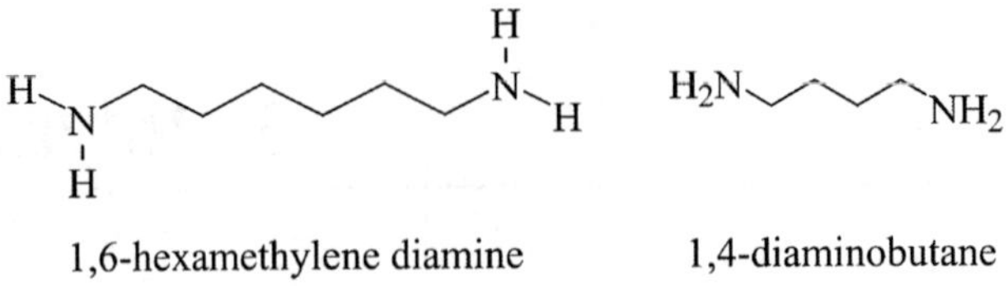

Figure 8.2 Chemical structures of diamines used to make polyamides.

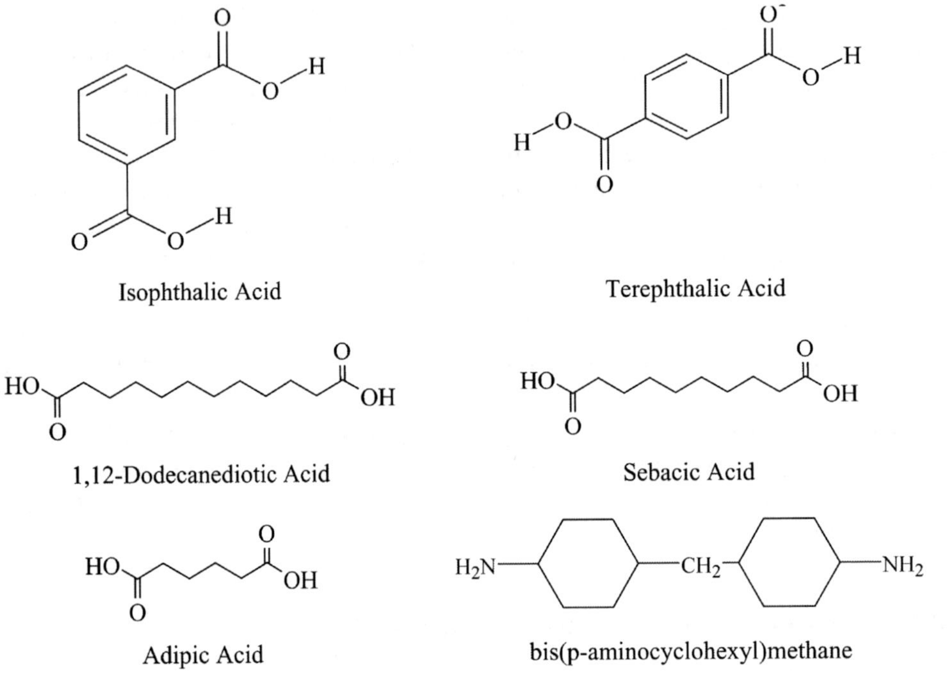

Figure 8.3 Chemical structures of diacids used to make polyamides.

Figure 8.4 Chemical structures of amino acids used to make polyamides.

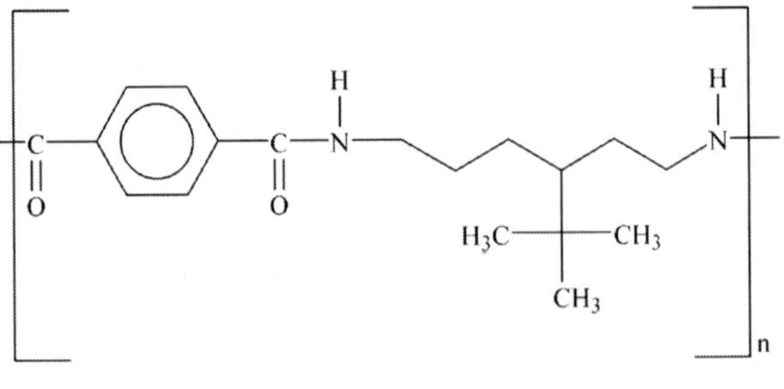

Figure 8.5 Chemical structure of amorphous nylon.

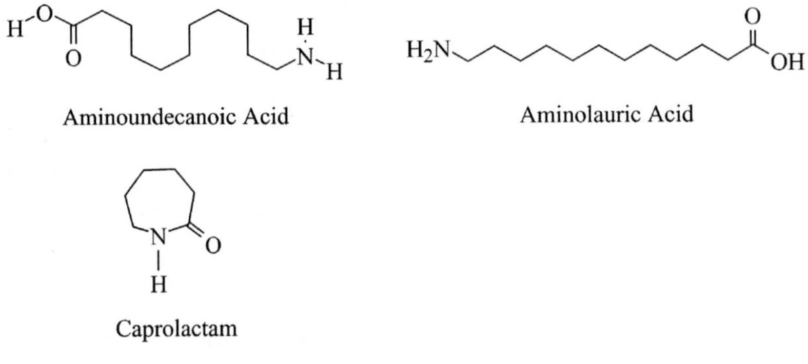

8.1 Amorphous Polyamide (Nylon)

Amorphous nylon is designed to give no crystallinity to the polymer structure. One such amorphous nylon is shown in Fig. 8.5.

The tertiary butyl group attached to the amine molecule is bulky and disrupts this molecule's ability to crystallize. This particular amorphous nylon is sometimes designated at nylon 6-3-T. Amorphous polymers can have properties that differ significantly from crystalline types, one of which is optical transparency.

Some of the amorphous nylon characteristics are as follows:

- Crystal-clear, high optical transparency
- High mechanical stability
- High heat deflection temperature
- High impact strength
- Good chemical resistance compared to other plastics
- Good electrical properties
- Low mold shrinkage

Blending even low percentages (20%) of Selar® PA with nylon 6, nylon 66, and nylon copolymers will result in a product that behaves like an amorphous polymer. These blends retain all of the advantages of the Selar® PA resin with some of the mechanical property advantages of semicrystalline nylon.

Manufacturers and trade names: DuPont™ Selar® PA, EMS Chemie Grivory® G 16, and Grivory® G21.

Applications and uses: Used as a monolayer or as a component of multilayer flexible in meat and cheese packages as well as rigid packaging; multilayer or monolayer are used in transparent hollow vessels (bottles), packaging films, and deep-drawn plates (Tables 8.2–8.7, Figs. 8.6–8.9).

Table 8.2 Permeation of Oxygen at 23 °C through EMS Chemie Grivory® G16 and Grivory® G21 Amorphous Nylon[1]

Grade	G16	G21	G16	G21
RH (%)	0	0	85	85
Test method	ASTM D3985	DIN 53380	ASTM D3985	DIN 53380
Source document units, gas permeability (cm^3/m^2 day bar)		30		
Normalized units, permeability coefficient (cm^3 mm/m^2 day atm)	1.54	1.5	0.512	0.4

Table 8.3 Water Vapor Permeation at 23 °C through EMS Chemie Grivory® G16 and Grivory® G21 Amorphous Nylon (0.05 mm)[1]

Grade	G16	G21
RH (%)	0	85
Test method	ASTM D3985	DIN 53122
Source document units, vapor transmission rate (g/100 in.^{2} day)	0.9	
Source document units, vapor transmission rate (g/m^2 day)		7
Normalized units, vapor transmission rate (g mm/m^2 day)	0.005	0.35

Table 8.4 Carbon Dioxide and Nitrogen at 23 °C through EMS Chemie Grivory® G16 Amorphous Nylon[1]

Penetrant	Carbon dioxide		Nitrogen
RH (%)	0	85	0
Test method	EMS method	EMS method	DIN 53380
Permeability coefficient (cm^3 mm/m^2 day atm)	4.57	2.05	0.512

Table 8.5 Permeation of Carbon Dioxide vs. Temperature and Humidity through DuPont™ Selar® PA Amorphous Nylon[2]

Temperature (°C)	0		30	
RH (%)	0–5	95–100	0–5	95–100
Source document units, permeability coefficient (cc mil/100 in.² day atm)	5.5	12.2	18	9.8
Normalized units, permeability coefficient (cm³ mm/m² day atm)	2.16	4.8	7.07	3.85

Table 8.6 Water Vapor at 90% RH through DuPont™ Selar® PA Amorphous Nylon[3]

Temperature (°C)	37.8	40
Source document units, vapor transmission rate (g mil/100 in.² day)	1.2	1.4
Normalized units, vapor transmission rate (g mm/m² day)	0.47	0.55

Table 8.7 Oxygen Permeation of DuPont™ Selar® Blends with Nylon 6[2]

Temperature	RH	% Selar® PA 3426 in Blend					
		0	20	30	50	80	100
Source Document Units; Permeability Coefficient (cm³ mil/100 in.² day atm)							
0	0–5	0.9	0.9	0.9	0.9	0.9	0.8
0	95–100	3.7	2.0	1.3	0.5	0.4	0.3
30	0–5	4.0	3.9	3.9	3.9	3.9	3.8
30	95–100	15.0	14.0	12.0	9.1	5.6	1.5
Normalized Units; Permeability Coefficient (cm³ mm/m² day atm)							
0	0–5	0.4	0.4	0.4	0.4	0.4	0.3
0	95–100	1.5	0.8	0.5	0.2	0.2	0.1
30	0–5	1.6	1.5	1.5	1.5	1.5	1.5
30	95–100	5.9	5.5	4.7	3.6	2.2	0.6

See also Figs. 8.6–8.10.

8: Polyamides (Nylons)

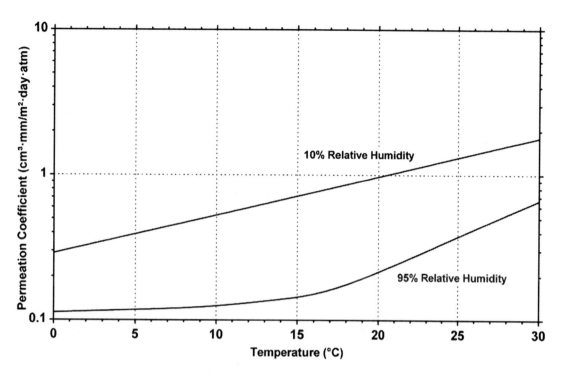

Figure 8.6 Permeation of oxygen vs. temperature at 10% and 95% RH through DuPont™ Selar® amorphous polyamide.[2]

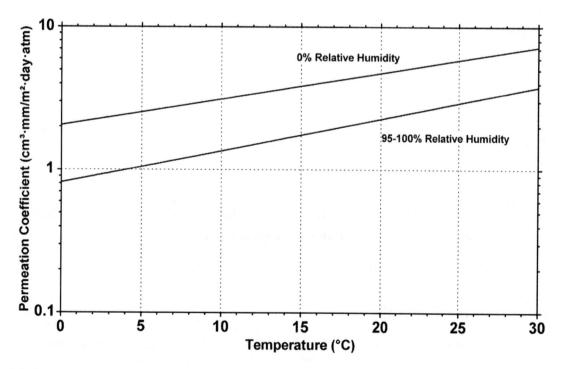

Figure 8.7 Permeation of carbon dioxide vs. temperature through DuPont™ Selar® PA and nylon 6 at 10% and 95% RH.[2]

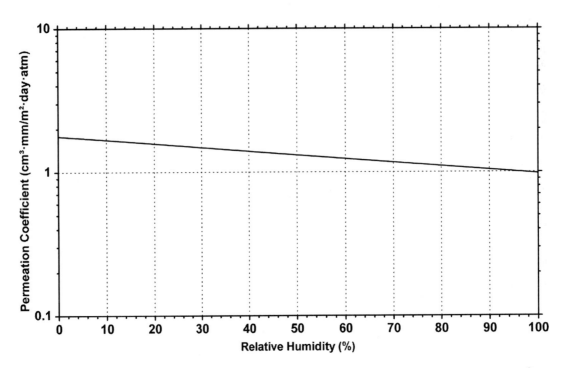

Figure 8.8 Permeation of carbon dioxide vs. RH through DuPont™ Selar® PA amorphous nylon.[3]

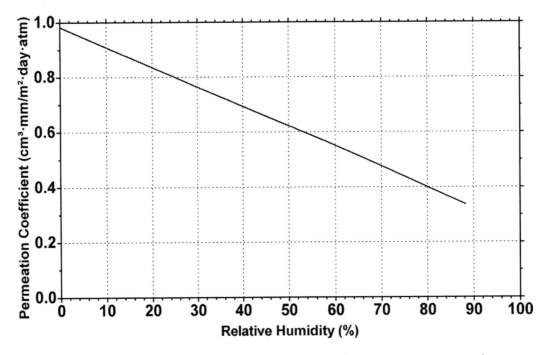

Figure 8.9 Permeation of oxygen vs. RH through DuPont™ Selar® PA amorphous nylon.[4]

8.2 Polyamide 6 (Nylon 6)

Nylon 6 begins as pure caprolactam, which is a ring-structured molecule. This is unique in that the ring is opened and the molecule polymerizes with itself. Since caprolactam has six carbon atoms, the nylon that it produces is called nylon 6, which is nearly the same as nylon 66 described in Section 8.5. The structure of nylon 6 is shown in Fig. 8.11 with the repeating unit in the brackets. The CAS Number is 628-02-4.

8: Polyamides (Nylons)

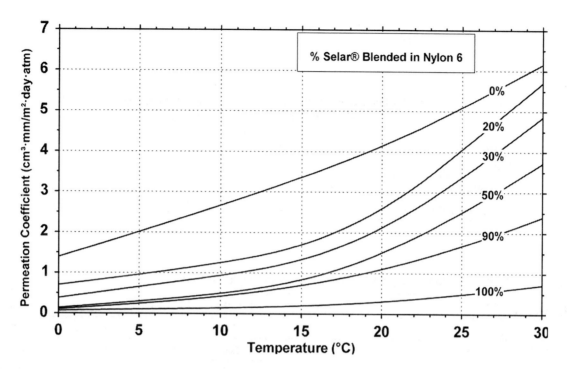

Figure 8.10 Permeation of oxygen vs. temperature through blends of DuPont™ Selar® with nylon 6.[2]

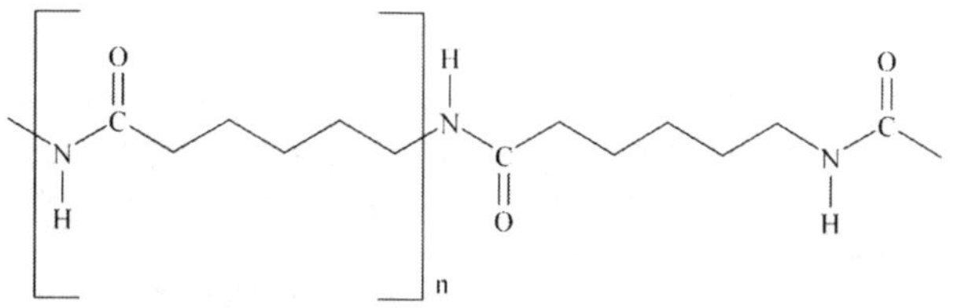

Figure 8.11 Chemical structure of nylon 6.

Some of the nylon 6 characteristics are as follows:

- Outstanding balance of mechanical properties.
- Outstanding toughness in equilibrium moisture content.
- Outstanding chemical resistance and oil resistance.
- Outstanding wear and abrasion resistance.
- Almost all grades are self-extinguishing. The flame-resistant grades are rated UL 94VO.
- Outstanding long-term heat resistance (at a long-term continuous maximum temperature ranging between 80 and 150 °C).
- Grades reinforced with glass fiber and other materials offer superior elastic modulus and strength.
- Offers low gasoline permeability and outstanding gas barrier properties.
- Highest rate of water absorption and highest equilibrium water content (8% or more).
- Excellent surface finish even when reinforced.
- Poor chemical resistance to strong acids and bases.

Manufacturers and trade names: BASF Ultramid® B, Honeywell Capran® and Aegis®, EMS Grilon® B, UBE Industries.

Applications and uses: Multilayer packaging, food and medical, industrial containers, and automotive underhood reservoirs (Tables 8.8–8.14, Fig. 8.10).

Table 8.8 Permeation of Gases at Various Temperatures through Honeywell Plastics Capron® Nylon 6 Films[5]

Permeant	Permeability Coefficient					
	Source Document Units (cm³/100 in.² day atm)			Normalized Units (cm³ mm/m² day atm)		
	0 °C	23 °C	50 °C	0 °C	23 °C	50 °C
Oxygen	0.5	2.6	14	0.2	1.02	5.5
Nitrogen	0.2	0.9	12	0.08	0.35	4.7
Carbon dioxide	0.6	4.7	44	0.24	1.8	17.3

Film thickness: 0.0254 mm; RH: 0%.

Table 8.9 Water Vapor through Honeywell Plastics Capron® Nylon 6 Films[6]

Temperature (°C)	23	23	37.8	37.8
Film thickness (mm)	0.019	0.0254	0.019	0.0254
Relative Humidity (%)	50	50	90	90
Source document units vapor permeation rate (g/day 100 in.²)	0.08	0.6	24–26	19–20
Normalized units vapor permeation rate WVTR (g mm/m² day)	0.24	0.24	7.1–7.7	5.6–5.9

Table 8.10 Permeation of Gases at 23 °C and 0% RH through Honeywell Plastics Aegis® Nylon 6 Films[6] (Applies to H73QP, H73ZP, H86MP, H85NP, H85QP, H100MP, H100QP, H100ZP, H135KQP, H135MP, H135QP, H135WP, H135ZP, H155MP, H155QP, H155WP, H155ZP, H205QP)

Permeate	Permeability Coefficient	
	Source Document Units (cm³ mil/m² day atm)	Normalized Units (cm³ mil/m² day atm)
Oxygen	40.3	1.02
Nitrogen	14	0.36
Carbon dioxide	72.8	1.85

Table 8.11 Permeation of Oxygen and Water Vapor through UBE Industries Nylon 6[7]

Grade	Oxygen	Water Vapor
	Source Document Units (cm³/m² day)	Normalized Units (g/m² day)
Test method	ASTM D3985	JIS Z-0208
1022B	41	125
1030B	41	125
1022FDX99	41	125
1022C2	25	65

8: POLYAMIDES (NYLONS)

Table 8.12 Permeation of Oxygen through Oriented and Un-Oriented Nylon 6[3]

Temperature °C	Oriented		Unoriented	
	Source Document Units; Permeability Coefficient (cm³ 25 μm/ m² day atm)	Normalized Units; Permeability Coefficient (cm³ mm/ m² day atm)	Source Document Units; Permeability Coefficient (cm³ 25 μm/ m² day atm)	Normalized Units; Permeability Coefficient (cm³ mm/ m² day atm)
5	7.59	0.19	22.3	0.57
23	25.6	0.7	78.7	2
35	51.2	1.3	155	3.9

Table 8.13 Permeation of Oxygen, Carbon Dioxide, and Nitrogen at 23 °C through EMS Grivory® Grilon® F 34 Type 6 Nylon[8]

Permeant	RH (%)	Permeability Coefficient	
		Source Document Units (cm³/m² day atm)	Normalized Units (cm³ mm/m² day atm)
Oxygen	0	25	1.26
	85	100	5.05
Nitrogen	0	10	0.5
Carbon dioxide	0	65	3.28

Thickness: 0.05 mm; test methods: DIN 53380, DIN 53122.

Table 8.14 Oxygen, Carbon Dioxide, Nitrogen, and Water Vapor through EMS Grilon® F 50 Type 6 Nylon[7]

Permeant	RH (%)	Permeability Coefficient	
		Source Document Units (cm³/m² day atm)	Normalized Units (cm³ mm/m² day atm)
Oxygen	0	25	1.26
	85	70	3.53
Nitrogen	0	10	0.5
Carbon dioxide	0	80	4.04
	85	250	12.6

Thickness: 0.05 mm; test methods: DIN 53380, DIN 53122. See also Figs. 8.12–8.14.

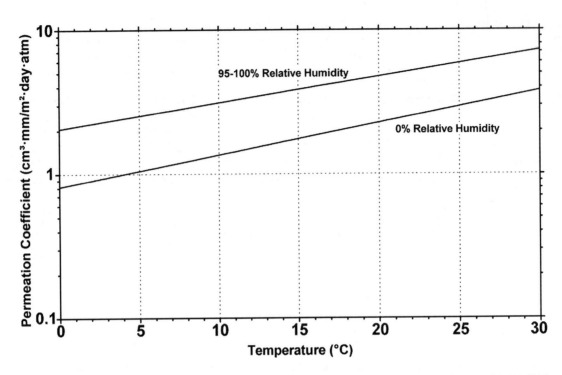

Figure 8.12 Permeation of carbon dioxide vs. temperature through nylon 6 at 10% RH and 95% RH.

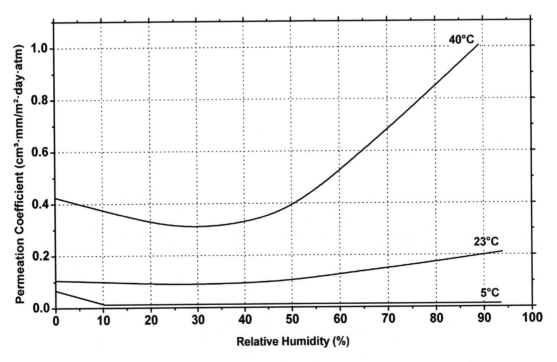

Figure 8.13 Permeation of hydrogen sulfide vs. temperature through DuPont™ nylon 6.[9]

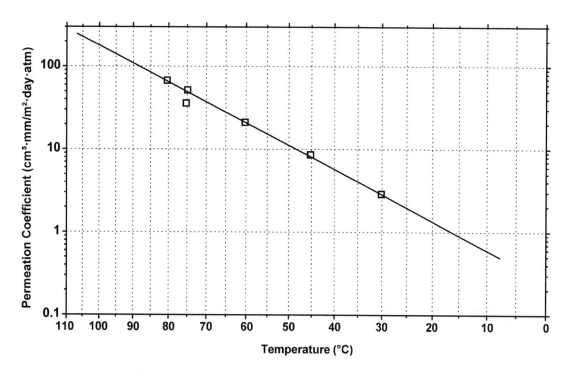

Figure 8.13 Continued

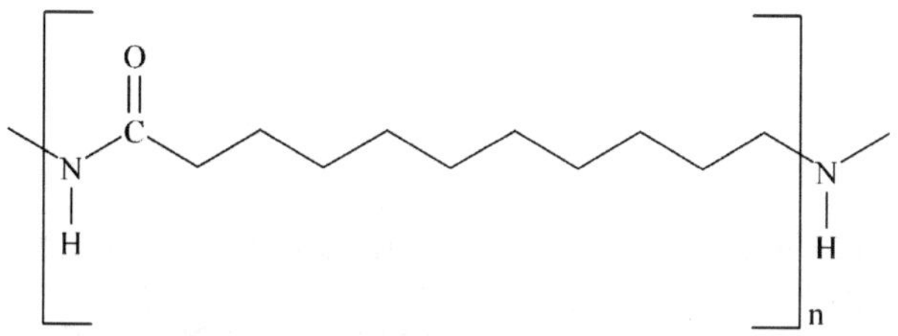

Figure 8.14 Chemical structure of nylon 11.

8.3 Polyamide 11 (Nylon 11)

Nylon 11 has only one monomer, aminoundecanoic acid. It has the necessary amine group at one end and the acid group at the other. It polymerizes with itself to produce the polyamide containing 11 carbon atoms between the nitrogen of the amide groups. Its structure is shown in Fig. 8.15 and it has a CAS number of 25035-04-5.

Rilsan® PA 11 is produced from a "green" raw material—castor beans.

Some of the nylon 11 characteristics are as follows:

- Low water absorption for nylon (2.5% at saturation)
- Reasonable UV resistance
- Higher strength
- Ability to accept high loading of fillers
- Better heat resistance than nylon 12
- More expensive than nylon 6 or nylon 6/6
- Relatively low impact strength

Manufacturers and trade names: Arkema Rilsan® B, Suzhou Hipro Polymers Hiprolon®.

Applications and uses: Automotive: fuel and brake lines, ski boots, tennis racquets, medical catheters, and tubing (Table 8.15, Figs. 8.15 and 8.16).

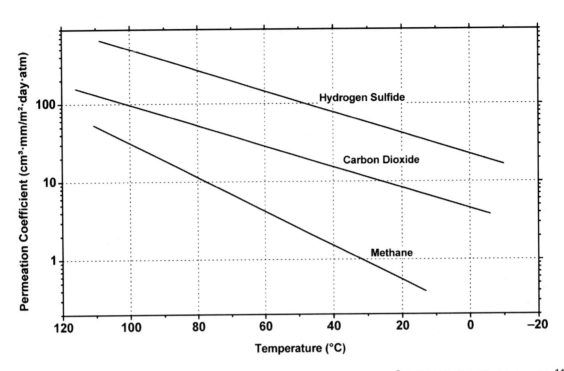

Figure 8.15 Permeability vs. temperature of various gases through Rilsan® BESNO P40TL Nylon 11.[11]

Table 8.15 Permeation of Various Gases at 20 °C through Arkema Rilsan® PA11[10]

Permeant Gas	Permeability Coefficient			
	Source Document Units, 10^{-9} (cm^3 cm/cm^2 s bar)		Normalized Units (cm^3 mm/m^2 day atm)	
Rilsan® Grade	BESNO TL	BESNO P40TL	BESNO TL	BESNO P40TL
Hydrogen	7	15	61	131
Nitrogen	0.15		1.3	
Oxygen	2		18	
Carbon dioxide	7	6	61	53
Water		0.04		0.35
Hydrogen sulfide		30		263
Methane	0.15	0.6	1	5
Ethane		2.3		20
Propane		0.75		7
Butane		5.4		47

See also Figs. 8.16 and 8.17.

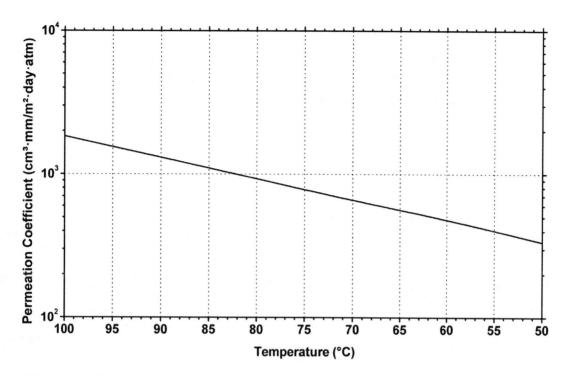

Figure 8.16 Permeability vs. temperature of natural gas (composition: 85% methane, 8% propane, and 2% butane) through Rilsan® BESNO P40TL Nylon 11.[11]

8.4 Polyamide 12 (Nylon 12)

Nylon 12 has only one monomer, aminolauric acid. It has the necessary amine group at one end and the acid group at the other. It polymerizes with itself to produce the polyamide containing 12 carbon atoms between the two nitrogen atoms of the two amide groups. Its structure is shown in Fig. 8.17.

The properties of semicrystalline polyamides are determined by the concentration of amide groups in the macromolecules. Polyamide 12 has the lowest amide group concentration of all commercially available polyamides thereby substantially promoting its characteristics:

- Lowest moisture absorption (~2%): Parts show largest dimensional stability under conditions of changing humidity.

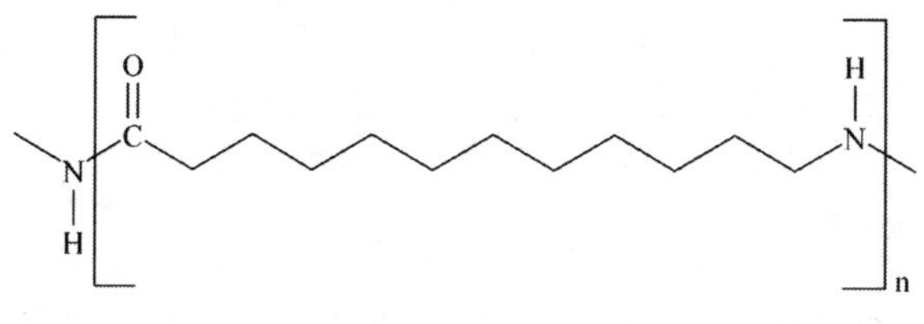

Figure 8.17 Chemical structure of nylon 12.

- Exceptional impact and notched impact strength, even at temperatures well below the freezing point.
- Good to excellent resistance against greases, oils, fuels, hydraulic fluids, various solvents, salt solutions, and other chemicals.
- Exceptional resistance to stress cracking, including metal parts encapsulated by injection molding or embedded.
- Excellent abrasion resistance.
- Low coefficient of sliding friction.
- Noise and vibration damping properties.
- Good fatigue resistance under high-frequency cyclical loading condition.
- High processability.
- Expensive.
- Lowest strength and heat resistance of any Polyamide unmodified generic.

Manufacturers and trade names: Arkema Rilsan® A, EMS-Grivory® Grilamid®, Exopack® Dartek® (Tables 8.16 and 8.17, Fig. 8.18).

Table 8.16 Permeation of UBE 303 XA Nylon 12 Resin[7]

Permeant	Oxygen	Water Vapor
Test method	ASTM D3985	JIS Z-0208
Source document units	cm^2/m^2 day	g/m^2 day
303XA	1050	50

Table 8.17 Permeation of EMS-Grivory® Grilamid L 25 Nylon 12 Resin[11]

Permeant	Conditions	Test Method	Source Document Units	Normalized Units
			(g/m^2 day)	($g\ mm/m^2$ day)
Water vapor	23 °C, 85% RH	DIS 15106-1/-2	8	0.4
			(cm^3/m^2 day bar)	($cm^3\ mm/m^2$ day bar)
Oxygen	23 °C, 0% RH	DIS 15105-1/-2	350	17.7
	23 °C, 85% RH	DIS 15105-1/-2	370	18.7
Carbon dioxide	23 °C, 0% RH	DIS 15105-1/-2	1500	76
	23 °C, 85% RH	DIS 15105-1/-2	1600	81

See also Fig. 8.19. Thickness: 0.05 mm.

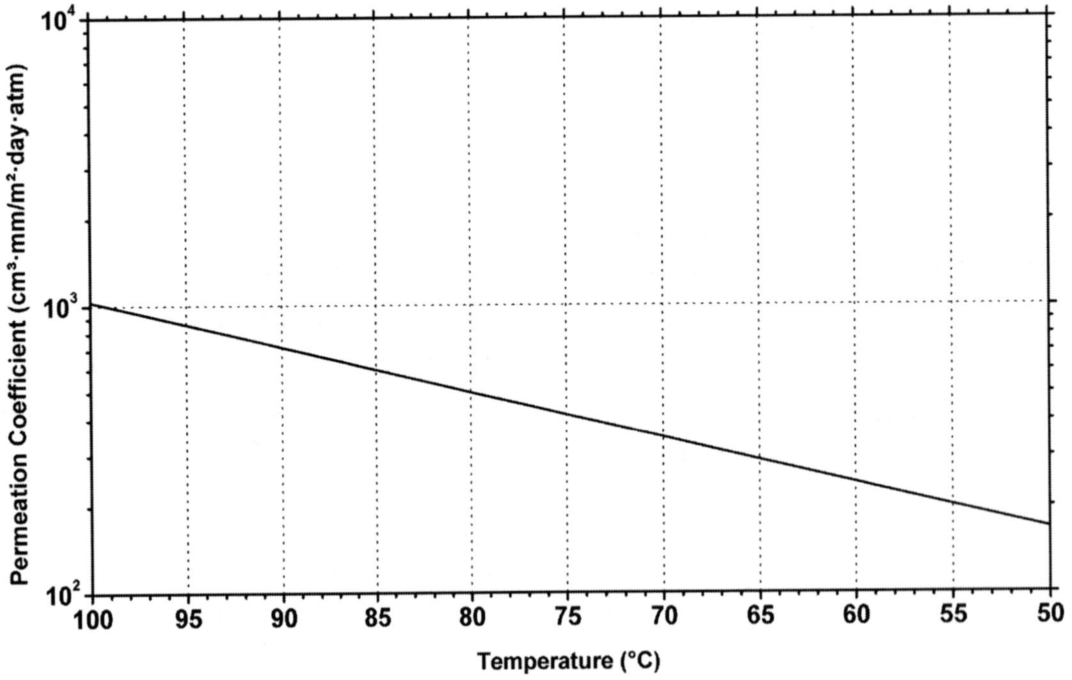

Figure 8.18 Permeability vs. temperature of natural gas (composition: 85% methane, 8% propane, and 2% butane) through Rilsan® AESNO P40TL Nylon 12.[10]

8.5 Polyamide 66 (Nylon 66)

The structure of nylon 66 is shown in Fig. 8.19. The CAS number is 32131-17-2.

Some of the nylon 66 characteristics are as follows:

- Outstanding balance of mechanical properties.
- Outstanding toughness in equilibrium moisture content.
- Outstanding chemical resistance and oil resistance.
- Outstanding wear and abrasion resistance.
- Almost all grades are self-extinguishing. The flame-resistant grades are rated UL 94 V0.
- Outstanding long-term heat resistance (at a long-term continuous maximum temperature ranging between 80 and 150 °C).
- Grades reinforced with glass fiber and other materials offer superior elastic modulus and strength.
- Offers low gasoline permeability and outstanding gas barrier properties.
- High water absorption.
- Poor chemical resistance to strong acids and bases.

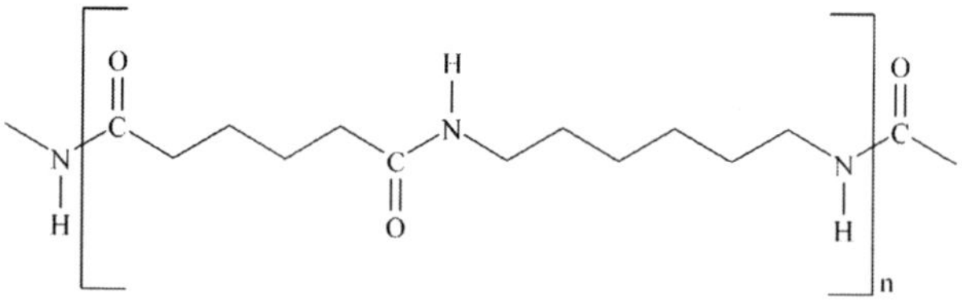

Manufacturers and trade names: Exopack Performance Films Inc. Dartek®, DuPont™ Zytel®.

Applications and uses: Packaging meat and cheese, industrial end uses, pouch and primal bag, stiff packages, snacks, condiments, shredded cheese, and coffee. Wrapping fine art, potable water, and electrical applications (Tables 8.18–8.22).

Figure 8.19 Chemical structure of nylon 66.

Table 8.18 Permeation of Oxygen and Water Vapor through Exopack® Performance Films Inc. Dartek® Nylon 6,6 Films[12]

Dartek® Product	Features	Thickness (mm)	Oxygen Permeability Coefficient (cm³ mm/ m² day atm)	Water Vapor Transmission Rate (cm³ mm/ m² day atm)
B-601	PVDF coated one side	0.025	0.19	0.23
B-602	PVDF coated one side	0.038	0.29	0.34
F-101	Cast film	0.025	1.4	7.38
N-201	For pouch and bag	0.025	1.4	7.38
O-401	Oriented in machine direction	0.015	0.59	2.18
SF-502	Super formable	0.076	3.6	
UF-412	Oriented with slip properties	0.015	0.59	2.18

Table 8.19 Permeation of Oxygen and Carbon Dioxide through BASF Ultramid A5 Nylon 66 Film

Permeant	RH (%)	Permeability Coefficient	
		Source Document Units (cm³ 100 μm/m² day bar)	Normalized Units (cm³ mm/m² day atm)
Oxygen	40	6–7	0.61–0.71
Carbon dioxide	0	45	4.6

Test method: DIN53380; thickness: 0.02 mm.

Table 8.20 Permeation of Water Vapor through BASF Ultramid A5 Nylon 66 Film

Film Type	Vapor Transmission Rate	
	Source Document Units (g 100 μm/m day)	Normalized Units (g mm/m² day)
Flat Film	11–12	1.1–1.2
Tubular Film	8	0.8

RH gradient: 85%–0%; standard test method: DIN 53122.

Table 8.21 Permeation of Various Gases at 23 °C and 50% RH through DuPont™ Zytel 42 Nylon 66 Film[13]

Permeant	Permeability Coefficient	
	Source Document Units (cm³ mil/100 in.² day atm)	Normalized Units (cm³ mm/m² day atm)
Oxygen	2	0.8
Carbon dioxide	9	3.5
Nitrogen	0.7	0.3
Helium	150	59.1

Table 8.22 Permeation of Liquids through DuPont™ Zytel 42 Nylon 66 Bottles[14]

Liquid	Vapor permeation rate (g mm/m² day)
Kerosene	0.08
Methyl salicylate	0.08
Motor oil (SAE 10)	0.08
Toluene	0.08
Fuel oil B	0.2
Water	1.2–2.4
Carbon tetrachloride	2.0
VMP naphtha	2.4

Thickness: 2.54 mm.

8.6 Polyamide 66/610 (Nylon 66/610)

Nylon 66/610 is a copolymer made from hexamethylenediamine, adipic acid, and sebacic acid. Its structure is represented in Fig. 8.20.

Manufacturers and trade names: EMS-Grivory® Grilon®.

Applications and uses: Flexible packaging for foodstuff and medical packaging such as IV bags (Table 8.23).

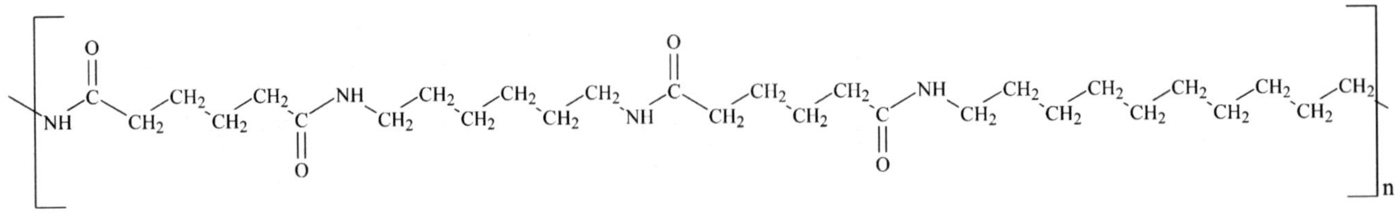

Figure 8.20 Structure of polyamide 66/610.

Table 8.23 Permeation of Oxygen, Carbon Dioxide, Nitrogen, and Water Vapor at 23 °C through EMS-Grivory® Grilon® BM 20 SBG[15]

Permeant	Test Method	RH (%)	Permeability Coefficient	
			Source Document Units (cm^3/m^2 day bar)	Normalized Units (cm^3 mm/m^2 day atm)
Oxygen	ISO 15105-1	0	25	1.3
Oxygen	ISO 15105-1	85	70	3.5
Carbon dioxide	ISO 15105-2	0	80	4.1
Carbon dioxide	ISO 15105-2	85	250	12.7
Nitrogen	DIN 53380	0	15	0.8
			Vapor Transmission Rate	
			Source Document Units (g/m^2 day)	Normalized Units (g mm/m^2 day)
Water vapor	ISO 15106-1	85	20	1

Thickness: 0.050 mm.

8.7 Polyamide 6/12 (Nylon 6/12)

The structure of nylon 6/12 is given in Fig. 8.21. The CAS number is 24936-74-1.

Some of the nylon 6/12 characteristics are as follows:

- High impact strength
- Very good resistance to greases, oils, fuels, hydraulic fluids, water, alkalis, and saline
- Very good stress cracking resistance, even when subjected to chemical attack and when used to cover metal parts
- Low coefficients of sliding friction and high abrasion resistance, even when running dry
- Heat deflection temperature (melting point nearly 40 °C higher than Nylon 12)
- Tensile and flexural strength
- Outstanding recovery at high wet strength

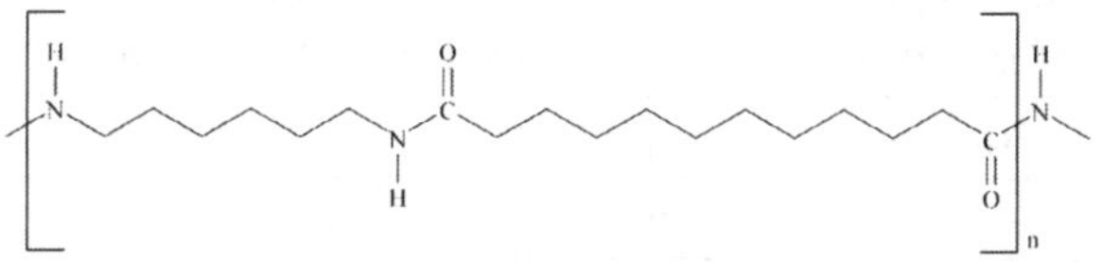

Figure 8.21 Chemical structure of nylon 6/12.

Manufacturers and trade names: EMS-Grivory®—Grilon® CF, CR, Ube Industries.

Applications: Multilayer food packaging and boil in bag (Tables 8.24–8.29).

Table 8.24 Permeation of Oxygen and Carbon Dioxide at 23 °C through EMS-Grivory® Grilon® CF 6 Nylon 6/12 Film[14]

Permeant	Test Method	RH (%)	Permeation Coefficient	
			Source Document Units (cm^3/m^2 day bar)	Normalized Units (cm^3 mm/m^2 day atm)
Oxygen	ISO 15105-1	0	120	6
	ISO 15105-1	85	200	10
Carbon dioxide	ISO 15105-2	0	400	20
	ISO 15105-2	85	800	41

Thickness: 0.050 mm.

Table 8.25 Permeation of Oxygen and Carbon Dioxide at 23 °C through EMS-Grivory® Grilon® CA 6 Nylon 6/12 Film[14]

Permeant	Test Method	RH (%)	Permeation Coefficient	
			Source Document Units (cm^3/m^2 day bar)	Normalized Units (cm^3 mm/m^2 day atm)
Oxygen	ISO 15105-1	0	150	8
	ISO 15105-1	85	250	13
Carbon dioxide	ISO 15105-2	0	450	23
	ISO 15105-2	85	850	43

Thickness: 0.050 mm.

Table 8.26 Permeation of Oxygen and Carbon Dioxide at 23 °C through EMS-Grivory® Grilon® CF 7 Nylon 6/12 Film[16]

Permeant	Test Method	RH (%)	Permeation Coefficient	
			Source Document Units (cm^3/m^2 day bar)	Normalized Units (cm^3 mm/m^2 day atm)
Oxygen	ISO 15105-1	0	110	6
	ISO 15105-1	85	130	7
Carbon dioxide	ISO 15105-2	0	400	20
	ISO 15105-2	85	800	41

Thickness: 0.050 mm.

Table 8.27 Permeation of Oxygen and Carbon Dioxide at 23 °C through EMS-Grivory® Grilon® CR 8 Nylon 6/12 Film[17]

Permeant	Test Method	RH (%)	Permeation Coefficient	
			Source Document Units (cm^3/m^2 day bar)	Normalized Units (cm^3 mm/m^2 day atm)
Oxygen	ISO 15105-1	0	80	4.1
	ISO 15105-1	85	90	4.6
Carbon dioxide	ISO 15105-2	0	300	15
	ISO 15105-2	85	800	41

Thickness: 0.050 mm.

Table 8.28 Permeation of Oxygen, Carbon Dioxide, and Nitrogen at 23 °C through EMS-Grivory® Grilon® CR 9 Nylon 6/12 Film[18]

Permeant	Test Method	RH (%)	Permeation Coefficient	
			Source Document Units (cm^3/m^2 day bar)	Normalized Units (cm^3 mm/m^2 day atm)
Oxygen	ISO 15105-1	0	55	2.8
	ISO 15105-1	85	75	3.8
Carbon dioxide	ISO 15105-2	0	200	10
	ISO 15105-2	85	350	18
Nitrogen	DIN 53380	0	15	0.8

Thickness: 0.050 mm.

Table 8.29 Permeation Water Vapor at 23 °C through EMS-Grivory® Grilon® Nylon 6/12 Films[19–21]

Grilon® Product Code	Vapor Transmission Rate	
	Source Document Units (g/m² day)	Normalized Units Rate (g mm/m² day)
CF 6	15	0.8
CA 6	20	1.0
CF 7	15	0.8
CR 8	15	0.8
CR 9	15	0.8

Thickness: 0.050 mm; RH: 85%.

8.8 Polyamide 666 (Nylon 666 or 6/66)

This is the name given to copolyamides made from PA 6 and PA 66 building blocks. A precise structure cannot be drawn.

Manufacturers and trade names: Honeywell Aegis™, UBE Industries, BASF Ultramid® (Tables 8.30–8.33).

Table 8.30 Permeation of Oxygen, Nitrogen, and Carbon Dioxide through Honeywell Aegis™ Nylon 6/66 Films[19]

Permeant	Permeation Coefficient	
	Source Document Units (cm³ mil/100 in.² day atm)	Normalized Units (cm³ mm/m² day atm)
Oxygen	2.40	1.0
Nitrogen	19.8	8.8
Carbon dioxide	287	113

Table 8.31 Permeation of Oxygen at 23 °C and Different RHs through UBE Industries LTD. UBE 5033B Nylon 6/66 Films[20]

RH (%)	Permeation Coefficient	
	Source Document Units (cm³ 25 μm/m² day atm)	Normalized Units (cm³ mm/m² day atm)
0	52	1.3
65	55	1.4
100	198	5.0

Table 8.32 Permeation of Oxygen, Carbon Dioxide, Nitrogen, and Water Vapor at 23 °C through BASF Ultramid® C35 Nylon 6/66 Film[21]

Permeant	Test Method	RH (%)	Permeation Coefficient	
			Source Document Units (cm³ 100 μm/m² day bar)	Normalized Units (cm³ mm/m² day atm)
Oxygen	DIN 53380	40	8–9	0.8–0.9
Carbon dioxide	DIN 53380	0	40–45	4.0–4.6
			Vapor Transmission Rate	
			Source Document Units (g 100 μm/m² day)	Normalized Units (g mm/m² day)
Water vapor		85%–0% Gradient	15–18	1.5–1.8

Thickness: 0.02–0.1 mm.

Table 8.33 Permeation of Oxygen, Carbon Dioxide, Nitrogen, and Water Vapor at 23 °C through Honeywell Capran® Nylon 6/66 Films[22]

Permeant	Test Method	RH (%)	Permeation Coefficient	
			Source Document Units (cm³/m² day)	Normalized Units (cm³ mm/m² day atm)
Oxygen	ASTM D3985	0	37.2	0.94
		90	232.5	5.91
Carbon dioxide	ASTM D1434	0	113.2	2.88
Nitrogen	ASTM D1434	0	7.75	0.2
			Vapor Transmission Rate	
			Source Document Units (g/m² day)	Normalized Units (g mm/m² day)
Water vapor		90	341	8.7

Thickness: 0.0254 mm.

8.9 Polyamide 6/69 (Nylon 6/6.9)

This resin is specifically suited for applications requiring superior toughness and abrasion resistance.

Manufacturers and trade names: Shakespeare Monofilaments and Specialty Polymers Isocor™, EMS-Grivory® Grilon®.

Applications and uses: Cable jacketing, film extrusions, monofilaments/bristles, impact modifier, electrical connectors, and trimmer line (Tables 8.34 and 8.35).

8: POLYAMIDES (NYLONS)

Table 8.34 Permeation of Oxygen, Carbon Dioxide, Nitrogen, and Water Vapor at 23 °C through EMS-Grivory® Grilon® BM 13 SBG or Grilon® BM 13 SBGX[23]

Permeant	Test Method	RH (%)	Permeation Coefficient	
			Source Document Units (cm^3/m^2 day bar)	Normalized Units (cm^3 mm/m^2 day atm)
Oxygen	ISO 15105-1	0	50	2.5
	ISO 15105-1	85	100	5.0
Carbon dioxide	ISO 15105-2	0	130	6.5
	ISO 15105-2	85	500	25
Nitrogen	DIN 53380	0	10	0.5
			Vapor Transmission Rate	
			Source Document Units (g/m^2 day)	Normalized Units (g mm/m^2 day)
Water vapor	ISO 15106-1	85	15	0.8

Thickness: 0.050 mm.

Table 8.35 Permeation of Oxygen, Carbon Dioxide, and Water Vapor at 23 °C through EMS-Grivory® Grilon® BM 17 SBG[24]

Permeant	Test Method	RH (%)	Permeation Coefficient	
			Source Document Units (cm^3/m^2 day bar)	Normalized Units (cm^3 mm/m^2 day atm)
Oxygen	ISO 15105-1	0	65	3.3
	ISO 15105-1	85	45	2.3
Carbon dioxide	ISO 15105-2	0	200	10.3
	ISO 15105-2	85	470	23.5
			Vapor Transmission Rate	
			Source Document Units (g/m^2 day)	Normalized Units (g mm/m^2 day)
Water vapor	ISO 15106-1	85	18	0.9

Thickness: 0.050 mm.

8.10 Polyarylamide

Another partially aromatic high-performance polyamide is polyarylamide (PAA). The primary commercial polymer, PAMXD6, is formed by the reaction of *m*-xylylenediamine and adipic acid giving the structure shown in Figs. 8.22 and 8.23. It is a semicrystalline polymer.

- Very high rigidity.
- High strength.
- Very low creep.

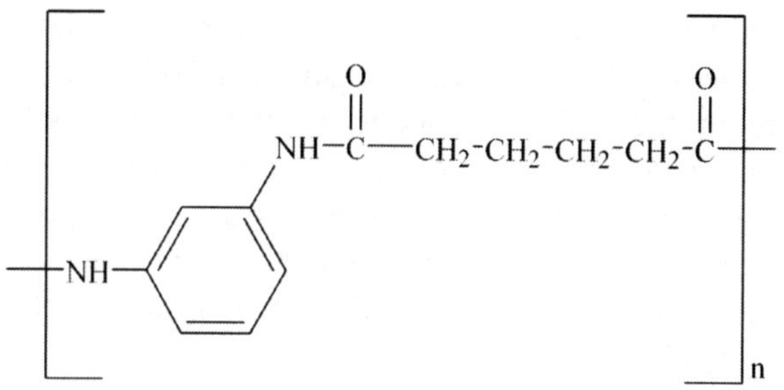

Figure 8.22 Chemical structure of PAMXD6 polyarylamide.

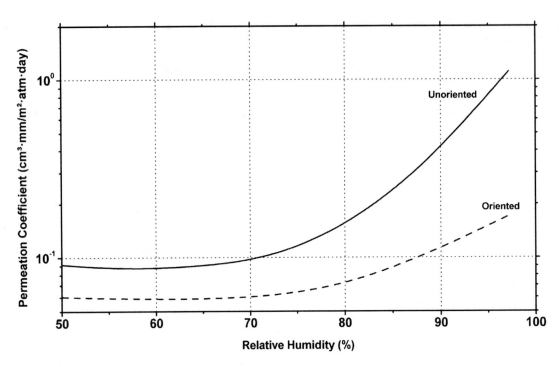

Figure 8.23 Permeation of oxygen vs. RH at 23 °C for Mitsubishi gas chemical nylon-MXD6 PAA films.[26]

Table 8.36 Permeation of CE 10 Fuel Components at 60 °C through Solvay Advanced Polymers Ixef® PAA[25]

Permeant	Transmission Rate (g mm/m² day)
Ethanol	0.83
Toluene	0.003
Isooctane	0.001

See also Fig. 8.24.

- Excellent surface finish even for a reinforced product even with high glass fiber content.
- Ease of processing.
- Good dimensional stability.
- Slow rate of water absorption.

Graphs of multipoint properties of polyamides as a function of temperature, moisture, and other factors are in the following sections. Because the polyamides do absorb water, and that affects the properties, some of the data are dry or better DAM. Some of the data are for conditioned specimen; they have reached equilibrium water absorption from 50% RH at 23 °C.

Manufacturers and trade names: Solvay Advanced Polymers Ixef®, Mitsuibishi Gas Chemical Co. Nylon-MXD6, Nanocor® Imperm®.

Applications and uses: Automotive fuel systems and packaging (Table 8.36).

8.11 Polyphthalamide/High Performance Polyamide

As a member of the nylon family, it is a semi-crystalline material composed from a diacid and a diamine. However, the diacid portion contains at least 55% terephthalic acid (TPA) or isophthalic acid (IPA). TPA or IPA are aromatic components that serve to raise the melting point, glass transition temperature, and generally improve chemical resistance vs. standard aliphatic nylon polymers. The structure of the polymer depends on the ratio of the diacid ingredients and the diamine used and varies from grade to grade. The polymer usually consists of

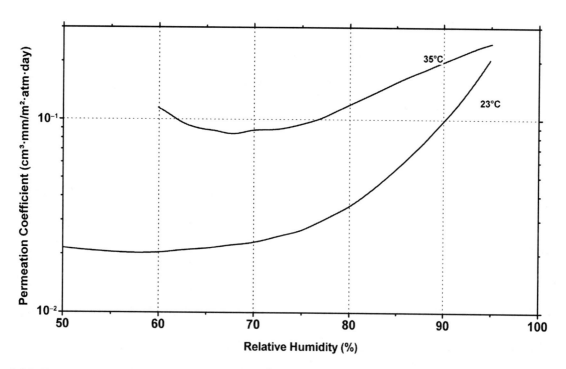

Figure 8.24 Permeation of oxygen vs. RH Nanocor® Imperm® 105 nanoclay-filled nylon-MXD6 PAA films.[27]

mixtures of blocks of two or more different segments, four of which are shown in Fig. 8.25.

Some of the polyphthalamide (PPA) characteristics are as follows:

- Very high heat resistance
- Good chemical resistance
- Relatively low moisture absorption
- High strength or physical properties over a broad temperature range
- Not inherently flame retardant
- Requires good drying equipment
- High processing temperatures

Manufacturers and trade names: Solvay Advanced Polymers Amodel®.

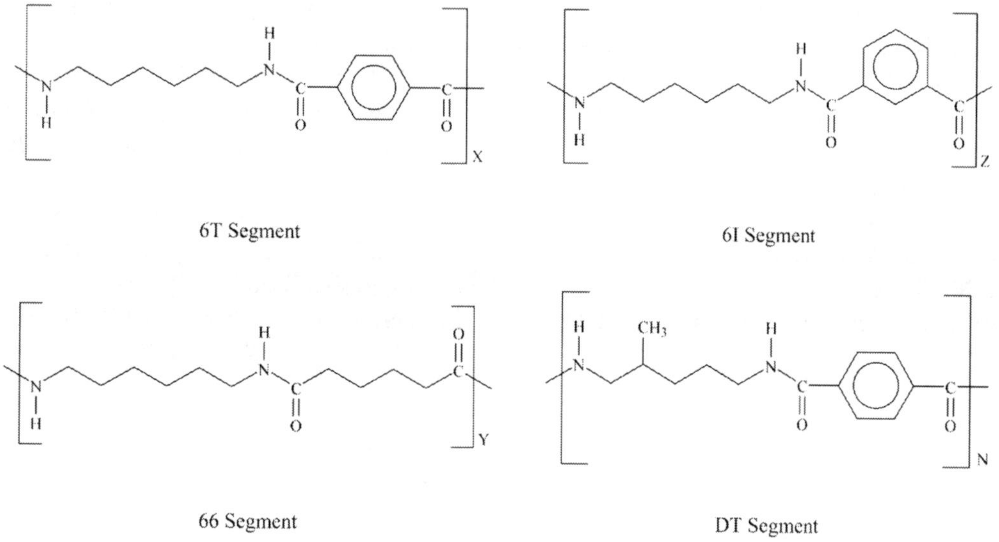

Figure 8.25 Chemical structures of block used to make PPA.

Table 8.37 Permeation of CE 10 Fuel at 60 °C Through Solvay Advanced Polymers Amodel® PPA (CE10 = 45% isooctane, 45% toluene, 10% ethanol.)

Grade	Transmission Rate (g mm/m² day)
A-1004	0.03
AT-1002	0.4

(CE10 = 45% isooctane, 45% toluene, 10% ethanol.)

Applications and uses: Automotive fuel systems (Table 8.37).

References

1. *Data Sheets, Grivory® G16 and G21*. EMS Chemie; 2010.
2. DuPont™ Selar® PA3426 Blends with Nylon 6 — General information; 2005.
3. Gas Barrier Properties of EVAL Resins—Technical Bulletin No. 110, supplier technical report. EVAL Company of America.
4. Capran Nylon Films, supplier technical report. Allied Signal Inc.
5. *Aegis® product specifications*. Honeywell; 2008–2009.
6. UBE nylon extrusion application; 2005.
7. *Data Sheets, Characteristics of Grivory® G16, Characteristics of Grivory® G21*. EMS Chemie; July 2000.
8. *Rilsan® PA11: Created from a renewable source*. Arkema; 2005.
9. *Specifications, campus database*. EMS Grivory; 2010.
10. Dartek® Product Specification Sheets, DuPont™ Packaging Polymers (Product line sold in 2007).
11. Zytel® /Minlon® design guide-module II, DuPont™ engineering polymers, 232409D; 1997.
12. *Technical Data Sheet-Grilon® BM 20 SBG*. EMS-Grivory; 2002.
13. *Technical Data Sheet-Grilon® CF 6*. EMS-Grivory; 2007.
14. *Technical Data Sheet-Grilon® CF 7*. EMS-Grivory; 2002.
15. *Technical Data Sheet-Grilon® CR 8*. EMS-Grivory; 2002.
16. *Technical Data Sheet-Grilon® CR 9*. EMS-Grivory; 2002.
17. Specification Sheet, Honeywell Aegis™ HCA73MP Nylon 6/6,6 Extrusion Grade Copolymer.
18. Specification Sheet, UBE 5033B Nylon 6/66 Film.
19. *Ultramid nylon resins product line, properties, processing, supplier design guide (B 568/1e/4.91)*. BASF Corporation; 1991.
20. *Capron Nylon Resins for Films—Operating Manual, supplier technical report (SFF-08)*. Allied Signal Inc.; 1992.
21. *Technical Data Sheet-Grilon® BM 13 SBG*. EMS-Grivory; 2002.
22. *Technical Data Sheet-Grilon® BM 17 SBG*. EMS-Grivory; 2002.
23. *Solvay advanced polymers unveils newly formulated Ixef® polyarylamide barrier material for automotive fuel systems. News release*. Solvay Advanced Polymers; 2007.
24. *Selar® PA 3426 Barrier Resin, supplier technical report (E-73974)*. DuPont™ Company; 1985.
25. Heilman W, Tammela V, Meyer J, Stannett V, Szwarc M. Permeability of polymer films to hydrogen sulfide gas. *Ind Eng Chem* 1956;**48**: 821–4.
26. Mitsubishi gas chemical website. Available from: http://www.gasbarriertechnologies.com/ds_gasbar.html; 2010.
27. *Technical Bulletin NC105-O1E, Imperm® Grade 105*. NANOCOR, INC.; 2008.

9 Polyolefins, Polyvinyls, and Acrylics

This chapter focuses on polymers made from monomers that contain a carbon—carbon double bond through which the polymer is made by addition polymerization as discussed in Section 2.1. An alkene, also called an olefin, is a chemical compound made of only carbon and hydrogen atoms containing at least one carbon-to-carbon double bond. The simplest alkenes, with only one double bond and no other functional groups, form a homologous series of hydrocarbons with the general formula C_nH_{2n}. The two simplest alkenes of this series are ethylene and propylene. When these are polymerized, they form polyethylene (PE) and polypropylene (PP), which are two of the plastics discussed in this chapter. A slightly more complex alkene is 4-methylpentene-1, the basis of poly(methylpentene) (PMP), known under the trade name of TPX™.

If one of the hydrogens on the ethylene molecule is changed to chlorine, the molecule is called vinyl chloride, the basis of polyvinyl chloride, commonly called PVC. There are many other vinyl monomers that substitute different functional groups onto the carbon—carbon double bond. Vinyl alcohol is particularly the important one.

Acrylic polymers are also polymerized through the carbon—carbon double bond. Methyl methacrylate is the monomer used to make poly(methyl methacrylate) (PMMA).

The structures of some of these monomers are shown in Fig. 9.1. Structures of the polymers may be found in the appropriate sections that contain the data for those materials.

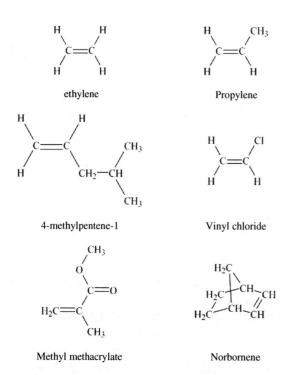

Figure 9.1 Chemical structures of some monomers used to make polyolefins, polyvinyls, and acrylics.

9.1 Polyethylene

Polyethylene can be made in a number of ways. The way it is produced can affect its physical properties. It can also have very small amounts of comonomers, which will alter its structure and properties.

The basic types or classifications of PE, according the ASTM 1248, are:

- Ultra-low-density polyethylene (ULDPE), polymers with densities ranging from 0.890 to 0.905 g/cm^3, contains comonomer
- Very-low-density polyethylene (VLDPE), polymers with densities ranging from 0.905 to 0.915 g/cm^3, contains comonomer
- Linear low-density polyethylene (LLDPE), polymers with densities ranging from 0.915 to 0.935 g/cm^3, contains comonomer
- Low-density polyethylene (LDPE), polymers with densities ranging from about 0.915 to 0.935 g/cm^3

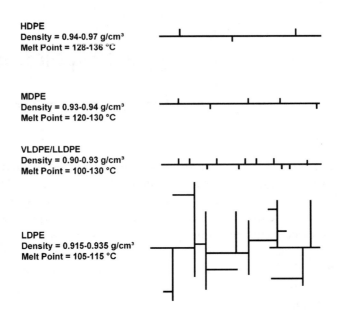

Figure 9.2 Graphical depictions of polyethylene types.

- Medium-density polyethylene (MDPE), polymers with densities ranging from 0.926 to 0.940 g/cm^3, may or may not contain comonomer
- High-density polyethylene (HDPE), polymers with densities ranging from 0.940 to 0.970 g/cm^3, may or may not contain comonomer

Figure 9.2 shows the differences graphically. The differences in the branches in terms of number and length affect the density and melting points of some of the types.

Branching affects the crystallinity. A representation of the crystal structure of PE is shown in Fig. 9.3. One can imagine how branching in the polymer chain can disrupt the crystalline regions. The crystalline regions are the highly ordered areas in the shaded rectangles of Fig. 9.3. A high degree of branching would reduce the size of the crystalline regions, which leads to lower crystallinity.

Manufacturers and trade names: There are many, some of which are Dow Attane™; Chevron Philips Marlex®; LlyondellBasell Polyolefins.

Applications and uses: ULDPE—Heavy duty sacks, turf bags, consumer bags, packaging for cheese, meat, coffee, and detergents, silage wrap, mulch films, extruded membranes, heating and water pipes, and injection-molded products. HDPE—Food packaging: dairy products and bottled water, cosmetics, medical products and household chemicals, automotive gas tanks, 55 gallon drums, sheets, pipes, recreational items, and geosynthetic materials.

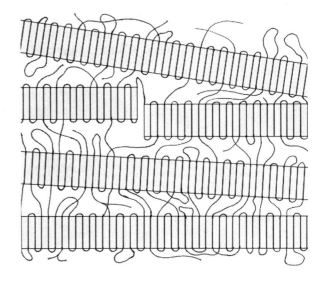

Figure 9.3 Graphical diagram of polyethylene crystal structure.

The data tables and graphs that follow will be in the order of the basic types or classifications of PE described in the first part of this section, except that data on unspecified PE and data that cover the range of PE molecular weights will be first.

9.1.1 Unclassified Polyethylene

See Figs 9.4–9.9.

9: Polyolefins, Polyvinyls, and Acrylics

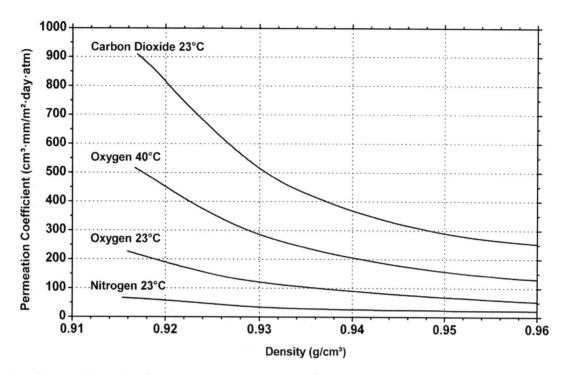

Figure 9.4 Permeability of gases vs. the density of LlyondellBasell polyolefins polyethylene films.[1]

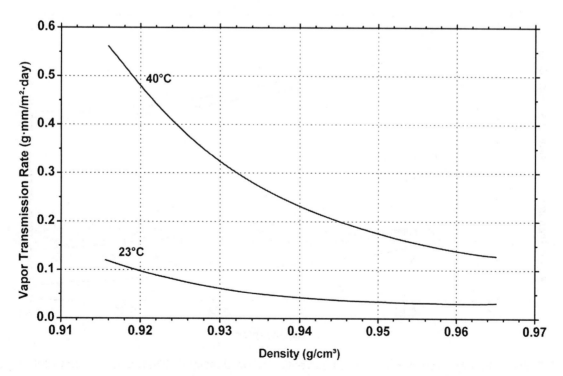

Figure 9.5 Water vapor transmission rate vs. the density of LlyondellBasell polyolefins polyethylene films.[2]

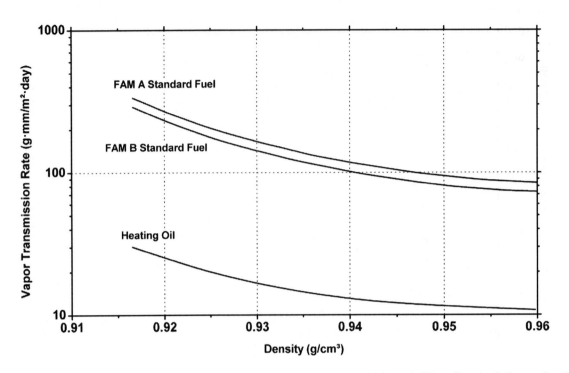

Figure 9.6 Vapor transmission rate of various fuels vs. the density of LlyondellBasell polyolefins polyethylene films.[2]

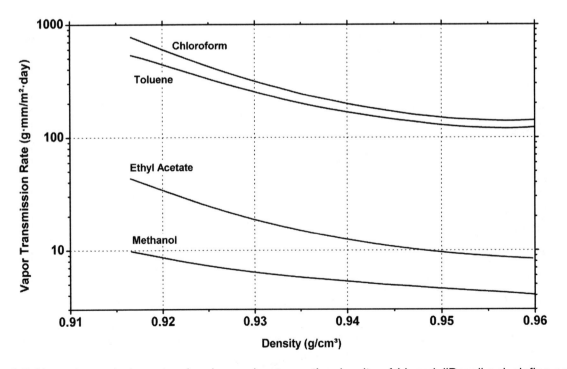

Figure 9.7 Vapor transmission rate of various solvents vs. the density of LlyondellBasell polyolefins polyethylene films.[2]

9: Polyolefins, Polyvinyls, and Acrylics

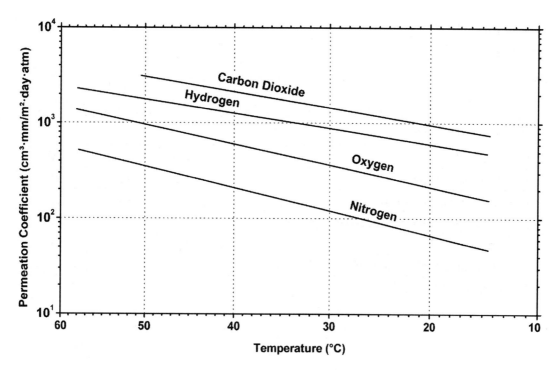

Figure 9.8 Permeability of gases vs. temperature of polyethylene film.[3]

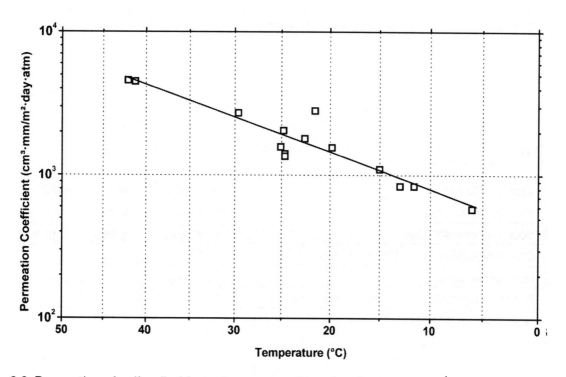

Figure 9.9 Permeation of sulfur dioxide vs. temperature through polyethylene film.[4]

9.1.2 Ultra-low-Density Polyethylene

See Tables 9.1–9.3.

Table 9.1 Permeation of Oxygen at 23 °C and 50% Relative Humidity through Dow Chemical Attane™ Blown Film[2]

Attane™ Grade	4201	4201	4202
Film thickness (mm)	0.02	0.05	0.02
Source document units Gas permeability (cm^3 mil/100 in.^{2} day atm)	716	711	695
Normalized units Permeability coefficient (cm^3 mm/m^2 day atm)	281	279	273

Test method: ASTM F1249.

Table 9.2 Permeation of Carbon Dioxide at 23 °C and 50% Relative Humidity through Dow Chemical Attane™ Blown Film[5]

Attane™ Grade	4201	4202
Film thickness (mm)	0.05	0.02
Source document units Gas permeability (cm^3 mil/100 in.^{2} day atm)	3128	3340
Normalized units Permeability coefficient (cm^3 mm/m^2 day atm)	1233	1312

Test method: ASTM F1249.

Table 9.3 Permeation of Water Vapor at 23 °C and 50% Relative Humidity through Dow Chemical Attane™ Blown Film[5]

Attane™ Grade	4201	4201	4202
Film thickness (mm)	0.02	0.05	0.02
Source document units Gas permeability (g mil/100 in.^{2} day atm)	1.58	1.39	2.15
Normalized units Permeability coefficient (g mm/m^2 day atm)	0.62	0.55	0.84

Test method: ASTM F1249.

9.1.3 Linear Low-Density Polyethylene

See Tables 9.4 and 9.5.

Table 9.4 Permeation of Water and Oxygen through Exopack® Sclairfilm® LX-1 (LLDPE) Film[6]

Film Thickness (mm)	Water Transmission Rate		Oxygen
	Source Document Units (g/m^2 day)	Normalized Units ($g \cdot mm/m^2$ day)	Permeability Coefficient ($cm^3 \cdot mm/m^2$ day atm)
0.0762	4.7	0.36	236
0.0508	9.3	0.47	193
0.0381	12.4	0.47	236

Table 9.5 Permeation of Carbon Dioxide and Nitrogen through Exopack® Sclairfilm® SL1 and SL3 LLDPE Films[7]

Permeant Gas	Permeability Coefficient	
	Source Document Units (cm^3/m^2 day)	Normalized Units ($cm^3 \cdot mm/m^2$ day $\cdot$ atm)
Oxygen	150	3.81
Carbon dioxide	1400	35.6

Thickness: 0.0254 mm. Test method: ASTM D3985.

9.1.4 Low-Density Polyethylene

See Tables 9.6 and 9.7. See also Fig. 9.10.

Table 9.6 Permeation of Gases at 24 °C through Dow Chemical Low-Density Polyethylene[5]

Penetrant	Vapor Transmission Rate	
	Source Document Units (g mil/100 $in.^2$ day)	Normalized Units ($g \cdot mm/m^2$ day)
Oxygen	250–350	98–138
Nitrogen	100–200	39–79
Carbon dioxide	1000–2000	394–787

Table 9.7 Permeation of Solvent Vapors at 24 °C through Dow Chemical Low-Density Polyethylene[8]

Penetrant	Vapor Transmission Rate	
	Source Document Units (g mil/100 in.² day)	Normalized Units (g mm/m² day)
Methyl alcohol	6–8	2.4–3.1
Ethyl alcohol	2–4	0.8–1.6
n-Heptane	300–500	118–197
Ethyl acetate	30–300	11.8–118
Formaldehyde	2–5	0.8–2.0
Tetrachloroethylene	500–750	197–295
Acetone	10–40	3.9–15.8
Benzene (35 °C)	600	236
Water	1–1.5	0.39–0.59

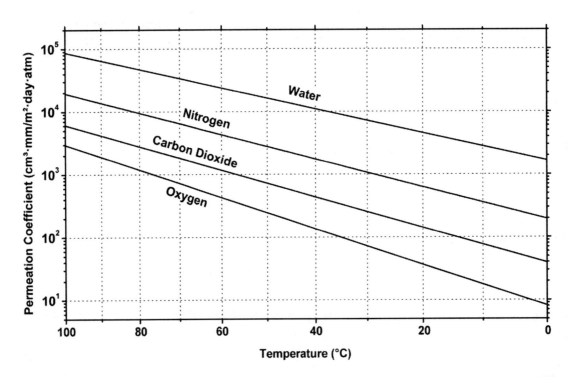

Figure 9.10 Permeation of oxygen, carbon dioxide, nitrogen, and water vs. temperature through LDPE polyethylene film.[9]

9.1.5 Medium-Density Polyethylene

See Table 9.8.

Table 9.8 Permeation of Carbon Dioxide and Nitrogen through Exopack® Sclairfilm® LWS-1 and LWS-2 Laminating MDPE Films[10]

Grade	Film Thickness (mm)	Permeation Coefficient	
		Water (cm^3 mm/ m^2 day atm)	Oxygen (cm^3 mm/ m^2 day atm)
Sclairfilm® LWS-1	0.0508	0.26	132
Sclairfilm® LWS-1	0.0381	0.27	133
Sclairfilm® LWS-2	0.0508	0.26	132
Sclairfilm® LWS-2	0.0381	0.27	133

9.1.6 High-Density Polyethylene

See Tables 9.9–9.15.
See also Fig. 9.11.

Table 9.9 Permeability Coefficient[a] for Pipes Made of LlyondellBasell Polyolefins HDPE at 20 °C[11]

Permeant Gas	P
	Permeability Coefficient[a] (cm^3/m bar day)
Nitrogen	0.018
Air	0.029
Carbon monoxide	0.036
Natural gas	0.056
Methane	0.056
Argon	0.066
Oxygen	0.072
Ethane	0.089
Helium	0.15
Hydrogen	0.22
Carbon dioxide	0.28
Sulfur dioxide	0.43

[a]This permeation coefficient in Table 9.9 is designed for pipe gas permeation calculations by Eqn (9.1).

where:
$$V = P\left(\frac{\pi d_e L p t}{e}\right) \quad (9.1)$$

V is the volume of permeating gas [cm^3 (STP)]

P is the permeability coefficient [cm^3 (STP)/m bar day]

d_e is outside diameter of the pipe (mm)

L is the length of the pipe (m)

p is the partial pressure of the gas in the pipe (bar)

t is time (days)

e wall thickness of the pipe (mm)

Table 9.10 Permeation of Gases through Chevron Philips Marlex® HDPE

Permeant Gas	Permeability Coefficient (cm^3 mm/m^2 day atm)
Carbon dioxide	136
Hydrogen	126
Oxygen	44
Helium	97
Ethane	93
Natural gas	44
Freon® 12	37
Nitrogen	21

Test method: ASTM D1434.[8]

Table 9.11 Permeation of Oxygen and Water through NOVA Chemicals Sclair® HDPE Films

Sclair® Code	Film Thickness (mm)	Water Transmission Rate Source Document Units (g/m^2 day)	Water Transmission Rate Normalized Units (g mm/m^2 day)	Oxygen Permeability Coefficient (cm^3 mm/m^2 day atm)
15A	0.038	5.4	0.21	91
19A	0.038	3.4	0.13	65
19C	0.038	2.6	0.10	57
19G	0.038	2.8	0.11	62
19H	0.038	6.2	0.24	65

Test methods: ASTM E96 and ASTM D3985.[12]

Table 9.12 Permeation of Hydrogen vs. Temperature and Pressure through HDPE[13]

Temperature (°C)	−15	25	68	−16	25	67	−18	25	67
Pressure gradient (kPa)	1724			3447			6895		
Source document units Permeability coefficient (cm^3 mm/cm^2 kPa s)	3.64×10^{-10}	1.78×10^{-9}	8.69×10^{-9}	3.49×10^{-10}	1.76×10^{-9}	8.54×10^{-9}	3.19×10^{-10}	1.84×10^{-9}	8.45×10^{-9}
Normalized units Permeability coefficient (cm^3 mm/m^2 day atm)	31.9	156	761	30.6	154	748	27.9	161	740

Thickness: 0.030 mm. Test method: mass spectrometry and calibrated standard gas leaks; developed by McDonnell Douglas Space Systems Company Chemistry Laboratory.

Table 9.13 Permeation of Nitrogen vs. Temperature and Pressure through HDPE[14]

Temperature (°C)	−10	25	72	−19	25	69	−17	25	68
Pressure gradient (kPa)	1724			3447			6895		
Source document units Permeability coefficient (cm^3 mm/cm^2 kPa s)	1.81×10^{-11}	1.77×10^{-10}	1.98×10^{-9}	1.08×10^{-11}	1.6×10^{-10}	1.46×10^{-9}	1.13×10^{-11}	1.68×10^{-10}	1.71×10^{-9}
Normalized units Permeability coefficient (cm^3 mm/m^2 day atm)	1.6	15.5	173	0.95	14.0	128	0.99	14.7	150

Thickness: 0.030 mm. Test method: mass spectrometry and calibrated standard gas leaks; developed by McDonnell Douglas Space Systems Company Chemistry Laboratory.

Table 9.14 Permeation of Oxygen vs. Temperature and Pressure through HDPE[14]

Temperature (°C)	−16	25	51	−15	25	52
Pressure gradient (kPa)	1724			3447		
Source document units Permeability coefficient (cm^3 mm/cm^2 kPa s)	5.75×10^{-11}	5.75×10^{-10}	2.49×10^{-9}	5.91×10^{-11}	5.64×10^{-10}	2.03×10^{-9}
Normalized units Permeability coefficient (cm^3 mm/m^2 day atm)	5.0	50.3	218	5.2	49.4	178

Thickness: 0.030 mm. Test method: mass spectrometry and calibrated standard gas leaks; developed by McDonnell Douglas Space Systems Company Chemistry Laboratory.

Table 9.15 Permeation of Ammonia vs. Temperature and Pressure through HDPE[14]

Temperature (°C)	−15	25	68
Pressure gradient (kPa)	965		
Source document units Permeability coefficient (cm^3 mm/cm^2 kPa s)	3.71×10^{-10}	1.4×10^{-9}	7.12×10^{-9}
Normalized units Permeability coefficient (cm^3 mm/m^2 day atm)	32.5	122.6	623

Thickness: 0.030 mm. Test method: mass spectrometry and calibrated standard gas leaks; developed by McDonnell Douglas Space Systems Company Chemistry Laboratory.

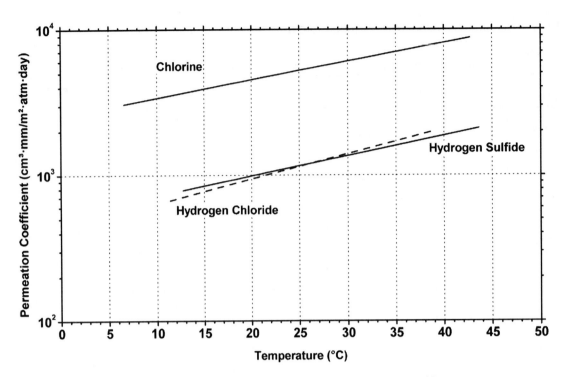

Figure 9.11 Permeation of gases vs. temperature of high-density polyethylene.[15]

9.2 Polypropylene

The three main types of PP generally available:

- Homopolymers are made in a single reactor with propylene and catalyst. It is the stiffest of the three propylene types and has the highest tensile strength at yield. In the natural state (no colorant added), it is translucent and has excellent see through or contact clarity with liquids. In comparison to the other two types, it has less impact resistance, especially below 0 °C.

- Random Copolymer (homophasic copolymer) is made in a single reactor with a small amount of ethylene (<5%) added, which disrupts the crystallinity of the polymer allowing this type to be the clearest. It is also the most flexible with the lowest tensile strength of the three. It has better room temperature impact than homopolymer but shares the same relatively poor impact resistance at low temperatures.

- Impact Copolymers (heterophasic copolymer), also known as block copolymers, are made in a two-reactor system where the homopolymer matrix is made in the first reactor and then transferred to the second reactor where ethylene and propylene are polymerized to create ethylene propylene rubber (EPR) in the form of microscopic nodules dispersed in the homopolymer matrix phase. These nodules impart impact resistance to the compound at both ambient and cold temperatures. This type has intermediate stiffness and tensile strength and is quite cloudy. In general, the more the ethylene monomer added, the greater the impact resistance with correspondingly lower stiffness and tensile strength.

Applications and uses:

- Homopolymer: Thermoforming, slit film and oriented fibers, high clarity, housewares, syringes, and closures.

- Random copolymer: Food, household chemicals, beauty aid products, clear containers, and hot fill applications.

- Impact copolymers: Automotive, housewares, film, sheet, profiles, high-pressure resistance, medical trays, and thin-wall parts (Tables 9.16–9.19).

Table 9.16 Permeation of Various Gases through Ineos Polypropylene[16]

Permeant	Temperature (°C)	Permeability Coefficient	
		Source Document Units, $\times 10^{-10}$ (cm²/s cm Hg)	Normalized Units (cm³ mm/m² day atm)
Water	25	5.1	335
Oxygen	30	2.3	151
Carbon dioxide	30	3.5	230
Hydrogen	20	41.0	2692
Nitrogen	30	0.27	18

Table 9.17 Permeability Coefficient[a] for LlyondellBasell Polyolefins Polypropylene at Different Temperatures[11]

Permeant Gas	P			
	Permeability Coefficient[a] (cm³/m bar day)			
	25 °C	30 °C	40 °C	50 °C
Air	0.028	0.038	0.072	0.144
Nitrogen	0.017	0.024	0.052	0.104
Oxygen	0.76	0.1	0.204	0.368
Carbon dioxide	0.244	0.336	0.6	1.08
Hydrogen	0.64	0.72	1.12	1.88
Helium	0.7	0.88	1.2	1.76
Argon	0.66	0.84	0.164	0.32

[a]Permeation coefficient is designed for pipe gas permeation calculations by Eqn (9.2).

$$V = P\left(\frac{\pi d_e L p t}{e}\right) \quad (9.2)$$

where:

V is the volume of permeating gas [cm³ (STP)]

P is the permeability coefficient [cm³ (STP)/m bar day]

d_e is the outside diameter of the pipe (mm)

L is the length of the pipe (m)

p is the partial pressure of the gas in the pipe (bar)

t is the time (days)

e is the wall thickness of the pipe (mm).

Table 9.18 Permeation of Carbon Dioxide through LlyondellBasell Adflex™ PP[17]

Adflex™ Grade	Source Document Units		Normalized Units	
	Permeability (cm³/100 in.² day)		Permeability Coefficient (cm³ mm/m² day atm)	
Film thickness (mm)	0.025	0.050	0.025	0.050
Q401F	2200	960	865	754
KS089P	2250	1060	884	833
KS353P	5760	2750	2240	2162

Table 9.19 Permeation of Oxygen through LlyondellBasell Adflex™ PP[18]

Adflex™ Grade	Source Document Units		Normalized Units	
	Permeability (cm³/100 in.² day)		Permeability Coefficient (cm³ mm/m² day atm)	
Film thickness (mm)	0.025	0.050	0.025	0.050
Q401F	100	45	39	35
KS089P	420	240	185	188
KS353P	980	475	377	373

9.3 Polybutadiene

1,3-Butadiene is a hydrocarbon molecule that has two carbon–carbon double bonds. PB can be formed from many 1,3-butadiene monomers undergoing free radical polymerization to make a much longer polymer chain molecule. But as discussed in Section 2.7, there are many ways for the monomers to react that gives slightly different isomeric structure. This is shown in Fig. 9.12.

Most of the time, adding monomer bonds to the #4 or terminal carbon of the previous butadiene results in a 1,4-addition of the previous butadiene unit. The new double bond may have either a cis- or a trans-isomer configuration. In a smaller fraction of the time (perhaps 20%), the new monomer bonds to the #2 carbon of the previous butadiene resulting in a 1,2-addition of the previous butadiene unit. The double bond between the #1 and #2 carbons turns into a single bond in the previous butadiene unit, and the double bond between the #3 and #4 carbons remains intact in a short vinyl side group available for branching or cross-linking. There are different catalysts available that can result in polymerization either in the cis or the trans configurations.

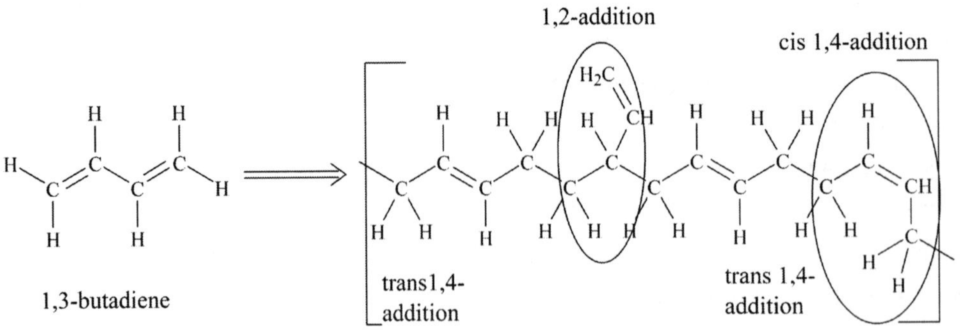

Figure 9.12 Polybutadiene structural isomers.

9: Polyolefins, Polyvinyls, and Acrylics

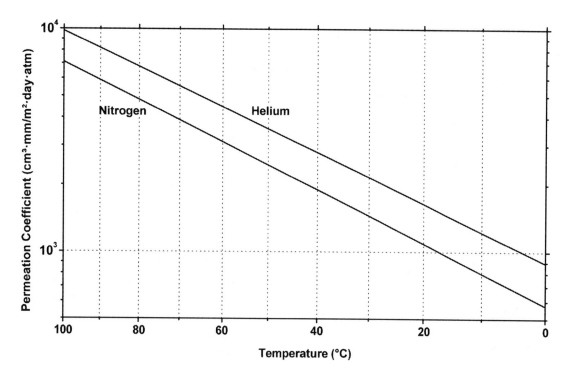

Figure 9.13 Permeation coefficients of nitrogen and helium vs. temperature through polybutadiene rubber.[9]

The CAS number of PB is 9003-29-6.

Manufacturers and trade names: There are many manufacturers including Firestone Polymers and LANXESS.

Applications and uses: PB is used synergistically as a blend component to improve and differentiate the properties of polyolefins in packaging films or nonwoven fabrics. Pressurized vessels, pressurized beverage tubing, seals such as beverage closure liners, architectural seals and gaskets, compression packaging films, peel seal, film modification, hot melt, and polyolefin modification applications. It is also used in various parts of automobile tires. Its use in the tread portion of giant truck tires, which helps to improve the abrasion, i.e., less wearing, and to run the tire comparatively cool. In the sidewall of truck tires, the use of PB rubber helps to improve fatigue to failure life due to the continuous flexing during run. As a result, tires will not blow out in extreme service conditions. PB rubber is blended with polystyrene to prepare high-impact polystyrene (HIPS) (Fig. 9.13).

9.4 Polymethylpentene

4-Methylpentene-1 based polyolefin is a lightweight, functional polymer that displays a unique combination of physical properties and characteristics due to its distinctive molecular structure, which includes a bulky side chain as shown in Fig. 9.14. PMP possesses many characteristics inherent in traditional polyolefins such as excellent electrical insulating properties and strong hydrolysis resistance. Moreover, it features low dielectric, superb clarity, transparency, gas permeability, heat and chemical resistance, and release qualities. Its CAS number is 89-25-8.

Manufacturers and trade names: Mitsui TPX™, Honeywell PMP; Chevron Philips Crystalor—discontinued.

Applications and uses: It can be used for extruded and film products, injection molded, and blow-molded application items, including:

- Paper coatings and baking cartons
- Release film and release paper

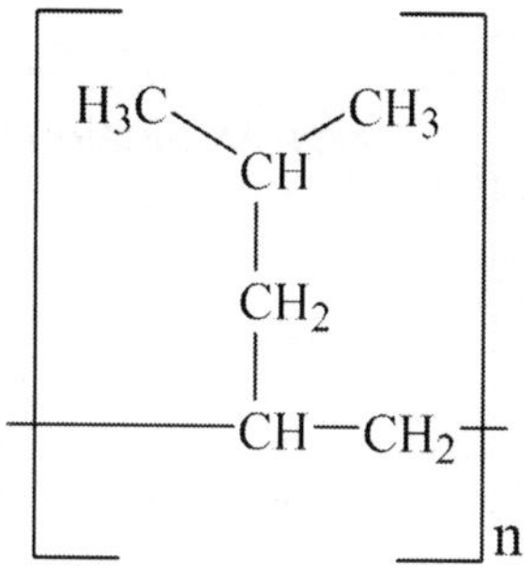

Figure 9.14 Structure of PMP.

Table 9.20 Water Transmission Rate at 38 °C (100 °F) and 100% Relative Humidity of Honeywell Series R PMP Films[14]

Product Code	Vapor Transmission Rate	
	Source Document Units (g/100 in.² day)	Normalized Units (g/m² day)
R100	5.85	91
R1180	0.5	7.8
R1500	3.9	60
R2000	3.0	47

Note: Film thickness note specified.

- High-frequency films
- Microwave cookware
- Food packaging such as gas permeable packages for fruit and vegetables
- LED molds (Table 9.20)

9.5 Cyclic Olefin Copolymer

COC is an amorphous polyolefin made by reaction of ethylene and norbornene in varying ratios. Its structure is given in Fig. 9.15. The properties can be customized by changing the ratio of the monomers found in the polymer. Being amorphous it is transparent. Other performance benefits include:

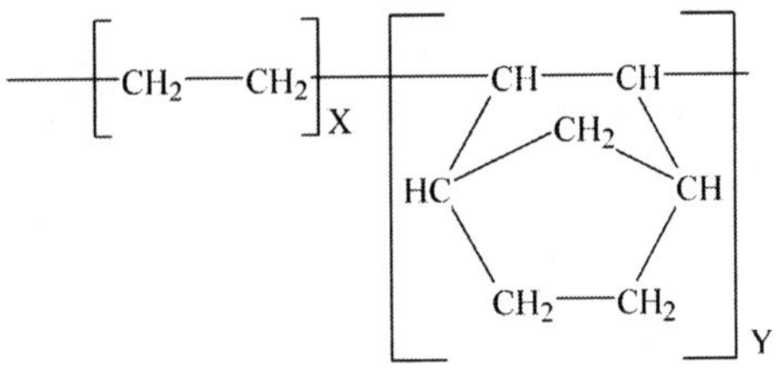

- Low density
- Extremely low water absorption
- Excellent water vapor barrier properties
- High rigidity, strength, and hardness
- Variable heat deflection temperature up to 170 °C
- Very good resistance to acids and alkalis

Manufacturers and trade names: Mitsui Chemical APEL, Topas advanced polymers, TOPAS®.

Applications and uses: Topas® COC is used as a core layer in push-through packaging (PTP), either in five layer coextruded or three layer laminated film structures; flexible and rigid packaging for food and consumer items; syringes, vials, and other prefillable containers (Tables 9.21 and 9.22).

Figure 9.15 Chemical structure of cyclic olefin copolymers.

Table 9.21 Water Vapor Permeation at 23 °C and 85% Relative Humidity through Topas Advanced Polymers TOPAS® COC[19]

Topas® Grade	Heat Deflection Temperature (°C)	Water Vapor Transmission Rate (g mm/m² day)
8077	75	0.023
6013	130	0.035
6015	130	0.035
5013	150	0.030
6017	170	0.045

Test method: DIN53122.

Table 9.22 Water Vapor Permeation at 38 °C and 90% Relative Humidity through Topas Advanced Polymers TOPAS® COC[20]

Film Layers	Layer Thickness (mm)	Vapor Transmission Rate (g²/day)
PVC/Topas®/PVDC	0.060/0.240/90g/m²	0.14
PP/Topas®/PP	0.030/0.300/0.030	0.23
PVC/Topas®/PVC	0.035/0.240/0.035	0.28
PP/Topas®/PP	0.030/0.240/0.030	0.28
PP/Topas®/PP	0.030/0.190/0.030	0.35

Test method: ASTM F1249.

9.6 Ethylene–Vinyl Acetate Copolymer

EVA is a copolymer of ethylene and vinyl acetate as shown in Figs 9.16 and 9.17. Its CAS number is 24937-78-8. Resins range in vinyl acetate content from 7.5 wt% to 33 wt%. Some grades are available with antiblock and slip additives. DuPont™ Elvax® grades vary by vinyl acetate content.

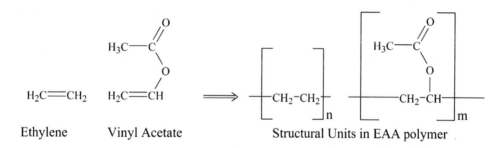

Figure 9.16 Structure of EVA polymers.

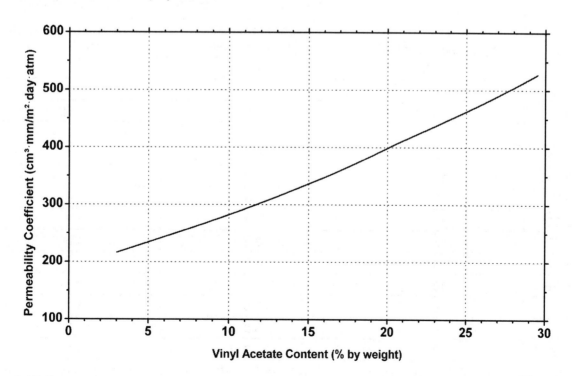

Figure 9.17 Permeation of oxygen vs. vinyl acetate content through BASF AG Lupolen V EVA.[21]

Table 9.23 Permeation of Oxygen through DuPont™ Elvax® EVA[22]

Grade	Vinyl Acetate Content (%)	Source Document Units Gas Permeability (cm^3/100 $in.^2$ day atm)	Normalized Permeability Coefficient (cm^3 mm/m^2 day atm)
3120	7.5	450	177
3121 A	7.5	580	228
3128	8.9	500	196
3130	12	400	157
3130 SB	12	570	224
3130 SBZ	12	570	224
3135 X	12	510	200
3135 SB	12	460	180
3150	15	500	196
3165	18	580	228
3165 SB	18	670	263
3169	18	500	196
3170	18	470	185
3170 SHB	18	535	210

Sample thickness (mm): 0.0254. Test method: ASTM D3985. Material note: SB: antiblock additive, SHB: slip additive and high antiblock additive.

Table 9.24 Permeation of Water Vapor through DuPont™ Elvax® EVA[23]

Grade	Vinyl Acetate Content (%)	Source Document Units Vapor Permeability (g/100 $in.^2$ day)	Normalized Units Vapor Transmission Rate (g mm/m^2 day)
3120	7.5	1.5	0.74
3121 A	7.5	1.5	0.74
3128	8.9	1.6	0.93
3130	12	2.3	1.1
3130 SB	12	2.2	1.1
3130 SBZ	12	2.2	1.1
3135 X	12	2.3	1.1
3135 SB	12	2.4	1.2
3150	15	3.3	1.6
3165	18	4.2	2.1
3165 SB	18	3.6	1.8
3169	18	3.4	1.7
3170	18	3.8	1.9
3170 SHB	18	3.7	1.8

Sample thickness (mm): 0.0254. Material note: SB: antiblock additive, SHB: slip additive and high antiblock additive; test method: ASTM E96.

Table 9.25 Permeation of Carbon Dioxide and Oxygen through Blown EVA Film[18]

Penetrant	Carbon Dioxide	Oxygen
Source document units	11,000	1800
Gas permeability (cm^3 100 μm/m^2 day atm)		
Normalized units	1100	180
Permeability coefficient (cm^3 mm/m^2 day atm)		

Features: 2.5 blow up ratio. Sample thickness (mm): 0.05. Vinyl acetate content (%): 12.0. Test method: ASTM D1434.

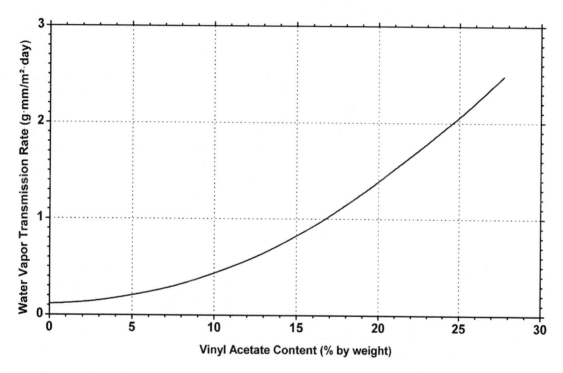

Figure 9.18 Permeation of water vapor vs. vinyl acetate content through EVA.[24]

Manufacturers and trade names: DuPont™ Appeel® and Elvax®, Celanese EVA Performance Polymers Ateva®, Lanxess Levapren®, Baymond® L and Levamelt®.

Applications and uses: Packaging, cap liners, pallet stretch wrapping, bundling, liquid packaging, and as a sealant in barrier bags for primal and sub-primal cuts of meat, medical packaging (Tables 9.23–9.25).

See also Fig. 9.18.

9.7 Ethylene–Vinyl Alcohol Copolymer

EVOH is a copolymer of ethylene and vinyl alcohol. The structure is shown in Fig. 9.19 and it has a CAS number of 26221-27-2. These materials are highly crystalline, and are produced with various levels of ethylene content. See Table 9.26 for EVAL™ Resins.

The predominant product line is Eval Company of America (Kuraray) EVAL™. The general classes of the EVAL™ product line are show in Table 9.26. The films are often heat treated and oriented. These processes can dramatically affect the properties.

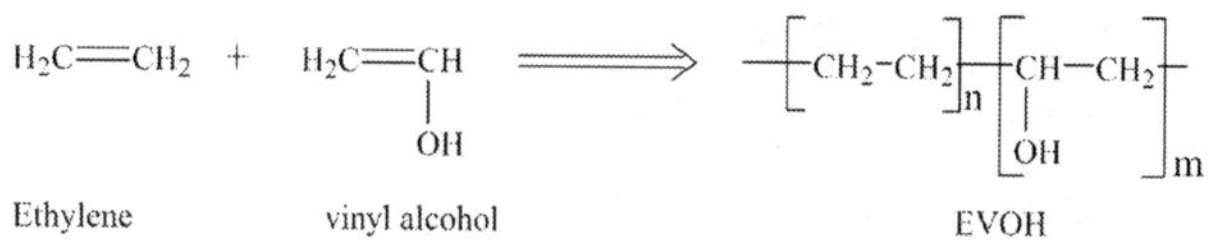

Figure 9.19 The structure of ethylene–vinyl alcohol copolymer (EVOH).

Table 9.26 EVAL™ Ethylene–Vinyl Alcohol Copolymer (EVOH) Polymer Grade Series[25]

EVAL™ Series	Ethylene Content (mol%)	General Characteristics
L Series	27	Has the lowest ethylene content of any EVOH and is suitable as an ultra-high-barrier grade in several applications.
F Series	32	Offers superior barrier performance and is widely used for automotive, bottle, film, tube, and pipe applications.
T Series	32	Specially developed to obtain good layer distribution in thermoforming, and has become the industry standard for multilayer sheet applications.
J Series	32	Offers thermoforming results even superior to those of T and can be used for unusually deep-draw or sensitive sheet-based applications.
H Series	38	Has a balance between high-barrier properties and long-term run stability. Especially suitable for blown film, special "U" versions exist to allow improved processing and longer running times even on less sophisticated machines.
E Series	44	Has a higher ethylene content which allows for greater flexibility and even easier processing. Different versions have been especially designed for cast and blown film as well as for pipe.
G Series	48	Has the highest ethylene content, making it the best candidate for stretch and shrink film applications.

Manufacturers and trade names: Eval Company of America (Kuraray) EVAL™, Soarus LLC Soarnol®.

Applications uses: Rigid packaging for entrees, edible oils, juice, cosmetics, pharmaceuticals, heating pipe, automotive plastic fuel tanks, packaging for condiments and toothpaste. Flexible packaging: Processed meats, bag-in-box, red meat, cereal, pesticides, and agrichemicals (Tables 9.27–9.57).

See also Figs. 9.20–9.28.

Table 9.27 Permeation of Oxygen at 20 °C and 65% Relative Humidity through Various EVAL™ Ethylene–Vinyl Alcohol Copolymer (EVOH) Resin Grades[26]

Grade	Application	Permeability Coefficient	
		Source Document Units (cm^3 20 μm/m^2 day atm)	Normalized Units (cm^3 mm/m^2 day atm)
F101B	Bottle, sheet, film	0.4	0.008
T101B	Thermoformed sheet	0.5	0.01
H101B	Bottle, sheet, film	0.7	0.014
E105B	Sheet, film	1.5	0.03
F101A	Without lubricant	0.4	0.008
F104B	High melt flow rate	0.4	0.008

Table 9.27 (Continued)

Grade	Application	Permeability Coefficient	
		Source Document Units (cm³ 20 μm/m² day atm)	Normalized Units (cm³ mm/m² day atm)
HU101B	Blown film	0.7	0.014
EU105B	Blown film	1.5	0.03
E151B	Low melt flow rate	1.5	0.03
FP101B	Pipe, with antioxidant	0.4	0.008
FP104B	Pipe, with antioxidant	0.4	0.008
EP105B	Pipe, with antioxidant	1.5	0.03
L101B	Ultrahigh barrier	0.2	0.004
J102B	Thermoforming	0.6	0.012
ES104B	Stretch/shrink film	4	0.08
G156B	Stretch/shrink film	3.2	0.064

Method: ASTM D3985.

Table 9.28 The Effect of Orientation, Heat Treatment, and Relative Humidity on the Oxygen Transmission Rate of EVAL™ F Series Ethylene–Vinyl Alcohol Copolymer (EVOH) Resins[27]

Chill Roll Temperature (°C)	Orientation	Heat Treatment	Permeability Coefficient			
			Source Document Units (cm³ 25 μm/m² day atm)		Normalized Units (cm³ mm/m² day atm)	
Relative humidity			0% RH	100% RH	0% RH	100% RH
50	None	None	0.126	40.9	0.00315	1.0225
110	None	None	0.118	33.8	0.00295	0.845
50	None	140	0.102	11	0.00255	0.275
50	Uniaxially 3 times	None	0.118	32.3	0.00295	0.8075
50	Uniaxially 3 times	140	0.094	3.9	0.00235	0.0975
50	Biaxially 3 × 3	None	0.118	31.5	0.00295	0.7875
50	Biaxially 3 × 3	140	0.094	2.3	0.00235	0.0575

Table 9.29 The Effect of Orientation, Heat Treatment, and Relative Humidity on the Oxygen Transmission Rate of EVAL™ E Series Ethylene–Vinyl Alcohol Copolymer (EVOH) Resins[28]

Chill Roll Temperature (°C)	Orientation	Heat Treatment	Permeability Coefficient			
			Source Document Units (cm^3 25 μ/m^2 day atm)		Normalized Units (cm^3 mm/m^2 day atm)	
Relative humidity			0% RH	100% RH	0% RH	100% RH
50	None	None	1.18	11.8	0.0295	0.295
110	None	None	1.02	9.4	0.0255	0.235
50	None	140	0.94	6.3	0.0235	0.1575
50	Uniaxially 3 times	None	1.02	10.2	0.0255	0.255
50	Uniaxially 3 times	140	0.94	3.1	0.0235	0.0775
50	Biaxially 3 × 3	None	1.02	10.2	0.0255	0.255
50	Biaxially 3 × 3	140	0.94	2.4	0.0235	0.06

Table 9.30 Permeation of Fluorocarbons through EVAL™ Ethylene–Vinyl Alcohol Copolymer (EVOH)[29]

Fluorocarbon	Temperature (°C)	Source Document Units		Normalized Units	
		Permeability Coefficient (cm^3 mil/m^2 day atm)		Permeability Coefficient (cm^3 mm/m^2 day atm)	
		Eval™ F Series	Eval™ E Series	Eval™ F Series	Eval™ E Series
HCFC-22 (R-22)	35		ND*		ND*
HCFC-22 (R-22)	50		ND*		ND*
HCFC-22 (R-22)	60		1.3		0.3
HCFC-22 (R-22)	65		4.1		0.1
CFC-12 (R-12)	70	0.24	8.0	0.006	0.2
HFC-134a (R-134a)	70		0.56		0.14

*None detected.

Table 9.31 Permeation of Oxygen vs. Temperature at 0% Relative Humidity through EVAL™ E Series EVOH[28]

Temperature (°C)	5	23	35	50
Source document units Permeability coefficient (cm^3 25 $\mu m/m^2$ day atm)	0.259	0.935	1.922	5.33
Normalized units Permeability coefficient (cm^3 mm/m^2 day atm)	0.01	0.02	0.05	0.14

Table 9.32 Permeation of Oxygen vs. Temperature at 0% Relative Humidity through EVAL™ G Series EVOH[28]

Temperature (°C)	5	23	35	50
Source document units Permeability coefficient (cm^3 25 μm/m^2 day atm)	1.034	1.8	2.7	6.11
Normalized units Permeability coefficient (cm^3 mm/m^2 day atm)	0.03	0.05	0.07	0.16

Table 9.33 Permeation of Oxygen vs. Temperature at 0% Relative Humidity through EVAL™ H Series EVOH[28]

Temperature (°C)	5	23	35	50
Source document units Permeability coefficient (cm^3 25 μm/m^2 day atm)	0.09	0.395	0.94	2.6
Normalized units Permeability coefficient (cm^3 mm/m^2 day atm)	0.0024	0.01	0.02	0.07

Table 9.34 Permeation of Oxygen vs. Temperature at 0% Relative Humidity through EVAL™ K Series EVOH[28]

Temperature (°C)	5	23	35	50
Source document units Permeability coefficient (cm^3 25 μm/m^2 day atm)	0.09	0.395	0.94	2.6
Normalized units Permeability coefficient (cm^3 mm/m^2 day atm)	0.0024	0.01	0.02	0.07

Table 9.35 Permeation of Oxygen vs. Temperature at 0% Relative Humidity through EVAL™ L Series EVOH[28]

Temperature (°C)	5	23	35	50
Source document units Permeability coefficient (cm^3 25 μm/m^2 day atm)	0.022	0.095	0.231	0.637
Normalized units Permeability coefficient (cm^3 mm/m^2 day atm)	0.00039	0.0024	0.01	0.02

Table 9.36 Permeation of Oxygen vs. Temperature at 0% Relative Humidity through EVAL™ F Series EVOH[28]

Temperature (°C)	5	23	35	50
Source document units Permeability coefficient (cm^3 25 μm/m^2 day atm)	0.045	0.2	0.48	1.34
Normalized units Permeability coefficient (cm^3 mm/m^2 day atm)	0.0012	0.01	0.01	0.03

Table 9.37 Permeation of Several Gases at 0% Relative Humidity through Various EVAL™ EVOH Grades[26]

Permeant	Temperature (°C)	Permeability Coefficient					
		Source Document Units (cm³ 20 μm/m² day atm)			Normalized Units (cm³ mm/m² day atm)		
		Eval™ F101B	Eval™ H101B	Eval™ H105B	Eval™ F101B	Eval™ H101B	Eval™ H105B
Nitrogen	25	0.017		0.13	0.00034		0.0026
Oxygen	25	0.27		1.23	0.0054		0.0246
Carbon dioxide	25	0.81		7.1	0.0162		0.142
Helium	25	160		410	3.2		8.2
Argon	35			1.6			0.032
Argon	50	0.5	3.5	7.0	0.01	0.07	0.14
Krypton	50	0.4	1.0	1.8	0.008	0.02	0.036

Table 9.38 Permeation of Carbon Dioxide, Nitrogen, and Helium at 0% Relative Humidity and Different Temperatures through EVAL™ E Series EVOH[28]

Penetrant	Carbon Dioxide			Nitrogen		Helium		
Temperature (°C)	5	23	35	23	35	5	23	35
Source document units Permeability coefficient (cm³ 25 μm/m² day atm)	0.87	3.32	7.72	0.124	0.232	102.3	368.9	551.8
Normalized units Permeability coefficient (cm³ mm/m² day atm)	0.02	0.08	0.2	0.0031	0.01	2.6	9.37	14.02

Table 9.39 Permeation of Carbon Dioxide, Nitrogen, and Helium at 0% Relative Humidity and Different Temperatures through EVAL™ F Series EVOH[28]

Penetrant	Carbon Dioxide			Nitrogen		Helium		
Temperature (°C)	5	23	35	23	35	5	23	35
Source document units Permeability coefficient (cm³ 25 μm/m² day atm)	0.155	0.496	1.023	0.015	0.031	41.8	144.1	212.3
Normalized units Permeability coefficient (cm³ mm/m² day atm)	0.0039	0.01	0.03	0.0004	0.0008	1.06	3.66	5.39

9: Polyolefins, Polyvinyls, and Acrylics

Table 9.40 Permeation of Carbon Dioxide, Nitrogen, and Helium at 0% Relative Humidity and Different Temperatures through EVAL™ H Series EVOH[28]

Penetrant	Carbon Dioxide			Nitrogen			Helium		
Temperature (°C)	5	23	35	5	23	35	5	23	35
Source document units	0.263	1.04	3.32	0.062	0.124	71.3	257.3	381.3	
Permeability coefficient (cm^3 25 μm/m^2 day atm)									
Normalized units	0.01	0.03	0.08	0.0016	0.0031	1.81	6.54	9.37	
Permeability coefficient (cm^3 mm/m^2 day atm)									

Note: columns 5/23/35 under Nitrogen show 0.062, 0.124; under Helium show 71.3, 257.3, 381.3. (Values read as reproduced above.)

Table 9.41 Permeation of Oxygen vs. Relative Humidity through EVAL™ EF-XL Biaxially Oriented EVOH Film[30]

Temperature (°C)	35			20		
Relative humidity (%)	0	65	85	65	85	100
Test method	JIS Z1707			ASTM D3985		
Source document units						
Permeability coefficient (cm^3 mil/100 $in.^2$ day)	0.03	0.02	0.08	0.07	0.39	
Normalized units						
Permeability coefficient (cm^3 mm/m^2 day atm)	0.01	0.01	0.03	0.03	0.15	

Sample thickness (mm): 0.015.

Table 9.42 Permeation of Oxygen vs. Relative Humidity at 20 °C through EVAL™ EF-XL, EVAL™ EF-F, and EF-E Series EVOH Film[23]

Eval™ Grade	EF-XL	EF-F	EF-E	EF-XL	EF-F	EF-E	EF-XL	EF-F	EF-E
Relative humidity (%)	0			65			100		
Source document units	0.02	0.02	0.16	0.01	0.02	0.08	0.23	1.0	0.52
Permeability coefficient (cm^3 mil/100 $in.^2$ day)									
Normalized	0.01	0.01	0.06	0.004	0.01	0.03	0.09	0.39	0.2
Permeability coefficient (cm^3 mm/m^2 day atm)									

Table 9.43 Oxygen Permeability at 20 °C and 0% RH vs. Orientation and Heat Treatment through EVAL™ E Series EVOH[28]

Chill Roll							
Temperature (°C)	50	110	50	50	50	50	50
Heat treatment (°C)	None	None	140	None	140	None	140
Orientation	None	None	None	Uniaxially (3 times)	Uniaxially (3 times)	Biaxially (3 × 3)	Biaxially (3 × 3)
Source document units Permeability coefficient (cm³ 25 μm/m² day atm)	1.18	1.02	0.94	1.02	0.94	1.02	0.94
Normalized units Permeability coefficient (cm³ mm/m² day atm)	0.03	0.026	0.024	0.028	0.024	0.028	0.024

Table 9.44 Oxygen Permeability at 20 °C and 100% RH vs. Orientation and Heat Treatment through EVAL™ E Series EVOH[28]

Chill Roll							
Temperature (°C)	50	110	50	50	50	50	50
Heat treatment (°C)	None	None	140	None	140	None	140
Orientation	None	None	None	Uniaxially (3 times)	Uniaxially (3 times)	Biaxially (3 × 3)	Biaxially (3 × 3)
Source document units Permeability coefficient (cm³ 25 μm/m² day atm)	11.8	9.4	6.3	10.2	3.1	10.2	2.4
Normalized units Permeability coefficient (cm³ mm/m² day atm)	0.299	0.24	0.16	0.28	0.079	0.28	0.06

Table 9.45 Oxygen Permeability at 20 °C and 0% RH vs. Orientation and Heat Treatment through EVAL™ F Series EVOH[28]

Chill Roll							
Temperature (°C)	50	110	50	50	50	50	50
Heat treatment (°C)	None	None	140	None	140	None	140
Orientation	None	None	None	Uniaxially (3 times)	Uniaxially (3 times)	Biaxially (3 × 3)	Biaxially (3 × 3)
Source document units Permeability coefficient (cm³ 25 μm/m² day atm)	0.008	0.0076	0.0066	0.0076	0.0006	0.0076	0.0061
Normalized units Permeability coefficient (cm³ mm/m² day atm)	0.126	0.118	0.102	0.118	0.094	0.118	0.094

9: Polyolefins, Polyvinyls, and Acrylics

Table 9.46 Oxygen Permeability at 20 °C and 100% RH vs. Orientation and Heat Treatment through EVAL™ F Series EVOH[28]

Chill Roll Temperature (°C)	50	110	50	50	50	50	50
Heat treatment (°C)	None	None	140	None	140	None	140
Orientation	None	None	None	Uniaxially (3 times)	Uniaxially (3 times)	Biaxially (3 × 3)	Biaxially (3 × 3)
Source document units Permeability coefficient (cm³ 25 μm/m² day atm)	40.9	33.8	11	32.3	3.9	31.5	2.3
Normalized units Permeability coefficient (cm³ mm/m² day atm)	1.02	0.87	0.28	0.83	0.1	0.79	0.06

Table 9.47 Organic Solvents at 20 °C through EVAL™ EF-E, EVAL™ EF-F, and EVAL™ EF-XL Series EVOH Film[31]

Grade	EF-F	EF-E	EF-XL	EF-F	EF-E	EF-XL	EF-F	EF-E	EF-XL
Penetrant	Chloroform			Xylene			Kerosene		
Source document units Vapor transmission rate (g mil/100 in.² day)	0.1	0.16	0.006	0.054	0.074	0.016	>0.001	0.0025	0.001
Normalized vapor transmission rate (g mm/m² day)	0.04	0.06	0.0024	0.02	0.03	0.01	>0.0004	0.00098	0.0004

Table 9.48 Organic Solvents at 20 °C and 65% Relative Humidity through Biaxially Oriented EVAL™ EF-XL Series EVOH Film[31]

Penetrant	Chloroform	Xylene	Methyl Ethyl Ketone	Kerosene
Source document units Vapor transmission rate (g/100 in.² day)	0.01	0.03	0.02	<0.003
Normalized vapor transmission rate (g mm/m² day)	0.002	0.007	0.005	<0.0007

Sample thickness (mm): 0.015.

Table 9.49 Organic Solvents at 20 °C and 65% Relative Humidity through EVAL™ E Series EVOH Film[31]

Penetrant	Chloroform		Xylene		Methyl Ethyl Ketone		Kerosene	
Sample thickness (mm)	0.02	0.032	0.02	0.032	0.02	0.032	0.02	0.032
Source document units Vapor transmission rate (g/100 in.^{2} day)	0.2	0.06	0.09	0.04	0.31	0.03	<0.003	<0.003
Normalized Vapor transmission rate (g mm/m^2 day)	0.06	0.03	0.03	0.02	0.12	0.01	<0.001	<0.002

Table 9.50 Organic solvents at 20 °C and 65% Relative Humidity through EVAL™ F Series EVOH Film[31]

Penetrant	Chloroform		Xylene		Methyl Ethyl Ketone		Kerosene	
Sample thickness (mm)	0.02	0.032	0.02	0.032	0.02	0.032	0.02	0.032
Source document Vapor transmission rate (g/100 in.^{2} day)	0.13	0.3	0.07	<0.003	0.25	0.02	<0.003	<0.003
Normalized Vapor transmission rate (g mm/m^2 day)	0.04	0.2	0.02	<0.002	0.08	0.01	<0.001	<0.002

Table 9.51 Permeation of Water Vapor at 40 °C and 90% Relative Humidity through EVAL™ EF-XL, EVAL™ EF-F, and EVAL™ EF-E Series EVOH Film[32]

Grade	EVAL™ EF-XL	EVAL™ EF-F	EVAL™ EF-E
Sample thickness (mm)	0.015	0.015	0.02
Source document Vapor transmission rate (g/100 in.^{2} day)	3	6	2
Normalized Vapor transmission rate (g mm/m^2 day)	1.2	2.4	0.8

Test method: JIS Z0208.

Table 9.52 Permeation of Water Vapor at 40 °C and 90% Relative Humidity through EVAL™ L, EVAL™ F, EVAL™ H, EVAL™ K, EVAL™ E, and EVAL™ G Series EVOH[28]

Grade	EVAL™ L	EVAL™ F	EVAL™ H	EVAL™ K	EVAL™ E	EVAL™ G
Source document Vapor transmission rate (g 25 μm/m^2 day)	124	58.9	32.6	32.6	21.7	21.7
Normalized Vapor transmission rate (g mm/m^2 day)	3.2	1.5	0.8	0.8	0.6	0.6

9: Polyolefins, Polyvinyls, and Acrylics

Table 9.53 Permeation of D-Limonene at 20 °C and 65% Relative Humidity through EVAL™ EVOH[24]

Material Grade	EVAL™ Series F	EVAL™ Series E	EVAL™ 5%	EVAL™ 7%
Source document Vapor permeability (g mil/100 in.² day)	0.002	0.003	98	113.5
Normalized Vapor permeability (g mm/m² day)	0.00098	0.0015	48	55.6

Table 9.54 Permeation of Various Gases at 20 °C through Nippon Gohse Soarnol® EVOH[33]

Product Grades	Ethylene Content (mol%)	Nitrogen	Oxygen	Carbon Dioxide	Helium	Hydrogen
		\multicolumn{5}{c}{Source Document Units (cm³ 20 μm/m² day atm)}				
D2908,[a] DT2903	29	0.018	0.23	0.49	110	27
DC3212,[a] DC3203F	32	0.024	0.3	0.62	120	32
E3808,[a] ET3803	38	0.041	0.53	1.3	180	
A4412,[a] AT4406,[a] AT4403	44	0.1	1.2	4.4	320	195
		\multicolumn{5}{c}{Normalized units (cm³ mm/m² day atm)}				
D2908,[a] DT2903	29	0.00036	0.0046	0.0098	2.2	0.54
DC3212,[a] DC3203F	32	0.00048	0.006	0.0124	2.4	0.64
E3808,[a] ET3803	38	0.00082	0.0106	0.026	3.6	
A4412,[a] AT4406,[a] AT4403	44	0.002	0.024	0.088	6.4	3.9

[a]Cast film, others blown film.

Table 9.55 Water Vapor Permeation Rate at 40 °C and 95% Relative Humidity through Nippon Gohse Soarnol® EVOH[26]

		Permeation Rate	
Product Grades	Ethylene Content (mol%)	Source Document Units (g 30 μm/m² day)	Normalized Units (g mm/m² day)
D2908,[a] DT2903	29	37–70	11–21
DC3212,[a] DC3203F	32	32–58	10–17
E3808,[a] ET3803	38	23–35	7–11
A4412,[a] AT4406,[a] AT4403	44	12–25	3.6–7.5

Film thickness: 0.03 mm.
[a]Cast films, others blown film.

Table 9.56 Liquid Permeation Rate of Chloroform and Kerosene at 20 °C through Nippon Gohse Soarnol® EVOH[28]

Product Grades	Ethylene Content (mol %)	Permeation Rate			
		Source Document Units (mg 20 μm/cm² day)		Normalized Units (g mm/m² day)	
		Chloroform	Kerosene	Chloroform	Kerosene
DC3212,[a] DC3203F	32	0.2	<0.005	0.0124	<0.001
A4412,[a] AT4406,[a] AT4403	44	0.31	<0.005	0	<0.001

[a]Cast films, others blown film.

Table 9.57 Liquid Permeation Rate of Chlorofluorocarbon R-22 at 30 °C and 50 °C through Nippon Gohse Soarnol® EVOH[34]

Product Grades	Ethylene Content (mol%)	Permeation Rate			
		Source Document Units (g 100 μm/m² day)		Normalized Units (g mm/m² day)	
		30 °C	50 °C	30 °C	50 °C
D2908,[a] DT2903	29	4.6	8.2	0.46	0.82
DC3212,[a] DC3203F	32	7.5	14.1	0.75	1.41
E3808,[a] ET3803	38	13.5	22.1	1.35	2.21
A4412,[a] AT4406,[a] AT4403	44	18.5	27.8	1.85	2.78

[a]Cast films, others blown film.

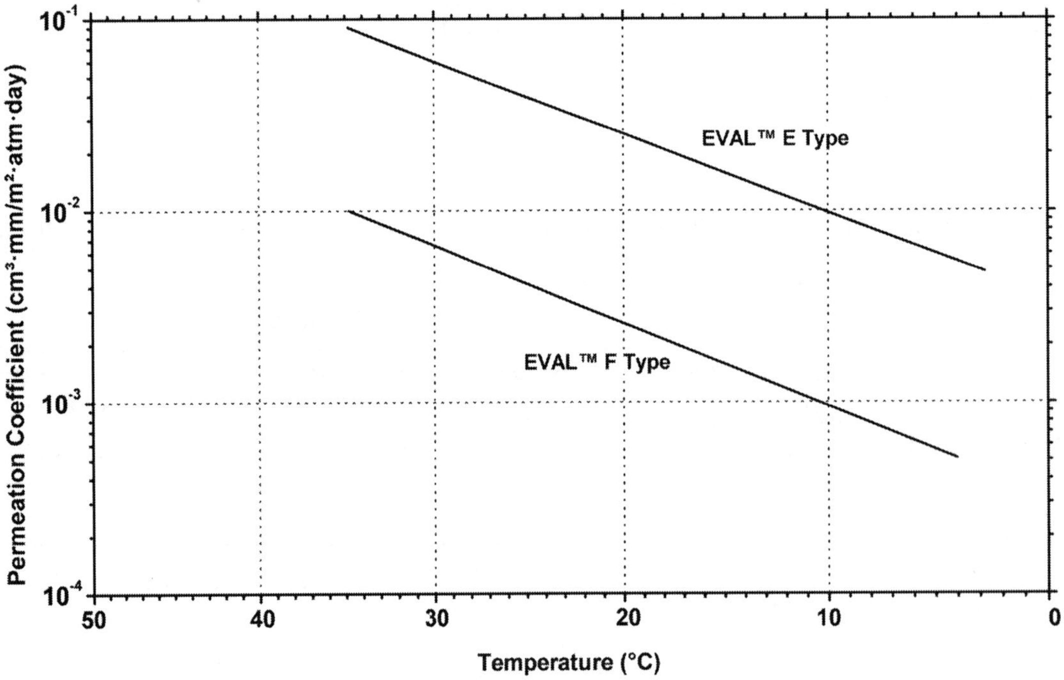

Figure 9.20 Permeation of oxygen vs. temperature through EVAL™ EVOH films.[26]

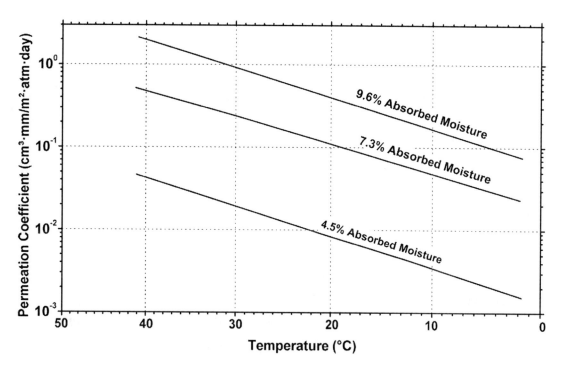

Figure 9.21 Permeation of oxygen vs. temperature at various moisture absorption conditions through EVAL™ EVOH films.[26]

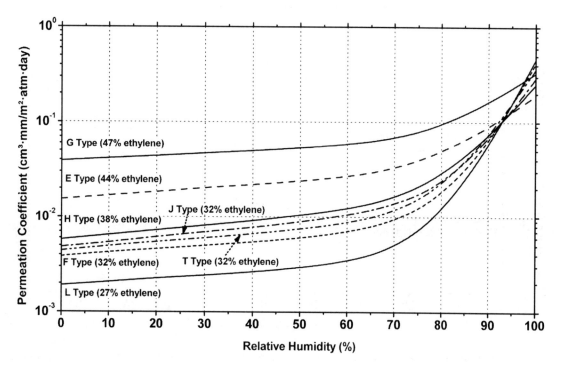

Figure 9.22 Permeation of oxygen vs. relative humidity at 20 °C through EVAL™ EVOH resins.[26]

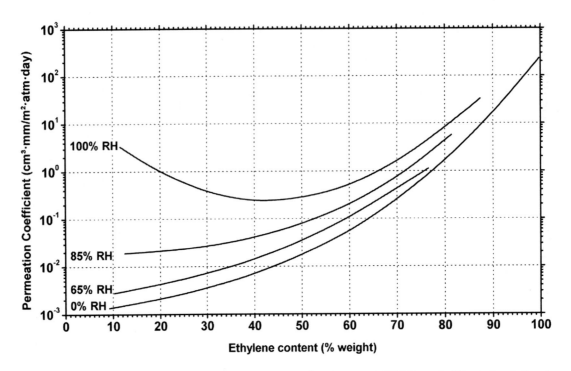

Figure 9.23 Permeation of oxygen vs. ethylene content of polymer at 20 °C and different relative humidities through EVAL™ EVOH films.[26]

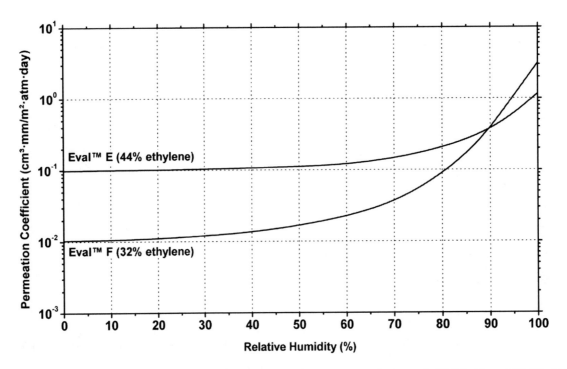

Figure 9.24 Permeation of carbon dioxide vs. relative humidity at 20 °C through EVAL F and EVAL E Series EVOH.[28]

9: Polyolefins, Polyvinyls, and Acrylics

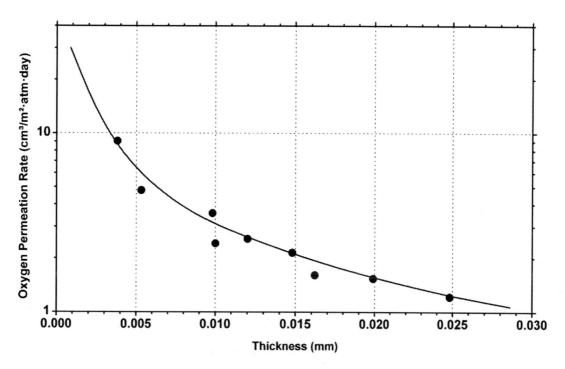

Figure 9.25 Permeation of oxygen vs. film thickness at 25 °C and 75% relative humidity through Nippon Gohsei Soarnol® E3808 and ET3803 EVOH.[35]

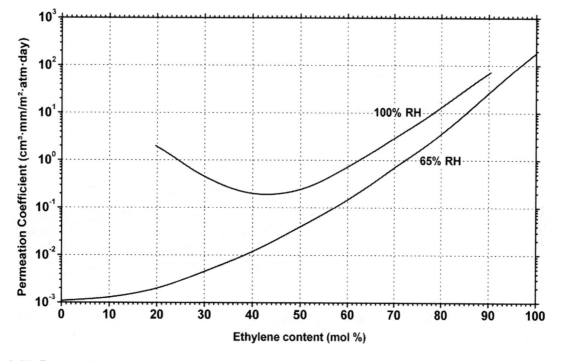

Figure 9.26 Permeation of oxygen vs. ethylene content at 20 °C through Nippon Gohsei Soarnol® EVOH.[36]

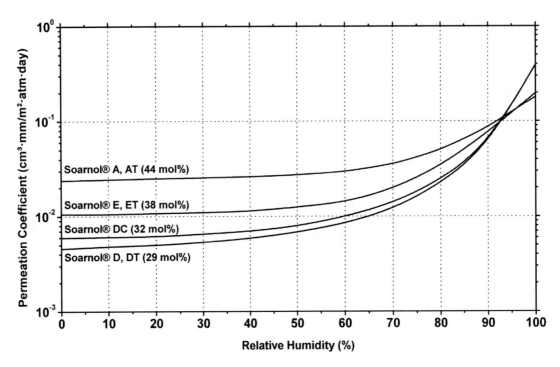

Figure 9.27 Permeation of oxygen vs. relative humidity at 20 °C through Nippon Gohsei Soarnol® EVOH.[37]

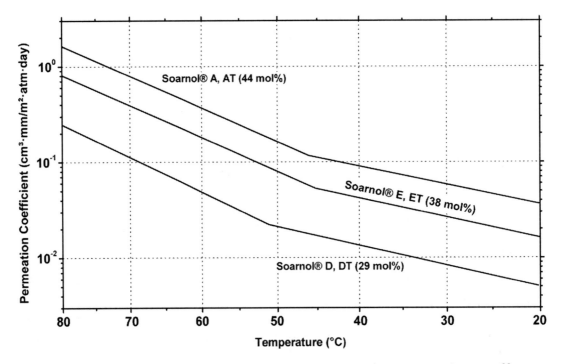

Figure 9.28 Permeation of oxygen vs. temperature through Nippon Gohsei Soarnol® EVOH.[38]

9.8 Polyvinyl Butyral

PVB is considered to be an acetal and is formed from the reaction of an aldehyde and alcohol. The structure of PVB is shown in Fig. 9.29, but it is generally not made in exactly this form. It is made in a way such that the polymer is a mixture of PVB, polyvinyl alcohol, and polyvinyl acetate segments as shown in the figure. The relative amounts of these segments are controlled but they are generally randomly distributed through the molecular chain. The properties of the polymers can be optimized by controlling the ratios of the three segments.

Manufacturers and trade names: GlasNovations KB®, DuPont™ Butacite®, Solutia Inc. Saflex®, Sekisui S-Lec, Kuraray TROSIFOL®.

Applications and uses: Safety glass, architectural glass (Table 9.58).

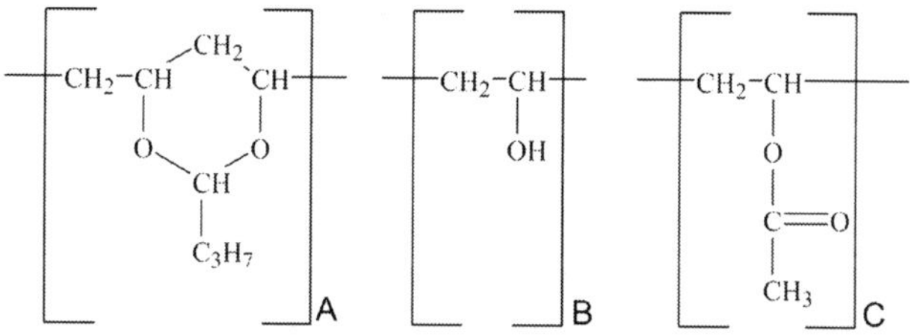

Figure 9.29 Structure of polyvinyl butyral.

9.9 Polyvinyl Chloride

PVC is a flexible or rigid material that is chemically nonreactive. Rigid PVC is easily machined, heat formed, welded, and even solvent cemented. PVC can also be machined using standard metal working tools and finished to close tolerances and finishes without great difficulty. PVC resins are normally mixed with other additives such as impact modifiers and stabilizers, providing hundreds of PVC-based materials with a variety of engineering properties.

There are three broad classifications for rigid PVC compounds: Type II, CPVC, and Type I. Type II differs from Type I due to greater impact values but lower chemical resistance. Chlorinated polyvinyl chloride (CPVC) is a thermoplastic produced by chlorination of polyvinyl chloride (PVC) resin and has greater high-temperature resistance. These materials are considered "unplasticized," because they are less flexible than the plasticized formulations. PVC has a broad range of applications, from high-volume construction related products to simple electric wire insulation and coatings.

Manufacturing and trade names: Polyone Geon™, Fiberloc™, VPI LLC Mirrex®.

Applications and uses: Packaging is a major market for PVC. Rigid grades are blown into bottles and made into sheets for thermoforming boxes and blister packs. Flexible PVC compounds are used in food packaging applications because of their strength, transparency, processability, and low raw material cost. Major markets for PVC are in building/construction, packaging, consumer and institutional products, and electrical/electronic uses (Tables 9.59 and 9.60).

See Figs. 9.30–9.32.

Table 9.58 Permeation of Gases at 25 °C through Polyvinyl Butyral Membranes[32]

Permeant Gas	Permeability Coefficient	
	Source Document units (Barrer)	Normalized Units (cm^3 mm/m^2 day atm)
Oxygen	0.771	50.6
Argon	0.463	30.4
Nitrogen	0.133	7.4

Table 9.59 Permeation of Water Vapor at 38 °C and 90% Relative Humidity through VPI Mirrex® 1025 PVC[31]

Film Thickness (mm)	Vapor Transmission Rate	
	Source Document (g mm/100 in.² day)	Normalized Units (g mm/m² day)
0.190	0.0608	0.94
0.254	0.0610	0.95
0.305	0.0610	0.95

Test method: ASTM F372.

Table 9.60 Permeation of Oxygen at 23 °C and 100% Relative Humidity through VPI Mirrex® PVC

Film Thickness (mm)	Permeability Coefficient	
	Source Document (cm³ mm/100 in.² day atm)	Normalized Units (cm³ mm/m² day atm)
0.190	0.304	4.7
0.254	0.305	4.7
0.305	0.305	4.7

Test method: Mocon Oxtran.

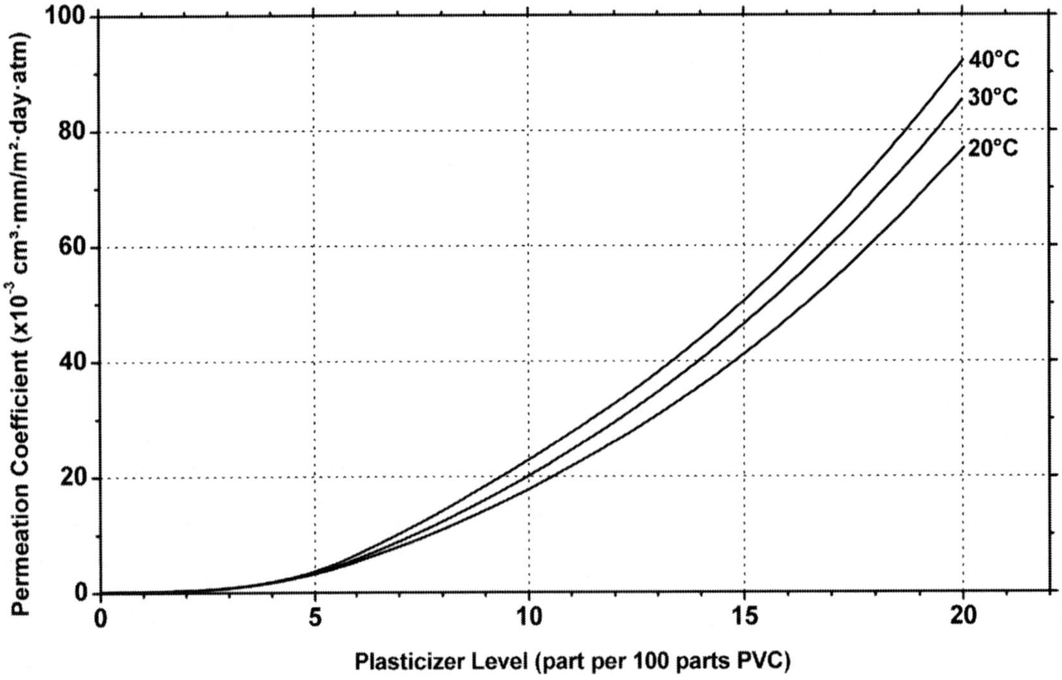

Figure 9.30 Permeation of carbon dioxide and hydrogen vs. temperature through Polyone Geon™ 101-EP-100 PVC film with varying levels of diethylhexyl phthalate plasticizer.[3]

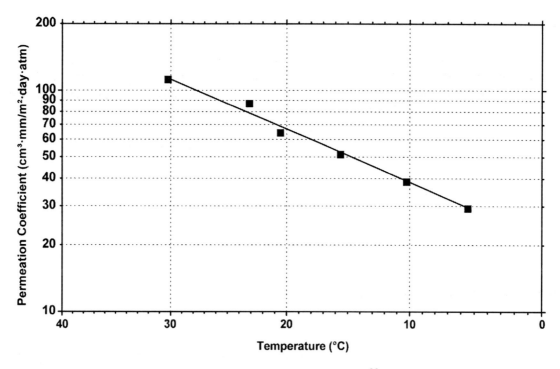

Figure 9.31 Permeation of oxygen vs. temperature through PVC film.[39]

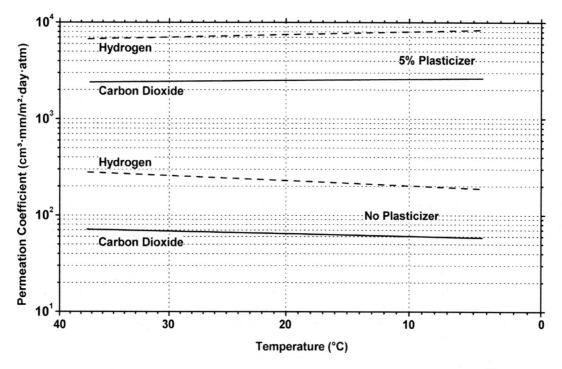

Figure 9.32 Permeation of oxygen vs. plasticizer level and temperature through PVC film.[15]

9.10 Polyvinylidene Chloride

PVDC resin, the structure of which is shown in Fig. 9.33, is usually a copolymer of vinylidene chloride with vinyl chloride or other monomers. Dow Plastics vinyl chloride and vinylidene chloride, Saran™, is usually supplied as a white, free flowing powder. PVDC has a CAS number of 9002-85-1.

Manufacturers and trade names: Dow Saran™.

Applications and uses: Monolayer films (Saran™) for food wrap and medical packaging, coextruded films and sheet structures as a barrier layer in medical, and packaging including fresh red meats, cheese, and sausages. Coatings are applied to containers to prevent gas transmission (Tables 9.61–9.66).

See also Figs. 9.34–9.36.

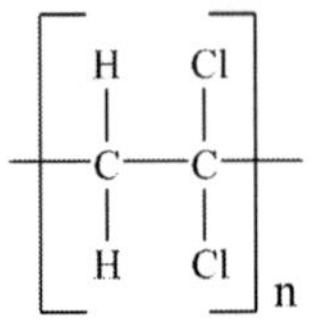

Figure 9.33 Structure of PVDC homopolymer.

Table 9.61 Permeation of Oxygen at 23 °C and 75% Relative Humidity through Dow Saran™ PVDC Films[40]

Saran™ Grade	Thickness (mm)	Permeability Coefficient	
		Source Document Units (cm³ mil/100 in.² day atm)	Normalized Units (cm³ mm/m² day atm)
469		0.10	0.04
516		0.10	0.04
525		0.10	0.04
MA 119		0.08	0.03
MA 123		0.08	0.03
MA 134		0.08	0.03
313		1.2	0.47
867		1.1	0.43
F239	0.01375	0.61	0.013
F239	0.025	0.35	0.0138
F278	0.01375	0.35	0.0075
F278	0.025	0.20	0.0079
F310	0.01375	1.5	0.032
F310	0.025	0.83	0.033
F271	0.01375	0.7	0.0149
F271	0.025	0.35	0.0138
F279	0.01375	0.35	0.0074
F279	0.025	0.20	0.00786
F281	0.01375	0.28	0.0059
F281	0.025	0.15	0.0059

Test method: ASTM D1434.

Table 9.62 Permeation of Oxygen at 23 °C and 50% Relative Humidity through Dow Saranex™ PVDC multilayer Films[35]

Saranex™ Grade	Thickness (mm)	Permeability Coefficient	
		Source Document Units (cm^3 mil/100 $in.^2$ day atm)	Normalized Units (cm^3 mm/m^2 day atm)
451	0.038	0.50	0.296
450	0.025	0.85	0.33
23P	0.051	0.75	0.612

Test method: ASTM D1434.

Table 9.63 Permeation of Gases through Dow Saranex™ 21 Films[35]

Permeant	Temperature (°C)	Relative Humidity (%)	Permeability Coefficient	
			Source Document Units (cm^3/m^2 day atm)	Normalized Units (cm^3 mm/m^2 day atm)
Carbon dioxide	23	10	21.7	0.95
Nitrogen	23	10	1.55	0.07
Oxygen	22	4	13	0.57
Air	23	10	1.6	0.07

Thickness: 0.044 mm. Test method: ASTM D1434.

Table 9.64 Permeation of Water Vapor at 38 °C and 90% Relative Humidity through Dow Saranex™ PVDC multilayer Films[35]

Saran™ Grade	Thickness (mm)	Vapor Transmission Rate	
		Source Document Units (g/100 $in.^2$ day)	Normalized Units (g mm/m^2 day)
451	0.038	0.33	0.194
450	0.025	0.35	0.140
553	0.076	0.33	0.194
23P	0.051	0.25	0.199

Test method: ASTM E96.

Table 9.65 Permeation of Water Vapor at 38 °C and 90% Relative Humidity through Dow Saran™ PVDC Films[35]

Saran™ Grade	Thickness (mm)	Vapor Transmission Rate	
		Source Document Units (g mil/100 in.² day)	Normalized Units (g mm/m² day)
469		0.13	0.06
516		0.13	0.06
525		0.13	0.06
MA 119		0.05	0.025
MA 123		0.05	0.025
MA 134		0.05	0.025
313		0.27	0.13
867		0.20	0.10
F239	0.01375	0.76	0.255
F239	0.025	0.38	0.228
F278	0.01375	0.43	0.140
F278	0.025	0.21	0.126
F310	0.01375	2.8	0.913
F310	0.025	1.3	0.786
F271	0.01375	0.6	0.197
F271	0.025	0.3	0.183
F279	0.01375	0.43	0.140
F279	0.025	0.21	0.126
F281	0.01375	0.41	0.134
F281	0.025	0.20	0.118

Test method: ASTM E96.

Table 9.66 Permeation of Gases at 23 °C and 90% Relative Humidity through Dow Saran™ Films[35]

Permeant Gas	Permeability Coefficient	
	Source Document Units (cm³ mil/100 in.² day atm)	Normalized Units (cm³ mm/m² day atm)
Oxygen	0.17	0.00425
Nitrogen	0.04	0.001
Carbon dioxide	0.25	0.00625
Air	0.07	0.00175

Thickness: 0.025 mm. Test methods: ASTM D1434 and ASTM F1249.

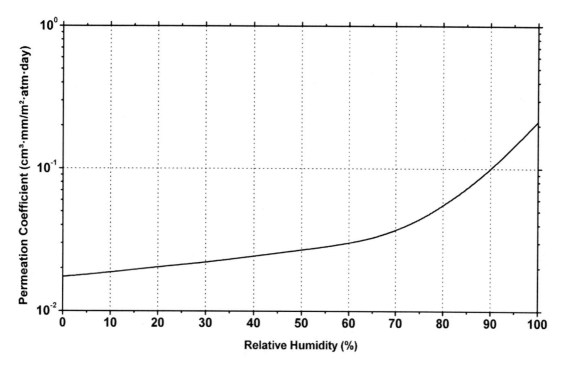

Figure 9.34 Oxygen transmission rate at 20 °C vs. relative humidity for Dow Saranex™ PVDC.[25]

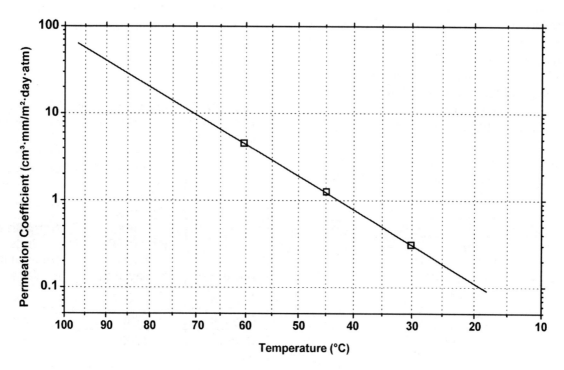

Figure 9.35 Permeation of hydrogen sulfide vs. temperature through Dow Saran™ PVDC.[41]

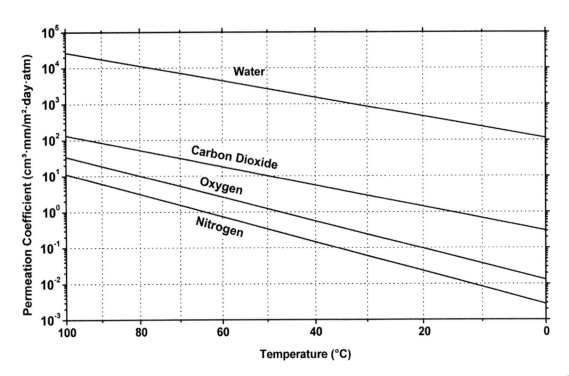

Figure 9.36 Permeation of water oxygen, carbon dioxide, and nitrogen vs. temperature through PVDC.[9]

9.11 Polyacrylics

While a large number of acrylic polymers are manufactured, PMMA is by far the most common. The structure of PMMA is shown in Fig. 9.37. Nearly everyone has heard of Plexiglas®. PMMA has two very distinct properties that set the products apart from others. First it is optically clear and colorless. It has a light transmission of 92%. The 4% reflection loss at each surface is unavoidable. Second its surface is extremely hard. They are also highly weather resistant. PMMA has a CAS number of 9011-14-7.

Manufacturers and trade names: Lucite International, Lucite Diakon®, and Perspex®, Evonik Industries LLC, Acrylite®, Europlex® and Rohaglas®, Arkema Oroglas.

Applications and uses: Optical parts, display items, tube and profile extrusion, automotive rear lights and dashboard lenses, extruded sheet, copying equipment, and lighting diffusers (Tables 9.67 and 9.68).

See also Figs. 9.38 and 9.39.

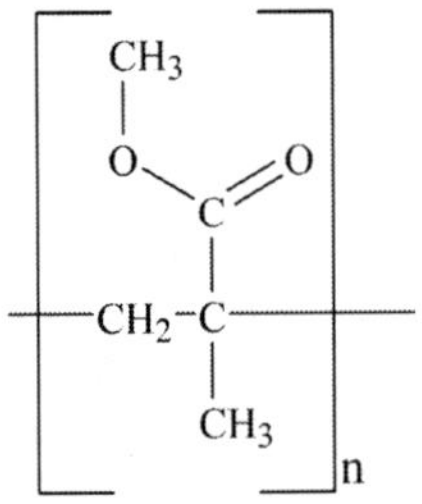

Figure 9.37 Structure of PMMA.

Table 9.67 Permeation of Oxygen, Nitrogen, and Carbon Dioxide at 25 °C through Lucite Diakon® Polymethyl Methacrylate[42]

Permeant	Permeability Coefficient	
	Source Document Units (cm³ (STP) 25 μm/ m² day atm)	Normalized Units (cm³ mm/ m² day atm)
Nitrogen	60	1.5
Oxygen	230	5.8
Carbon dioxide	1700	43

Film thickness: 25 μm.

9: Polyolefins, Polyvinyls, and Acrylics

Table 9.68 Permeation of Water Vapor at 25 °C and 75% Relative Humidity through Lucite Diakon® Polymethyl Methacrylate[3]

Vapor Transfer Rate	
Source Document Units (g 25 μm/m² day)	Normalized Permeability Units (g mm/m² day)
68	1.7

Film thickness: 25 μm.

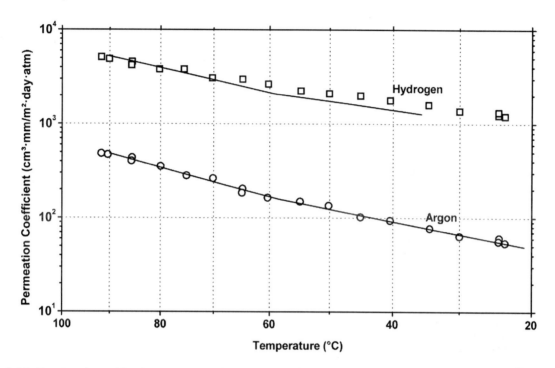

Figure 9.38 Permeation of hydrogen and argon vs. temperature through polyethylmethacrylate.[43]

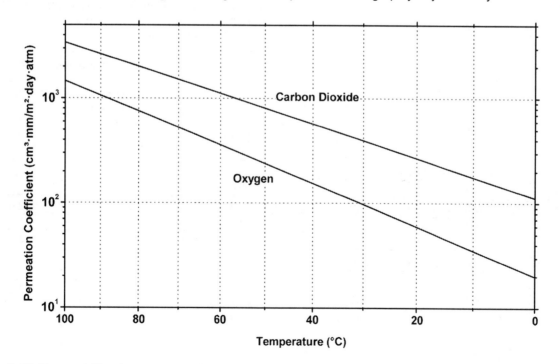

Figure 9.39 Permeability data vs. temp from Wood-Adams polyethylmethacrylate.

9.12 Acrylonitrile–Methyl Acrylate Copolymer

AMA is a copolymer of acrylonitrile and methyl acrylate. The generalized structure of AMA is shown in Fig. 9.40. It is a mixture of the acrylonitrile group and methyl group, with the ratio of the two groups affecting the properties. AMA has a CAS number of 24968-79-4.

Manufacturers and trade names: INEOS Barex®.
Applications and uses:

- Food packaging. Processed meats, fish, cheese, spices, sauces, extracts, and juice concentrates
- Medical packaging. Pharmaceutical, transdermal patches
- Personal care. Cosmetic packs, mouthwash, perfume (Tables 9.69 and 9.70).

See also Figs. 9.41 and 9.42.

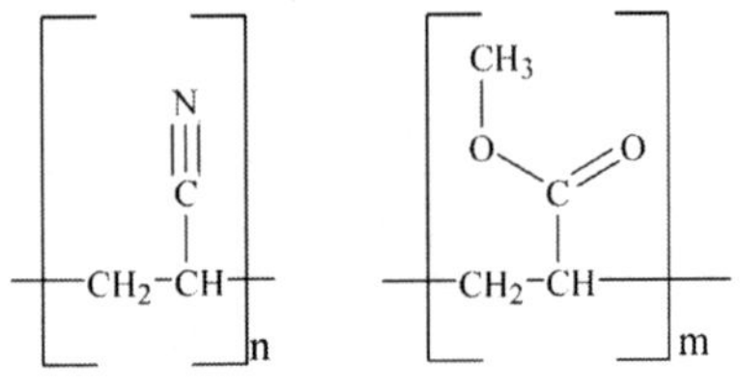

Figure 9.40 Structure of AMA segments.

Table 9.69 Permeation of Gases at 73 °F (23 °C) and 100% Relative Humidity through INEOS Barex® Acrylonitrile–Methyl Acrylate Copolymers[44]

Grade/Permeant	Permeability Coefficient	
	Source Document Units (cm^3 mil/100 in.^{2} day atm)	Normalized Units (cm^3 mm/m^2 day atm)
Barex® 210 extrusion grade		
Oxygen	0.8	0.3
Nitrogen	0.2	0.08
Carbon dioxide	1.2	0.45
Barex® 210 film grade		
Oxygen	0.8	0.3
Nitrogen	0.2	0.08
Carbon dioxide	1.2	0.45
Barex® 214 calendar grade		
Oxygen	1.2	0.45
Nitrogen	0.3	0.12
Carbon dioxide	1.4	0.52
Barex® 218 extrusion grade impact modified		
Oxygen	1.6	0.6
Nitrogen	0.4	0.16
Carbon dioxide	1.6	0.6

Test method: ASTM D3985.

Table 9.70 Permeation of Water Vapor at 100 °F (38 °C) and 90% Relative Humidity through INEOS Barex® Acrylonitrile–Methyl Acrylate Copolymers (per ASTM F1249-90)[45]

Grade	Vapor Transmission Rate	
	Source Document Units (g mil/100 in.^{2} day atm)	Normalized Units (g mm/m^2 day)
Barex® 210 extrusion grade	5.0	2.0
Barex® 210 film grade	5.0	2.0
Barex® 214 calendar grade	6.3	2.5
Barex® 218 extrusion grade impact modified	7.5	3.0

9: Polyolefins, Polyvinyls, and Acrylics

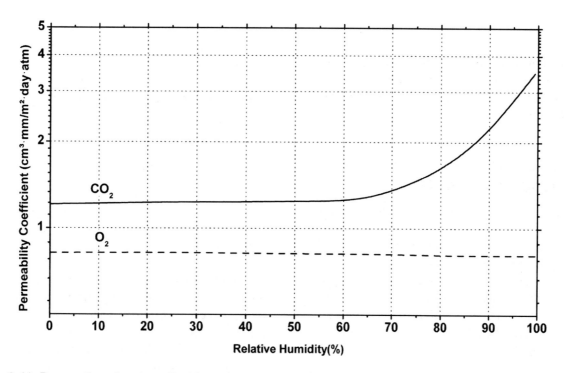

Figure 9.41 Permeation of carbon dioxide and oxygen vs. relative humidity through acrylonitrile–methyl acrylate copolymer.[42]

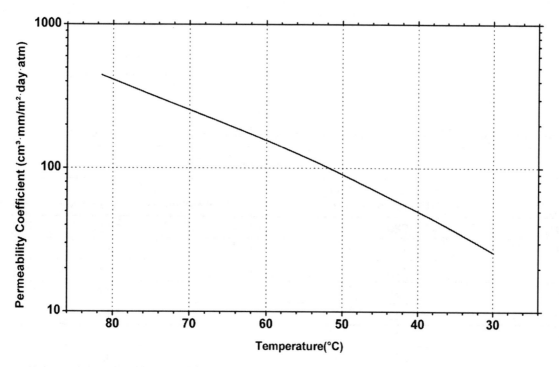

Figure 9.42 Permeation of oxygen vs. temperature through acrylonitrile–methyl acrylate copolymer.[46]

9.13 Ionomers

An ionomer is a polymer that comprises repeat units of both electrically neutral repeating units and a fraction of ionized units. Two types are discussed in the following sections.

9.13.1 Ionomer—Ethylene Acrylic Acid Copolymer (EAA)

Starting with selected various grades of copolymers such as ethylene/methacrylic acid, manufacturers add zinc, sodium, lithium, or other metal salts. Acid neutralization (for instance of the methacrylic acid in an ethylene methacrylic acid copolymer) results in the formation of ion clusters (hence the general term "ionomer") within the resulting polymer matrix. The chemical structure of this process is shown in Fig. 9.43.

The ionomer resins incorporate many of the performance features of the original ethylene-based copolymers, such as chemical resistance, melting range, density, and basic processing characteristics. However, with the alteration forming the ionomer resin, the performance is significantly enhanced in such areas as:

- Low temperature impact toughness
- Abrasion/scuff resistant
- Chemical resistance
- Transparency/clarity
- Melt strength
- Direct adhesion of epoxy and polyurethane finishes, to metal, glass, and natural fibers by heat lamination.

Manufacturers and trade names: DuPont™ Surlyn® and Bexloy® (ethylene methacrylic acid); Exxon Iotek™ (Ethylene–acrylic acid); Goodrich Hycar® (butadiene–acrylic acid)—discontinued; Dow Amplify™ (Ethylene–acrylic acid).

Applications and uses: Packaging films and sealants, glass coatings, and abrasion resistant surfaces (particularly golf ball covers) (Tables 9.71–9.74).

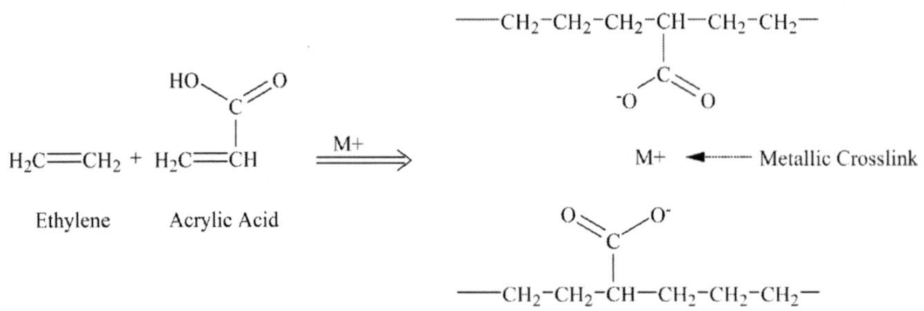

Figure 9.43 Structure of ethylene acrylic acid copolymer ionomers.

Table 9.71 Permeation of Oxygen through DuPont™ Surlyn® Zinc Ion Type Ionomer Film[47]

Grade	Permeability Coefficient	
	Source Document Units (cm^3/100 $in.^2$ day atm)	Normalized Units (cm^3 mm/m^2 day atm)
1650	220	174
1652	180	142
1702	175	138
1705	170	134
F1706	185	146
F1801	215	170
F1855	295	233

Blown film, thickness (mm): 0.051.

Table 9.72 Permeation of Oxygen through DuPont™ Surlyn® Sodium Ion Type Ionomer Film[9]

Grade	Permeability Coefficient	
	Source Document Units (cm^3/100 in.^{2} day atm)	Normalized Units (cm^3 mm/m^2 day atm)
1601	265	209
1603	190	150
F1605	200	158
1707	165	130
F1856	290	229

Blown film, thickness (mm): 0.051.

Table 9.73 Permeation of Water Vapor through DuPont™ Surlyn® Sodium Ion Type Ionomer Film[9]

Grade	Vapor Transmission Rate	
	Source Document Units (g/day 100 in.^{2})	Normalized Units (g mm/m^2 day)
1601	0.8	0.63
1603	0.65	0.51
F1605	0.8	0.63
1707	0.8	0.63
F1856	1.2	0.95

Blown film, thickness (mm): 0.051.

Table 9.74 Permeation of Water Vapor through DuPont™ Surlyn® Zinc Ion Type Ionomer Film[9]

Grade	Vapor Transmission Rate	
	Source Document Units (g/day 100 in.^{2})	Normalized Units (g mm/m^2 day)
1650	0.75	0.59
1652	0.6	0.47
1702	0.7	0.55
1705	0.7	0.55
F1706	0.7	0.55
F1801	0.7	0.55
F1855	1	0.79

Blown film, thickness (mm): 0.051.

9.13.2 Ionomer—Perfluorosulfonic Acid (PFSA)

Two types of fluoroionomers are common:

- Copolymer of tetrafluoroethylene and sulfonyl fluoride vinyl ether (SFVE) and
- Copolymer of tetrafluoroethylene and PFSA CAS number 66796-30-3.

Nafion® is a perfluorinated polymer that contains small proportions of sulfonic or carboxylic ionic functional groups. Its chemical structure is shown in Fig. 9.44. In this case, the acid end group is sulfonic functional group, but it may also be a carboxylic functional group. M^+ is either a metal cation in the neutralized form or an H^+ in the acid form.

Manufacturers and trade names: Solvay Solexis AQUIVION™, DuPont™ Nafion®.

Applications and uses: Chlor-alkali production cell membrane, proton exchange membrane (PEM) for fuel cells and water electrolyzer applications, ion-selective sensors (Table 9.75).

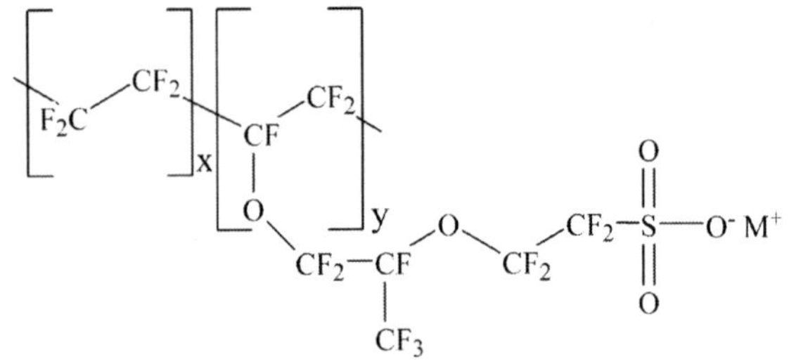

Figure 9.44 The chemical structure of DuPont™ Nafion®.

Table 9.75 Permeation of Nitrogen through Solvay Solexis AQUIVION™[48] PFSA Membranes

Grade	Temperature (°C)	Permeation Coefficient (cm³ mm/ m² day atm)
E87	80	110
	120	455
E79	80	65
	120	430

Membrane thickness (mm): 0.135.

References

1. *Polyethylene products and properties*. Basell Polyolefins; 2002.
2. *Attane™ data sheets*. Dow Chemical Company; May 2000.
3. Brubaker DW, Kammermeyer K. Separation of gases by means of permeable membranes. Permeability of plastic membranes to gases. *Ind Eng Chem* 1952;**44**:1465−74. Available from: http://pubs.acs.org/cgi-bin/doilookup/?10.1021/ie50510a071.
4. Felder RM, Spence RD, Ferrell JK. Permeation of sulfur dioxide through polymers. *J Chem Eng Data* 1975;**20**:235−42.
5. *Permeability of polymers to gases and vapors*. Supplier technical report (P302-335-79, D306-115-79). Dow Chemical Company; 1979.
6. Sclairfilm® specifications sheets. Exopack®; 2010.
7. *Sclairfilm® polyolefin film—SL-1 and SL-3 laminating film*. Supplier technical report (H-27763). Canada: DuPont; 1990.
8. *Engineering properties of Marlex® resins, PE TSM-1*. Chevron Philips; 2002.
9. Wood-Adams P. Permeation: essential factors for permeation; course notes, physical chemistry of polymers, *Polymer*. Concordia University; 2006. p. 1−16.
10. *Sclairfilm® polyolefin film—LWS-1 and LWS-2 film*. Supplier technical report. Canada: DuPont; 1990.
11. *Technical manual—materials used in pipe extrusion Hostalen, Lupolen and Hostalen PP products processing and applications*. Basell Polyolefins; 2005.
12. *Sclair® specification sheets*. NOVA Chemicals; 2010.
13. Adam SJ, David CE. Permeation measurement of fluoropolymers using mass spectrometry and calibrated standard gas leaks. In: *23rd International SAMPE Technical Conference. Conference proceedings*. SAMPE; 1991.
14. Honeywell R Series PMP datasheet; 12/2006.
15. *Halar® ECTFE, fluoropolymer coatings for cleanroom exhaust duct systems*. Solvay Solexis; 2003.
16. Ineos typical engineering properties of polypropylene; 2010.
17. *BOPP film, polypropylene, Form No. 022 PPe 10/01*. Basell Polypropylene; 2001.
18. Japan synthetic rubber JSR RB, supplier design guide. Japan Synthetic Rubber Company.

19. *Topas® cyclic olefin copolymer (COC) product brochure*; 2006.
20. Zürcher, Jörg, Topas COC in pharmaceutical blister packs; 2004.
21. *Lupolen, Lucalen product line, properties, processing. Supplier design guide [B 581e/(8127) 10.91]*. BASF Aktiengesellschaft; 1991.
22. *Elvax® selector guide, Form No. H-42042*. DuPont™ Packaging; 2001.
23. EVAL™ films the ultimate laminating film for barrier packaging applications—Technical bulletin No. 160. Supplier technical report. EVAL™ Company of America, 2007.
24. Flavor and aroma barrier properties of EVAL™ resins technical bulletin no. 190. Supplier technical report, rev. 07–00. EVAL™ Company of America, 2007.
25. *EVAL™∗ ethylene–vinyl alcohol copolymer explained*. EVAL™ Europe; 2002.
26. *Water vapor permeability and moisture absorption/water absorption. Soarnol® technical note*. Nippon Gohsei; 2003.
27. Gas barrier properties of EVAL™ resins—technical bulletin No. 110. Supplier technical report. EVAL™ Company of America, Rev 07–00, 2007.
28. *Organic solvent barrier property of "Soarnol®". Soarnol® technical note*. Nippon Gohsei; 2003.
29. *Chemical and solvent barrier properties of EVAL™ resins technical bulletin No. 180. Supplier technical report, rev.07/00*. EVAL™ Company of America; 2001.
30. EVAL™ film properties comparison. Supplier technical report. Kuraray Co., Ltd.
31. *Mirrex® product data sheet*. VPI; 2010.
32. Haraya K, Hwang S. Permeation of oxygen, argon and nitrogen through polymer membranes. *J Memb Sci* 1992;**71**:13–27.
33. *Various gases barrier property of Soarnol®. Soarnol® technical note*. Nippon Gohsei; 2003.
34. *Chlorofluorocarbon (R-22;CHCLF2) barrier property. Soarnol® technical note*. Nippon Gohsei; 2003.
35. *Thickness and OTR. Soarnol® technical note*. Nippon Gohsei; 2003.
36. *Ethylene content and OTR. Soarnol® technical note*. Nippon Gohsei; 2003.
37. *Relative humidity and OTR. Soarnol® technical note*. Nippon Gohsei; 2003.
38. *Temperature and OTR. Soarnol® technical note*. Nippon Gohsei; 2003.
39. Aiba S, Ohashi M, Huang S. Rapid determination of oxygen permeability of polymer membranes. *Ind Eng Chem* 1968;**7**(3):497–502. Available from: http://pubs.acs.org/doi/abs/10.1021/i160027a022.
40. *Saran™ F resins superior barrier for packaging, Form No. 190-00305-199X SMG*. Dow Plastics; 1999.
41. Heilman W, Tammela V, Meyer J, Stannett V, Szwarc M. Permeability of polymer films to hydrogen sulfide gas. *Ind Eng Chem* 1956;**48**:821–4.
42. Lucite® Diakon® technical manual (DTM/E/2Ed/Dec02), http://www.lucitesolutions.com; 2002.
43. Stannett V. The transport of gases in synthetic polymeric membranes—an historic perspective. *J Memb Sci* 1978;**3**(2):97–115. Available from: http://linkinghub.elsevier.com/retrieve/pii/S0376738800830161.
44. Ineos Barex specification sheets; June 2006.
45. *621 Ways to succeed – 1993–1994 materials selection guide. Supplier technical report (304-00286-1292X SMG)*. Dow Chemical Company; 1992.
46. *Barex barrier resins—barrier properties. Supplier technical report (Bx-555)*. BP Chemicals Inc.; 1992.
47. *Surlyn ionomer resin selector guide. Supplier technical report [E-48623 (1/90)]*. DuPont Company; 1986.
48. *AQUIVION™ Low-EW PFSA Membranes Product Description*. Yl01E004. Solvay Solexis; 2008.

10 Fluoropolymers

Traditionally, a fluoropolymer or a fluoroplastic is defined as a polymer consisting of carbon (C) and fluorine (F). Sometimes these are referred to as perfluoropolymers to distinguish them from partially fluorinated polymers, fluoroelastomers, and other polymers that contain fluorine in their chemical structure. For example, fluorosilicone and fluoroacrylate polymers are not referred to as fluoropolymers. The monomers used to make the various fluoropolymers are shown in Fig. 10.1.

Details of each of the fluoropolymers are in the following sections. The fluoroelastomers data are included in Chapter 12 covering elastomers and rubbers. The melting points are all compared in Table 10.1.

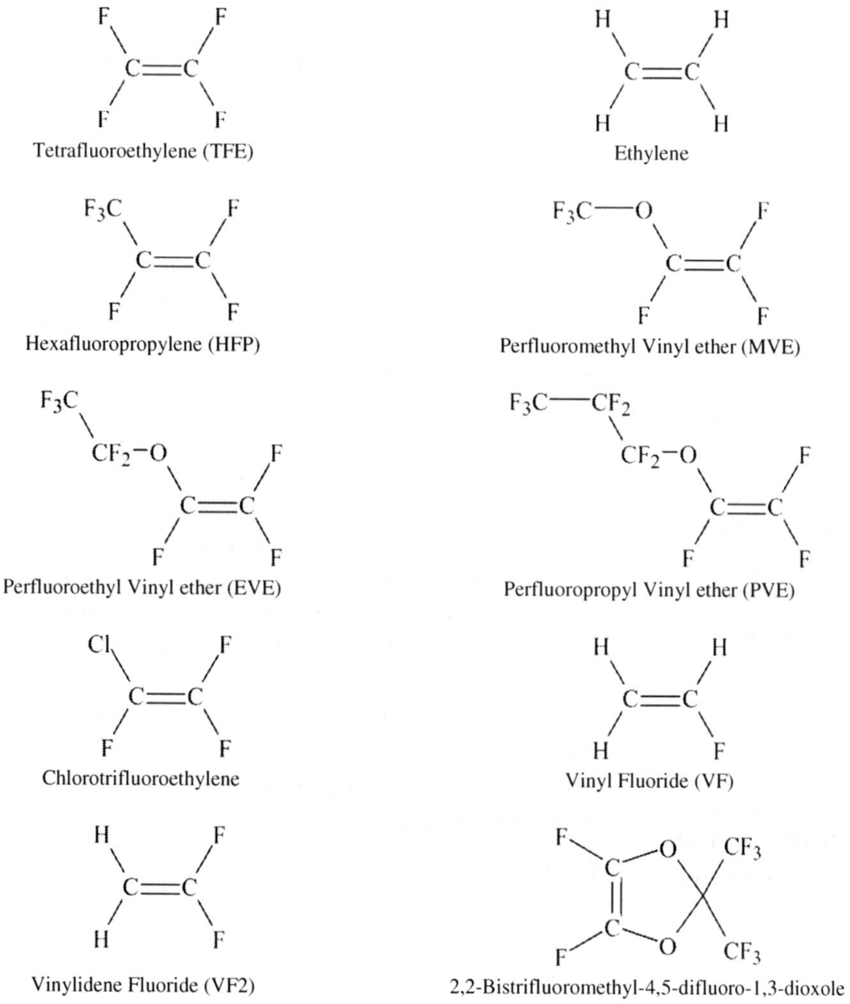

Figure 10.1 Structures of many monomers used to make fluoropolymers.

Tetrafluoroethylene (TFE)

Ethylene

Hexafluoropropylene (HFP)

Perfluoromethyl Vinyl ether (MVE)

Perfluoroethyl Vinyl ether (EVE)

Perfluoropropyl Vinyl ether (PVE)

Chlorotrifluoroethylene

Vinyl Fluoride (VF)

Vinylidene Fluoride (VF2)

2,2-Bistrifluoromethyl-4,5-difluoro-1,3-dioxole

Permeability Properties of Plastics and Elastomers. DOI: 10.1016/B978-1-4377-3469-0.10010-4
Copyright © 2012 Elsevier Inc. All rights reserved.

Table 10.1 Melting Point Ranges of Various Fluoroplastics

Fluoroplastic	Melting Point (°C)
Polytetrafluoroethylene (PTFE)	320–340
Polyethylene Chlorotrifluoroethylene (ECTFE)	240
Polyethylene Tetrafluoroethylene (ETFE)	255–280
Fluorinated Ethylene Propylene (FEP)	260–270
Perfluoroalkoxy (PFA)	302–310
Perfluoroalkoxy (PFA)	280–290
Polychlorotrifluoroethylene (PCTFE)	210–212
Polyvinylidene Fluoride (PVDF)	155–170
THV™	115–235
HTE	155–215

10.1 Polytetrafluoroethylene (PTFE)

PTFE polymer is an example of a linear fluoropolymer. Its structure in simplistic form is shown in Fig. 10.2. The CAS number for PTFE is 9002-84-0.

Formed by the polymerization of tetrafluoroethylene (TFE), the ($-CF_2-CF_2-$) groups repeat many thousands of times. The fundamental properties of fluoropolymers evolve from the atomic structure of fluorine and carbon and their covalent bonding in specific chemical structures. The backbone is formed of carbon–carbon bonds and the pendant groups are carbon–fluorine bonds. Both are extremely strong bonds. The basic properties of PTFE stem from these two very strong chemical bonds. The size of the fluorine atom allows the formation of a uniform and continuous covering around the carbon–carbon bonds and protects them from chemical attack, thus imparting chemical resistance and stability to the molecule. PTFE is rated for use up to 260 °C. PTFE does not dissolve in any known solvent. The fluorine sheath is also responsible for the low surface energy (18 dyn/cm) and low coefficient of friction (0.05–0.8, static) of PTFE. Another attribute of the uniform fluorine sheath is the electrical inertness (or nonpolarity) of the PTFE molecule. Electrical fields impart only slight polarization in this molecule, so volume and surface resistivity are high.

The PTFE molecule is simple and is quite ordered and so it can align itself with other molecules or other portions of the same molecule. Disordered regions are called *amorphous* regions. This is important because polymers with high crystallinity require more energy to melt. In other words they have higher melting points. When this happens it forms what is called a crystalline region. Crystalline polymers have a substantial fraction of their mass in the form of parallel, closely packed molecules. High-molecular weight PTFE resins have high crystallinity and therefore high melting points, typically as high as 320 °C–342 °C (608 °F–648 °F). The crystallinity of as-polymerized PTFE is typically 92–98%. Further, the viscosity in the molten state (called melt creep viscosity) is so high that high-molecular weight PTFE particles do not flow even at temperatures above its melting point. They sinter much like powdered metals; they stick to each other at the contact points and combine into larger particles.

PTFE is called a *homopolymer*, a polymer made from a single monomer. Recently many PTFE manufacturers have added minute amounts of other monomers to their PTFE polymerizations to produce alternate grades of PTFE designed for specific applications. Fluoropolymer manufacturers continue to call these grades modified homopolymer at below 1% by weight of comonomer. DuPont™ grades of this type are called *Teflon*® *NXT* resins. Dyneon™ TFM™ modified PTFE incorporates less than 1% of a comonomer perfluoropropyl vinyl ether (PPVE). Daikin's modified grade is Polyflon™ M-111. These modified granular PTFE materials retain the exceptional chemical, thermal, antistick, and low-friction properties of conventional PTFE resin, but offer some improvements:

Weldability
Improved permeation resistance
Less creep
Smoother, less porous surfaces
Better high-voltage insulation

The copolymers described in the next sections contain significantly more of the non-TFE monomers.

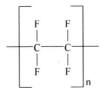

Figure 10.2 Chemical structure of PTFE.

10.1.1 PTFE Homopolymer

Manufacturers and trade names: DuPont™ Teflon® PTFE, Dyneon PTFE, Daikin Polyflon™, and many others.

Applications and uses: Pipe liners, fittings, valves, pumps, and other components used for transferring aggressive, ultrapure fluids (Tables 10.2–10.18). See also Fig. 10.3.

Table 10.2 Permeation of Hydrogen vs. Temperature and Pressure through DuPont™ Teflon® PTFE[1]

Pressure Gradient (kPa)	1724			3447			6895		
Temperature (°C)	−16	25	68	−17	25	67	−18	25	63
Source document units, permeability coefficient (cm³ mm/ cm² kPa s)	1.7×10^{-9}	6.34×10^{-9}	1.88×10^{-8}	1.63×10^{-9}	5.9×10^{-9}	1.86×10^{-8}	1.59×10^{-9}	5.94×10^{-9}	1.64×10^{-8}
Normalized units, permeability coefficient (cm³ mm/ m² day atm)	149	555	1646	143	516	1628	139	520	1436

Sample thickness: 0.03 mm. Test method: mass spectrometry and calibrated standard gas leaks; developed by McDonnell Douglas Space Systems Company Chemistry Laboratory.

Table 10.3 Permeation of Nitrogen versus Temperature and Pressure through DuPont™ Teflon® PTFE[1]

Pressure Gradient (kPa)	1724			3447			6895		
Temperature (°C)	−23	25	71	−25	25	70	−23	25	68
Source document units, permeability coefficient (cm³ mm/ cm² kPa s)	9.46×10^{-11}	7.87×10^{-10}	2.9×10^{-9}	8.89×10^{-11}	7.88×10^{-10}	2.89×10^{-9}	9.47×10^{-11}	7.84×10^{-10}	2.87×10^{-9}
Normalized units, permeability coefficient (cm³ mm/ m² day atm)	8.3	68.9	254	7.8	69	253	8.3	68.6	251

Sample thickness: 0.03 mm. Test method: mass spectrometry and calibrated standard gas leaks; developed by McDonnell Douglas Space Systems Company Chemistry Laboratory.

Table 10.4 Permeation of Ammonia vs. Temperature and Pressure through DuPont™ Teflon® PTFE[1]

Pressure Gradient (kPa)	965		
Temperature (°C)	−3	25	63
Source document units, permeability coefficient (cm^3 mm/cm^2 kPa s)	4.71×10^{-10}	1.73×10^{-9}	8.62×10^{-9}
Normalized units, permeability coefficient (cm^3 mm/m^2 day atm)	41.2	151	755

Sample thickness: 0.03 mm. Test method: mass spectrometry and calibrated standard gas leaks; developed by McDonnell Douglas Space Systems Company Chemistry Laboratory.

Table 10.5 Permeation of Oxygen vs. Temperature and Pressure through DuPont™ Teflon® PTFE[1]

Pressure Gradient (kPa)	1724			3447		
Temperature (°C)	−17	25	51	−17	25	51
Source document units, permeability coefficient (cm^3 mm/cm^2 kPa s)	5.27×10^{-10}	2.55×10^{-9}	5.38×10^{-9}	4.55×10^{-10}	2.54×10^{-9}	5.46×10^{-9}
Normalized units, permeability coefficient (cm^3 mm/m^2 day atm)	46.1	223	471	39.8	222	478

Sample thickness: 0.03 mm. Test method: mass spectrometry and calibrated standard gas leaks; developed by McDonnell Douglas Space Systems Company Chemistry Laboratory.

Table 10.6 Permeation of Hydrogen vs. Temperature and Pressure through Carbon-Filled DuPont™ Teflon® PTFE[1]

Pressure Gradient (kPa)	1724			3447			6895		
Temperature (°C)	−15	25	68	−11	25	67	−14	25	65
Source document units, permeability coefficient (cm^3 mm/cm^2 kPa s)	3.95×10^{-9}	1.34×10^{-8}	3.53×10^{-8}	4.51×10^{-9}	1.27×10^{-8}	3.42×10^{-8}	4.17×10^{-9}	1.23×10^{-8}	3.32×10^{-8}
Normalized units, permeability coefficient (cm^3 mm/m^2 day atm)	346	1173	3090	395	1112	2994	365	1077	2906

Sample thickness: 0.05 mm. Test method: mass spectrometry and calibrated standard gas leaks; developed by McDonnell Douglas Space Systems Company Chemistry Laboratory.

Table 10.7 Permeation Nitrogen vs. Temperature and Pressure through Carbon-Filled DuPont™ Teflon® PTFE[1]

Pressure Gradient (kPa)	1724			3447			6895		
Temperature (°C)	−14	25	68	−17	25	71	−17	25	67
Source document units, permeability coefficient (cm³ mm/ cm² kPa s)	2.5×10^{-10}	1.46×10^{-9}	5.28×10^{-9}	2.34×10^{-10}	1.52×10^{-9}	5.32×10^{-9}	2.34×10^{-10}	1.42×10^{-9}	4.78×10^{-9}
Normalized units, permeability coefficient (cm³ mm/ m² day atm)	21.9	128	462	20.5	133	466	20.5	124	418

Sample thickness: 0.05 mm. Test method: mass spectrometry and calibrated standard gas leaks; developed by McDonnell Douglas Space Systems Company Chemistry Laboratory.

Table 10.8 Permeation of Oxygen through Dyneon™ TF 1750 PTFE[2]

Temperature (°C)	20	40	80
Source document units, permeability coefficient (cm³ 200 μm/m² day bar)	1259	2054	4685
Normalized units, permeability coefficient (cm³ mm/m² day atm)	255	416	849

Sample thickness: 0.2 mm. Test method: DIN 53380 Part 4.1.2.

Table 10.9 Permeation of Carbon Dioxide through Dyneon™ TF 1750 PTFE[2]

Temperature (°C)	20	40	80
Source document units, permeability coefficient (cm³ 200 μm/m² day bar)	3551	4982	8490
Normalized units, permeability coefficient (cm³ mm/m² day atm)	720	1010	1721

Sample thickness: 0.2 mm. Test method: DIN 53380 Part 4.1.2.

Table 10.10 Permeation of Nitrogen through Dyneon™ TF 1750 PTFE[2]

Temperature (°C)	20	40	80
Source document units, permeability coefficient (cm³ 200 μm/m² day bar)	437	814	2086
Normalized units, permeability coefficient (cm³ mm/m² day atm)	89	165	423

Sample thickness: 0.2 mm. Test method: DIN 53380 Part 4.1.2.

Table 10.11 Comparative Permeation Rates of Various Liquids, Vapors, and Gases through DuPont™ Teflon® NXT and Conventional PTFE[3]

Permeant	Specimen Thickness (mm)	Vapor		Liquid		Gas	
		PTFE	Teflon® NXT	PTFE	Teflon® NXT	PTFE	Teflon® NXT
Perchloroethylene	1	5.5	2	13	4	—	—
	2	1.4	0.1	0.019	0.005	—	—
	4	0.08	0.05	0.006	0	—	—
	5	0.055	0.05	—	—	—	—
Hexane	2	3.4	0.2	23.4	0	—	—
	5	0.045	0.015	—		—	—
Methyl ethyl ketone	2	36.3	23.3	49.4	34.2	—	—
	5	22.6	20.8	35.5	25.2		—
HCl (20%)	1	0.4	0.1	—	—	—	—
Helium	2	—	—	—	—	93	1
	5	—	—	—	—	0.18	0.12

Table 10.12 Permeation of Oxygen, Carbon Dioxide, and Nitrogen through Dyneon™ TFM™ 1700 PTFE[2]

Penetrant	Oxygen			Carbon Dioxide			Nitrogen		
Temperature (°C)	20	40	80	20	40	80	20	40	80
Source document units, permeability coefficient (cm^3 200 μm/ m^2 day bar)	879	1557	3550	2405	3653	6698	316	637	1676
Normalized units, permeability coefficient (cm^3 mm/ m^2 day atm)	178	316	720	487	740	1358	64	129	340

Sample thickness: 0.2 mm. Test method: DIN 53380 Part 4.1.2.

Table 10.13 Permeation of Sulfur Dioxide, Hydrogen Chloride and Chlorine Gas through Dyneon™ TFM™ 1700 and TF 1750 PTFE[4]

		Permeability Coefficient			
		Source Document Units (cm^3 mm/m^2 day bar)		Normalized Units (cm^3 mm/m^2 day atm)	
Penetrant Gas	Temperature (°C)	TFM™ 1700	TF™ 1750	TFM™ 1700	TF™ 1750
Sulfur dioxide	23	210	310	213	314
Hydrogen chloride	54	460	640	466	648
Chlorine	54	160	320	162	324

Table 10.14 Permeation of Water Vapor through Dyneon™ TFM™ 1700 PTFE[2]

Temperature (°C)	20	40	80
Source document units, vapor transmission rate (g 200 μm/m² day)	0.09	0.348	4.827
Normalized units, vapor transmission rate (g mm/m² day)	0.0045	0.0174	0.241

Test method: DIN 53122 Part 2.

Table 10.15 Permeation of Water Vapor through Dyneon™ TF™ 1750 PTFE[2]

Temperature (°C)	20	40	80
Source document units, vapor transmission rate (g 200 μm/m² day)	0.085	0.435	6.01
Normalized units, vapor transmission rate (g mm/m² day)	0.0425	0.022	0.30

Test method: DIN 53122 Part 2.

Table 10.16 Permeation of Hydrogen Chloride through Dyneon™ PTFE[5]

Temperature (°C)	23	100
Source document units, permeability coefficient (cm³ 100 μm/m² day bar)	630	1600
Normalized units, permeability coefficient (cm³ mm/m² day atm)	64	162

Thickness: 0.1 mm. Test method: DIN 53380 Part 4.1.2.

Table 10.17 Permeation of Hydrogen Chloride through Dyneon™ TFM™ PTFE[5]

Temperature (°C)	23	100
Source document units, permeability coefficient (cm³ 100 μm/m² day bar)	445	1150
Normalized units, permeability coefficient (cm³ mm/m² day atm)	45	117

Thickness: 0.1 mm. Test method: DIN 53380 Part 4.1.2.

Table 10.18 Fuel Vapor Permeation at 60 °C through Dyneon™ PTFE[6]

Fuel	Description	Vapor Transmission Rate (g mm/m² day)	
		Dyneon™ TFM™ PTFE	Dyneon™ PTFE
CM15	15% Methanol	2.2	1.5
CE10	10% Ethanol	1.1	0.8
RFC	Reference fuel C	0.6	0.5

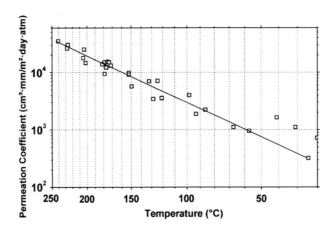

Figure 10.3 Permeation of sulfur dioxide vs. temperature through PTFE.[7]

10.1.2 Modified PTFE

Manufacturers and trade names: DuPont™ Teflon® NXT, Dyneon™ TFM™, Daiken M-111.

Applications and uses: Chemical process industry equipment linings, seals, gaskets, and other parts; electrical connectors and capacitor films (Tables 10.19 and 10.20).

Table 10.19 Comparative Permeation Rates for Teflon® NXT and Conventional PTFE[8]

Permeant	Specimen Thickness, mm	Vapor PTFE	Vapor Teflon® NXT	Liquid PTFE	Liquid Teflon® NXT	Gas PTFE	Gas Teflon® NXT
Perchloroethylene	1	5.5	2	13	4	—	—
	2	1.4	0.1	0.019	0.005	—	—
	4	0.08	0.05	0.006	0	—	—
	5	0.055	0.05	—	—	—	—
Hexane	2	3.4	0.2	23.4	0	—	—
	5	0.045	0.015	—	—	—	—
Methyl ethyl ketone	2	36.3	23.3	49.4	34.2	—	—
	5	22.6	20.8	35.5	25.2	—	—
HCl, 20%	1	0.4	0.1	—	—	—	—
Helium	2	—	—	—	—	93	1
	5	—	—	—	—	0.18	0.12

Table 10.20 Permeation of Gases through Dyneon™ TFM™ 1700 Modified PTFE and Unmodified PTFE[9]

Permeant Gas	Temperature (°C)	Dyneon™ TFM™ 1700 Source Document Units, Permeability Coefficient (cm³ mm/m² day bar)	Dyneon™ TF 1750 PTFE Source Document Units, Permeability Coefficient (cm³ mm/m² day bar)	Dyneon™ TFM™ 1700 Normalized Units, Permeability Coefficient (cm³ mm/m² day atm)	Dyneon™ TF 1750 PTFE Normalized Units, Permeability Coefficient (cm³ mm/m² day atm)
Sulfur dioxide	23	210	310	213	314
Hydrogen chloride	54	460	640	466	648
Chlorine	54	160	320	162	324

10.2 Fluorinated Ethylene Propylene (FEP)

If one of the fluorine atoms on TFE is replaced with a trifluoromethyl group ($-CF_3$) then the new monomer is called hexafluoropropylene (HFP). Polymerization of monomers HFP and TFE yield a fluoropolymer, fluorinated ethylene propylene, called FEP. The number of HFP groups is typically 13% by weight or less and its structure is shown in Fig. 10.4. The CAS number for FEP is 25067-11-2.

The effect of using HFP is to put a "bump" along the polymer chain. This bump disrupts the crystallization of the FEP, which has atypical as-polymerized crystallinity of 70% versus 92–98% for PTFE. It also lowers its melting point. The reduction of the melting point depends on the amount of trifluoromethyl groups added and secondarily on the molecular weight. Most FEP resins melt around 274 °C (525 °F), although lower melting points are possible. Even high-molecular weight FEP will melt and flow. The high chemical resistance, low surface energy, and good electrical insulation properties of PTFE are retained.

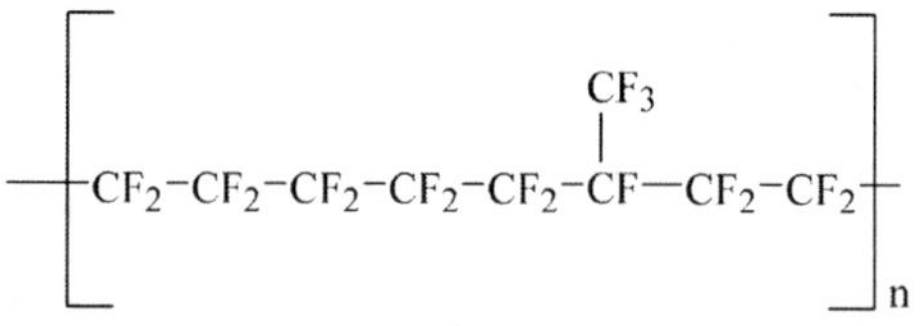

Manufacturers and trade names: DuPont™ Teflon® FEP, Dyneon™ THV FEP, Daikin Neoflon™.

Applications and uses: Applications requiring excellent chemical resistance, superior electrical properties, and high-service temperatures; and release films, tubing, cable insulation, and jacketing (Tables 10.21–10.28).

See also Figs. 10.5–10.8.

Figure 10.4 Chemical structure of FEP.

Table 10.21 Permeation of Carbon Dioxide, Hydrogen, Nitrogen, and Oxygen at 25 °C through DuPont Fluorocarbon FEP Film[10]

Permeant Gas	Source Document Units, Permeation Coefficient (cm^3 25 μm/m^2 day atm)	Normalized Units, Permeation Coefficient (cm^3 mm/m^2 day atm)
Carbon dioxide	25,900	648
Hydrogen	34,100	853
Nitrogen	5000	125
Oxygen	11,600	290

Test method: ASTM D1434. Film thickness: 25 μm.

Table 10.22 Vapor Permeation of Many Liquids through DuPont Fluorocarbon FEP Film[11]

	Vapor Transmission Rate					
	Source Document Units (g mil/100 in.² day)			Normalized Units (g mm/m² day)		
Vapor Permeant						
Temperature (°C)	23	35	>50	23	35	50
Acetic acid		0.42			0.165	
Acetone	0.13	0.95	3.29	0.051	0.374	1.295
Acetophenone	0.47			0.185		
Benzene	0.15	0.64		0.059	0.252	
N-Butyl ether	0.08		0.65	0.031		0.256
Carbon tetrachloride	0.11	0.31		0.043	0.122	
Decane	0.72		1.03	0.283		0.406
Freon® 12	24			9.4		
Limonene	0.17		0.35	0.067		0.138

(Continued)

Table 10.22 (Continued)

Vapor Permeant	Vapor Transmission Rate					
	Source Document Units (g mil/100 in.² day)			Normalized Units (g mm/m² day)		
Ethyl acetate	0.06	0.77	2.9	0.024	0.303	1.142
Ethanol	0.11	0.69		0.043	0.272	
Fuming nitric acid	10.5			5.6		
Hexane		0.57			0.224	
Hydrochloric acid (20%)	<0.01			<0.004		
Methanol			5.61			2.209
Piperdine	0.04			0.016		
Skydrel hydraulic fluid	0.05			0.020		
Sodium hydroxide (50%)	4.00×10^{-5}			1.6×10^{-5}		
Sulfuric acid (98%)	8.00×10^{-6}			3.1×10^{-6}		
Toluene	0.37		2.93	0.146		1.154
Water	0.09	0.45	0.89	0.035	0.177	0.350

Test method: ASTM E-96-53T.

Table 10.23 Permeation of Hydrogen vs. Temperature and Pressure through DuPont™ Teflon® FEP Copolymer[1]

Pressure Gradient (kPa)	1724			3447			6895		
Temperature (°C)	−15	25	68	−13	25	67	−16	25	67
Source document units, permeability coefficient (cm³ mm/cm² kPa s)	9.06×10^{-10}	4.41×10^{-9}	1.87×10^{-8}	9.64×10^{-10}	4.35×10^{-9}	1.77×10^{-8}	8.77×10^{-10}	4.4×10^{-9}	1.8×10^{-8}
Normalized units, permeability coefficient (cm³ mm/m² day atm)	79.3	386	1637	84.4	381	1550	76.8	385	1576

Test method: mass spectrometry and calibrated standard gas leaks; developed by McDonnell Douglas Space Systems Company Chemistry Laboratory.

Table 10.24 Permeation of Nitrogen vs. Temperature and Pressure through DuPont™ Teflon® FEP Copolymer[2]

Pressure Gradient (kPa)	1724			3447			6895		
Temperature (°C)	−9	25	71	−7	25	66	−5	25	68
Source document units, permeability coefficient (cm³ mm/cm² kPa s)	5.06×10^{-11}	3.8×10^{-10}	3.79×10^{-9}	5.64×10^{-11}	3.86×10^{-10}	3.85×10^{-9}	6.39×10^{-11}	3.85×10^{-10}	3.8×10^{-9}
Normalized units, permeability coefficient (cm³ mm/m² day atm)	4.4	33.3	332	4.9	33.8	337	5.6	33.7	333

Test method: mass spectrometry and calibrated standard gas leaks; developed by McDonnell Douglas Space Systems Company Chemistry Laboratory.

Table 10.25 Permeation of Ammonia vs. Temperature and Pressure through DuPont™ Teflon® FEP Copolymer[2]

Pressure Gradient (kPa)	965	1724	3447
Temperature (°C)	0	25	66
Source document units, permeability coefficient (cm^3 mm/cm^2 kPa s)	3.31×10^{-10}	1.15×10^{-9}	6.3×10^{-9}
Normalized units, permeability coefficient (cm^3 mm/m^2 day atm)	29	101	552

Test method: mass spectrometry and calibrated standard gas leaks; developed by McDonnell Douglas Space Systems Company Chemistry Laboratory.

Table 10.26 Permeation of Oxygen vs. Temperature and Pressure through DuPont™ Teflon® FEP Copolymer[2]

Pressure Gradient (kPa)	1724			3447		
Temperature (°C)	−16	25	52	−16	25	53
Source document units, permeability coefficient (cm^3 mm/cm^2 kPa s)	1.04×10^{-10}	1.33×10^{-9}	5.16×10^{-9}	1.03×10^{-10}	1.15×10^{-9}	5.31×10^{-9}
Normalized units, permeability coefficient (cm^3 mm/m^2 day atm)	9.1	116	452	9	101	465

Test method: mass spectrometry and calibrated standard gas leaks; developed by McDonnell Douglas Space Systems Company Chemistry Laboratory.

Table 10.27 Permeation of Gases at 25 °C through Daikin Neoflon™ FEP film[12]

Permeant Gas	Source Document Units, Permeation Coefficient $\times 10^{-8}$ [cm^3(STP) cm/cm^2 s atm]	Normalized Units, Permeation Coefficient (cm^3 mm/m^2 day atm)
Nitrogen	1.2	104
Oxygen	3.7	320
Carbon dioxide	9.7	838
Methane	0.66	57
Ethane	0.33	29
Propane	0.11	10
Ethylene	0.48	41

Table 10.28 Fuel Vapor Permeation at 60 °C through Dyneon™ FEP[6]

Fuel	Description	Vapor Transmission Rate (g mm/m² day)
CM15	15% Methanol	1.3
CE10	10% Ethanol	0.6
RFC	Reference fuel C	0.3

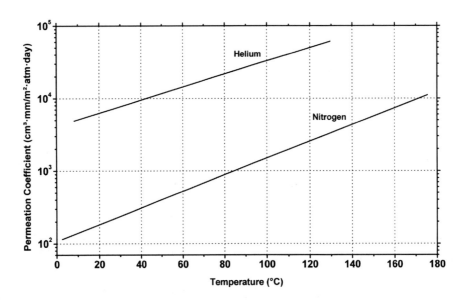

Figure 10.5 Permeation of nitrogen and helium vs. time after retort through FEP copolymer.

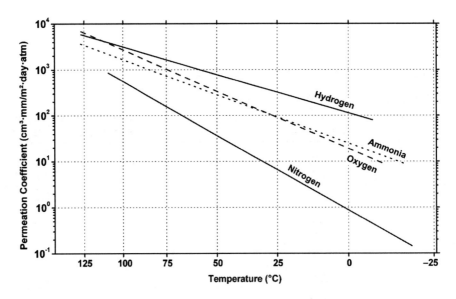

Figure 10.6 Permeation of gases vs. temperature through DuPont™ Teflon® FEP copolymer.[10]

10: Fluoropolymers

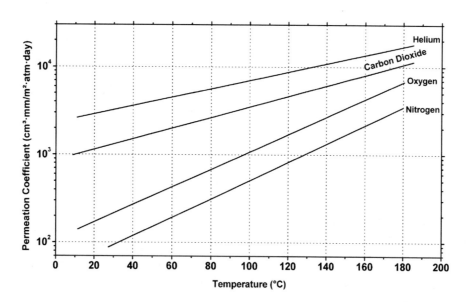

Figure 10.7 Permeation of gases vs. temperature through Daikin Neoflon™ FEP film.[1]

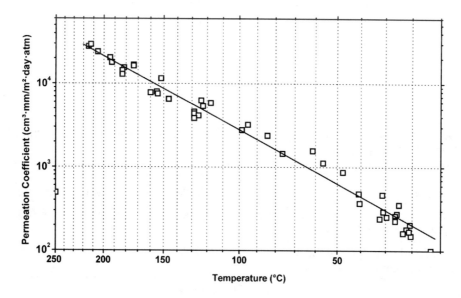

Figure 10.8 Permeation of sulfur dioxide vs. temperature through FEP.[7]

10.3 Perfluoroalkoxy (PFA)

Making a more dramatic change in the side group than that done in making FEP, chemists put a PFA group on the polymer chain. This group is signified as $-O-R_f$, where R_f can be any number of totally fluorinated carbons. The most common comonomer is perfluoropropyl ($-O-CF_2-CF_2-CF_3$). However, other comonomers are shown in Table 10.29.

Table 10.29 PFA Comonomers

Comonomer	Structure
Perfluoromethyl vinyl ether (MVE)	$CF_2=CF-O-CF_3$
Perfluoroethyl vinyl ether (EVE)	$CF_2=CF-O-CF_2-CF_3$
PPVE	$CF_2=CF-O-CF_2-CF_2-CF_3$

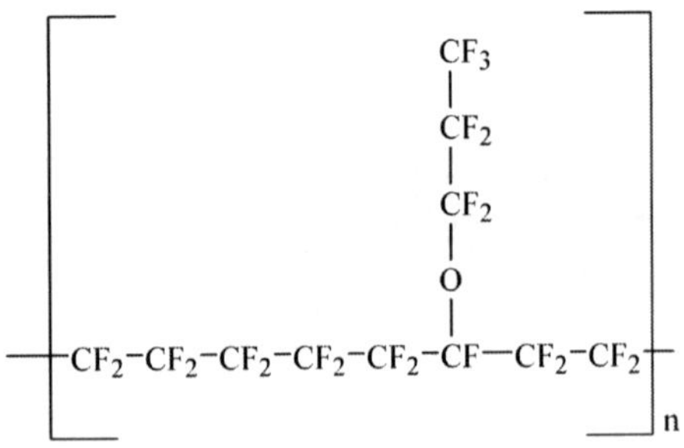

Figure 10.9 Chemical structure of PFA.

The polymers based on PVE are called PFA and the perfluoroalkylvinylether group is typically added at 3.5% or less. When the comonomer is MVE the polymer is called MFA. A structure of PFA is shown in Fig. 10.9. The CAS number of PFA using PPVE as comonomer is 26655-00-5.

The large side group reduces the crystallinity drastically. The melting point is generally between 305 °C and 310 °C (581 °F–590 °F) depending on the molecular weight. The melt viscosity is also dramatically dependent on the molecular weight. Since PFA is still perfluorinated as with FEP the high chemical resistance, low surface energy, and good electrical insulation properties are retained.

Solvay Solexis Hyflon® MFA and PFA are semi-crystalline fully fluorinated melt-processable fluoropolymers. Hyflon® PFA belongs to the class of PFA having a lower melting point than standard PFA grades.

Manufacturers and trade names: DuPont™ Teflon®, Solvay Solexis Hyflon®, Dyneon™ (a 3 M company), Daikin.

Applications and uses: Lined and coated processing equipment, vessels and housings; high-purity chemical storage and transport; and downhole components in harsh oil and gas well environments.

10.3.1 PFA

See Tables 10.30–10.35. See also Figs. 10.10 and 10.11.

Table 10.30 Permeation of Carbon Dioxide, Nitrogen, and Oxygen at 25 °C through DuPont™ Teflon® PFA Film[14]

Penetrant	Carbon Dioxide	Nitrogen	Oxygen
Source document units, permeability coefficient (cm^3/m^2 day atm)	14,000	2000	6700
Normalized units, permeability coefficient (cm^3 mm/m^2 day atm)	700	100	335

Sample thickness: 0.05 mm. Test method: ASTM D1434.

Table 10.31 Permeation of Oxygen, R22, and Chlorine through Solvay Solexis Hyflon® PFA 420

Penetrant	Oxygen				R22	Chlorine
Film thickness (mm)	0.05	0.1	0.1	0.1	0.1	0.7
Temperature (°C)	23	23	40	50	10	50
Permeability coefficient (cm^3 mm/m^2 day atm)	380	280	450	570	40	625

Test method: Swedish Corrosion Institute.

10: Fluoropolymers

Table 10.32 Permeation of Oxygen, Carbon Dioxide, and Nitrogen through Dyneon 6510N PFA[3]

Penetrant	Oxygen			Carbon Dioxide			Nitrogen		
Temperature (°C)	20	40	80	20	40	80	20	40	80
Source document units, permeability coefficient (cm^3 100 μm/m^2 day·bar)	2740	4910	15,100	8650	12,600	29,400	792	2010	4780
Normalized units, permeability coefficient (cm^3 mm/m^2 day atm)	277	497	1530	876	1276	2978	80	204	484

Test method: DIN 53380 Part 4.1.2. Sample thickness: 0.1 mm.

Table 10.33 Permeation of Water Vapor through Dyneon 6510N PFA[1]

Temperature (°C)	20	40	80
Source document units, vapor transmission rate (g 100 μm/m^2 day)	0.223	1.02	12.3
Normalized units, vapor transmission rate (g mm/m^2 day)	0.002	0.102	1.23

Test method: DIN 53122 Part 2.

Table 10.34 Permeation of Hydrogen Chloride through Dyneon™ PFA[5]

Temperature (°C)	23	100
Source document units, permeability coefficient (cm^3 100 μm/m^2 day bar)	160	1400
Normalized units, permeability coefficient (cm^3 mm/m^2 day atm)	16	142

Test method: DIN 53380 Part 4.1.2. Thickness: 0.1 mm.

Table 10.35 Permeation of solvents at 50 °C through Solvay Solexis Hyflon® PFA[15]

Solvent	Normalized units, Vapor Transmission Rate (g mm/m^2 day)
Hexane	0.904
Methylene chloride	13.010
Dimethylacetamide	0.120
Methanol	1.595

Figure 10.10 Permeation of gases vs. temperature of Solvay Solexis Hyflon® PFA.[13]

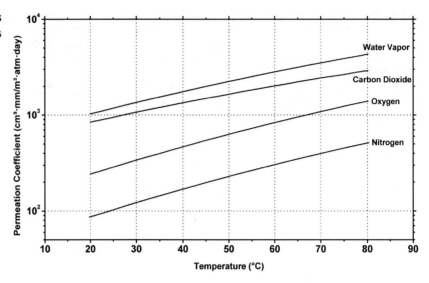

Figure 10.11 Permeation of more gases vs. temperature of Solvay Solexis Hyflon® PFA.[11]

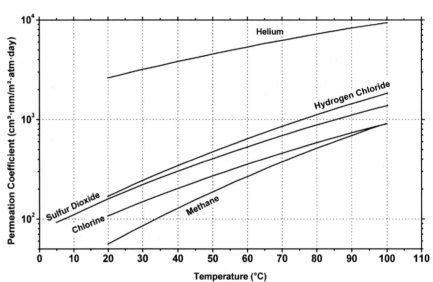

10.3.2 MFA

See Tables 10.36 and 10.37. See also Fig. 10.12.

Table 10.36 Vapor Permeation Rates for Various Liquids at 23 °C through Solvay Solexis Hyflon® MFA[16]

Chemical	T (°C)	Vapor Permeation Rate (g mm/m² day)
Acetone	30	<0.3[a]
	50	1.6
Methanol	30	<0.3[a]
	50	0.8
Toluene	30	<0.3[a]
	50	0.5
	80	8.4
5% Sodium hypochlorite	70	2.0

(*Continued*)

Table 10.36 (*Continued*)

Chemical	T (°C)	Vapor Permeation Rate (g mm/m² day)
96% Sulfuric acid	23	<0.3[b]
Hydrogen chloride from a 37% aqueous solution	23	0.1
Sulfuric acid from a 96% aqueous solution	23	<0.001[c]

[a]No permeation detected after 1000 h on 0.15 mm thick samples.
[b]No permeation detected after 2000 h on 0.3 mm thick samples.
[c]No permeation detected after 1000 h on 0.3 mm thick samples.

Table 10.37 Oxygen vs. Temperature, R22, and Chlorine through Solvay Solexis Hyflon® MFA 620[18]

Sample thickness (mm)	0.05	0.1	0.1	0.1	0.1	0.7
Penetrant	Oxygen	Oxygen	Oxygen	Oxygen	R22	Chlorine
Temperature (°C)	23	23	40	50	10	50
Source document units, permeability coefficient (cm³ mm/m² day atm)	300	270	380	540	36	567
Normalized units, permeability coefficient (cm³ mm/m² day atm)	300	270	380	540	36	567

Test method: Swedish Corrosion Institute.

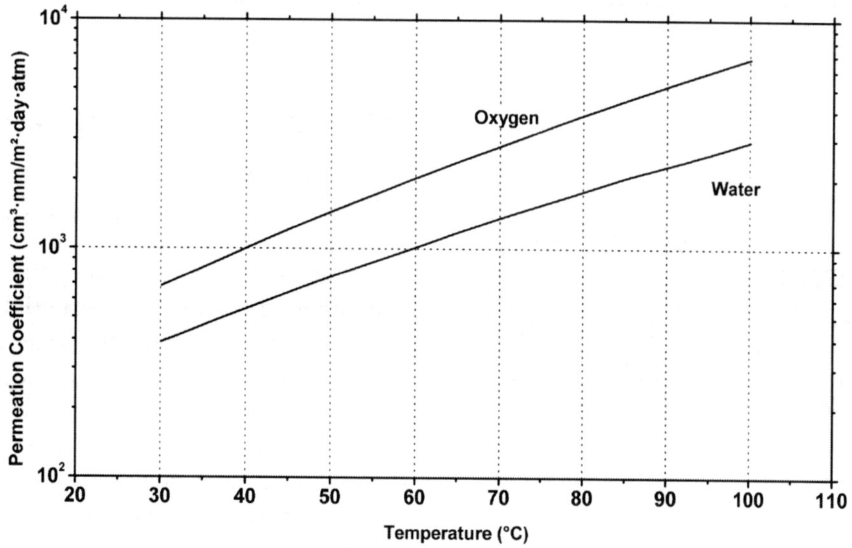

Figure 10.12 Permeability of oxygen and water through Solvay Solexis Hyflon® MFA.[11]

10.4 Hexafluoropropylene, Tetrafluoroethylene, Ethylene Terpolymer (HTE)

HTE is a terpolymer of HFP, TFE, and ethylene.
Manufacturers and trade names: Dyneon™ HTE.

Applications and uses: Pipe, tube, film, sheet, tank lining (Tables 10.38–10.40).

Table 10.38 Permeation of Oxygen, Carbon Dioxide, and Nitrogen through Dyneon™ 1700 HTE[3]

Penetrant	Oxygen			Carbon Dioxide			Nitrogen		
Temperature (°C)	20	40	80	20	40	80	20	40	80
Source document units, permeability coefficient (cm³ 100 μm/m² day bar)	801	1540	6990	4270	7400	35,900	194	453	2920
Normalized units, permeability coefficient (cm³ mm/m² day atm)	81	156	708	433	750	3637	20	46	296

Test method: DIN 53380 Part 4.1.2.

Table 10.39 Water Vapor Transmission through Dyneon™ 1700 HTE

Temperature (°C)	20	40	80
Source document units, vapor transmission rate (g 100 μm/m² day)	1.17	2.56	36.8
Normalized units, vapor transmission rate (g mm/m² day)	0.117	0.256	3.68

Test method: DIN 53122 Part 2.

Table 10.40 Permeation of Hydrogen Chloride through Dyneon™ 1706 HTE[5]

Temperature (°C)	23	100
Source document units, permeability coefficient (cm³ 100 μm/m² day bar)	190	4250
Normalized units, permeability coefficient (cm³ mm/m² day atm)	19	431

Test method: DIN 53380 Part 4.1.2. Thickness (mm): 0.1.

10.5 Tetrafluoroethylene, Hexafluoropropylene, Vinylidene Fluoride Terpolymer (THV™)

Dyneon™ THV™ is a polymer of TFE, HFP, and vinylidene fluoride. This material has some advantages over other fluoropolymer materials, which include low processing temperature, ability to bond to elastomers and hydrocarbon-based plastics, flexibility, and optical clarity. These combined advantages create new opportunities to make multilayer hoses, tubing, film, sheet, seals, and containers.

Manufacturers and trade names: Dyneon™ THV™.

10: Fluoropolymers

Applications and uses: Multilayer hoses, tubing, film, sheet, seals, and containers. These products are used in a variety of markets and applications such as automotive (low-permeation fuel systems), chemical processing industry, semiconductor, solar energy, polymer optical fiber, and architectural and protective coatings (Tables 10.41–10.46).

Table 10.41 Permeation of Oxygen through Dyneon™ 500 THV™[3]

Temperature (°C)	20	40	80
Source document units, permeability coefficient (cm^3 100 μm/m^2 day bar)	696	1930	13,100
Normalized units, permeability coefficient (cm^3 mm/m^2 day atm)	71	196	1327

Test method: DIN 53380 Part 4.1.2. Thickness (mm): 0.1.

Table 10.42 Permeation of Carbon Dioxide through Dyneon™ 500 THV™[18]

Temperature (°C)	20	40	80
Source document units, permeability coefficient (cm^3 100 μm/m^2 day bar)	2060	5680	29,800
Normalized units, permeability coefficient (cm^3 mm/m^2 day atm)	209	575	3019

Test method: DIN 53380 Part 4.1.2. Thickness (mm): 0.1.

Table 10.43 Permeation of Nitrogen through Dyneon™ 500 THV™[18]

Temperature (°C)	20	40	80
Source document units, permeability coefficient (cm^3 100 μm/m^2 day bar)	217	675	5280
Normalized units, permeability coefficient (cm^3 mm/m^2 day atm)	22	68	535

Test method: DIN 53380 Part 4.1.2. Thickness (mm): 0.1.

Table 10.44 Permeation of Hydrogen Chloride through Dyneon™ 500 THV™[5]

Temperature (°C)	23	100
Source document units, permeability coefficient (cm^3 100 μm/m^2 day bar)	180	4250
Normalized units, permeability coefficient (cm^3 mm/m^2 day atm)	18	431

Test method: DIN 53380 Part 4.1.2. Thickness (mm): 0.1.

Table 10.45 Water Vapor Permeation through Dyneon™ 500 THV™[18]

Temperature (°C)	20	40	80
Source document units, vapor transmission rate (g 100 μm/m^2 day)	1.73	7.38	137
Normalized units, vapor transmission rate (g mm/m^2 day)	0.173	0.738	13.7

Test method: DIN 53122 Part 2.

Table 10.46 Fuel Vapor Permeation at 60 °C through Dyneon™ THV™[6]

Fuel	Description	Vapor Transmission Rate (g mm/m² day)			
		THV™ 510 ESD	THV™ 510 ESD	THV™ X610	THV™ X815G
CM15	15% Methanol	35.3	22.1	11.2	6.2
CE10	10% Ethanol	18.1	9.9	5.3	2.9
RFC	Reference fuel C	6.3	4.1	2.4	1.5

10.6 Amorphous Fluoropolymer (AF)—Teflon AF®

A perfluorinated polymer made by DuPont™ called Teflon AF® breaks down the crystallinity completely, hence its designation AF. It is a copolymer made from 2,2-bistrifluoromethyl-4,5-difluoro-1,3-dioxole (PPD) and TFE. The structure of Teflon AF® is shown in Fig. 10.13.

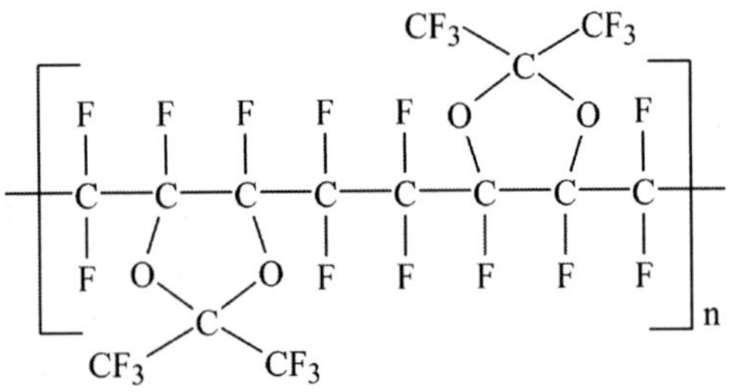

The Teflon AF® family of AFs is similar to Teflon® PTFE and PFA in many of the usual properties but is unique in the following ways. It is:

1. A true amorphous fluoropolymer.
2. Somewhat higher coefficient of friction than Teflon® PTFE and PFA.
3. Excellent mechanical and physical properties at end-use temperature up to 300 °C (572 °F).
4. Excellent light transmission from ultraviolet (UV) through a good portion of infrared (IR).
5. Very low refractive index.
6. Lowest dielectric constant of any plastic even at gigahertz frequencies.
7. Solubility to a limited extent in selected perfluorinated solvents.

Figure 10.13 Structure of Teflon® AF.

Table 10.47 Pure Gas Permeability Coefficients of Solution-Cast[19] and Melt-Pressed[20] DuPont™ Teflon® AF 2400 Films

Permeant Gas	Source Document Units, Permeability Coefficient, ×10⁻¹⁰ [cm³(STP) cm/cm² s cm Hg]		Normalized Units, Permeability Coefficient (cm³ mm/m² day atm)	
	Solution-Cast Film	Melt-Pressed Film	Solution-Cast Film	Melt-Pressed Film
Carbon dioxide	3900	2800	256,090	183,859
Helium	3600	2700	236,391	177,293
Hydrogen	3400	2200	223,258	144,461
Oxygen	1600	990	105,062	65,007
Nitrogen	780	490	51,218	32,175
Methane	600	340	39,398	22,326
Ethane	370	180	24,296	11,820

10: Fluoropolymers

Table 10.48 Permeability of Gases through DuPont™ Teflon® AF Films[21]

Permeant Gas	AF Grade			
	1600	2400	1600	2400
	Source Document Units, Permeability Coefficient, Barrer [cm³(STP) cm/ cm² s cm Hg]		Normalized Units, Permeability Coefficient (cm³ mm/m² day atm)	
Water	1142	4026	74,988	264,363
Oxygen	340	990	22,326	65,007
Nitrogen	130	490	8536	32,175
Carbon dioxide		2800		183,859

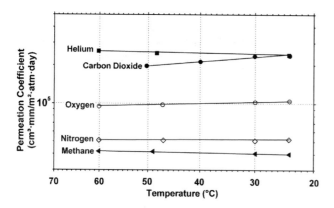

Figure 10.14 Permeability of gases vs. temperature through DuPont™ Teflon® AF 2400 films.[6]

Teflon® AF can be designed to have some solubility in selected perfluorinated solvents but remains chemically resistant to all other solvents and process chemicals. The solubility is typically only 3–15% by weight, but this allows you to solution cast extremely thin coatings in the submicron thickness range.

Manufacturers and trade names: DuPont™ Teflon AF®.

Applications and uses: Developmental applications are currently under evaluation for optical fiber cladding, photolithography, and as an electronic dielectric for the next generation of high-speed computer circuits (Tables 10.47 and 10.48).

See also Fig. 10.14.

10.7 Polyvinyl Fluoride (PVF)

PVF is a homopolymer of vinyl fluoride. The molecular structure of PVF is shown in Fig. 10.15.

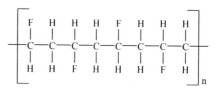

Figure 10.15 Structure of PVF.

DuPont™ is the only known manufacturer of this polymer they call Tedlar®. The structure above shows a head-to-tail configuration of the CF monomer; there are no fluorines on adjacent carbons. But in reality vinyl fluoride polymerizes in both head-to-head and head-to-tail configurations. DuPont's commercial PVF contains 10–12% of head-to-head and tail-to-tail units, also called *inversions*.[22] Its CAS is 24981-14-4.

PVF has excellent resistance to weathering, staining, and chemical attack (except ketones and esters). It exhibits very slow burning and low permeability to vapor. Its most visible use is on the interiors of the passenger compartments of commercial aircraft.

General description: PVF is available only in film form. DuPont™ Tedlar® films are available in clear, or translucent, or opaque white film and in several surface finishes.

Applications and uses: Release films for epoxies, phenolics, polyesters, and rubber compounds. Printed circuit boards, molded parts, resin overflow containment, surfacing of rubber laminating and printing rolls, and photovoltaic module back sheets (Tables 10.49 and 10.50).

See also Figs. 10.16 and 10.17.

Table 10.49 Permeation of Oxygen, Nitrogen, and Carbon Dioxide at 24 °C through DuPont™ Tedlar® PVF[23]

Penetrant	Carbon Dioxide	Helium	Hydrogen	Nitrogen	Oxygen
Source document units, permeability coefficient (cm^3 mil/100 in^2 day)	11.1	150	58.1	0.25	3.2
Normalized units, permeability coefficient (cm^3 mm/m^2 day atm)	4.4	59.1	22.9	0.1	1.3

Test method: ASTM D 1434-75.

Table 10.50 Vapor Permeability of DuPont™ SP TTR10AH9 Transparent High Gloss PVF Film[24]

Permeant	Source Document Units, Vapor Permeation Rate (g mil/100 in.^{2} h)	Normalized Units, Vapor Permeation Rate (g mm/cm^2 day)
Acetic acid	45	18
Acetone	10,000	3937
Benzene	90	35
Carbon tetrachloride	50	20
Ethyl acetate	1000	394
Ethyl alcohol	35	14
Hexane	55	22
Water	22	9

Thickness (mm) 0.0254. Test method: ASTM E96-80.

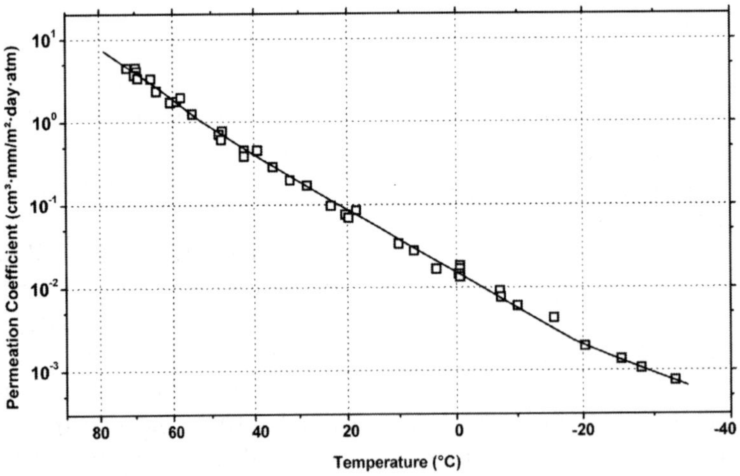

Figure 10.16 Permeability of argon vs. temperature through PVF.[17]

10: Fluoropolymers

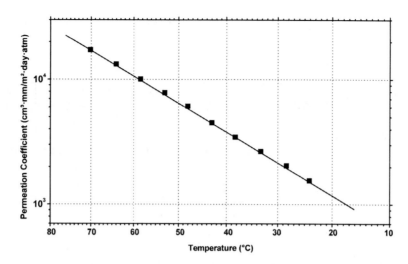

Figure 10.17 Permeability of hydrogen vs. temperature through PVF.[22]

10.8 Polychlorotrifluoroethylene (PCTFE)

CTFE is a homopolymer of chlorotrifluoroethylene, characterized by the following structure shown in Fig. 10.18. The CAS number is #9002-83-9.

The addition of the one chlorine atom contributes to lowering the melt viscosity to permit extrusion and injection molding. It also contributes to the transparency, the exceptional flow, and the rigidity characteristics of the polymer. Fluorine is responsible for its chemical inertness and zero moisture absorption. Therefore, PCTFE has unique properties. Its resistance to cold flow, dimensional stability, rigidity, low gas permeability, and low moisture absorption is superior to any other fluoropolymer. It can be used at low temperatures. Some products contain a small amount of a comonomer.

Manufacturers and trade names: Honeywell Aclar®, Arkema VOLTALEF®, Daikin Industries Neoflon® CTFE.

Applications and uses: Industrial and electronics packaging, pharmaceutical packaging, and blister packages; encapsulating film for clean room packaging and electroluminescent lamps, military, and industrial packaging as either a monolayer film or as a chemical and moisture barrier in laminate structures (Tables 10.51–10.55).

Figure 10.18 Chemical structure of PCTFE.

Table 10.51 Permeation of Water Vapor at 37.8 °C and 100% Relative Humidity through Honeywell Aclar® PCTFE Film[25]

Grade	11A	11A	11A	22A	22C	22C	33C	33C
Sample thickness (mm)	0.015	0.0225	0.05	0.0375	0.05	0.125	0.0187	0.195
Source document units, vapor transmission rate (g/100 in.² day)	0.027	0.017	0.008	0.022	0.019	0.007	0.027	0.003
Normalized units, vapor transmission rate (g mm/m² day)	0.0064	0.006	0.0063	0.013	0.015	0.014	0.008	0.009

Test method: ASTM F1249.

Table 10.52 Permeation of Water Vapor at Different Temperatures and Relative Humidity through Honeywell Aclar® PCTFE Film[27]

Temperature (°C)	Relative Humidity (%)	Aclar® Grade					
		Rx20e	Rx160	SupRx 900	UltRx 2000	UltRx 3000	UltRx 4000
Thickness (mm)		0.020	0.015	0.023	0.051	0.076	0.102
Source Document Units, Vapor Transmission Rate (g/m day)							
25	60	0.0367	0.0510	0.0279	0.0188	0.0124	0.0093
30	60	0.0620	0.0930	0.0465	0.0388	0.0248	0.0155
40	75	0.2330	0.3260	0.2020	0.1020	0.0620	0.0481
37.8	100	0.2950	0.4190	0.2640	0.1190	0.0844	0.0651
Normalized Units, Vapor Transmission Rate (g mm/m² day)							
25	60	0.000734	0.000765	0.000642	0.000959	0.000942	0.000949
30	60	0.001240	0.001395	0.001070	0.001979	0.001885	0.001581
40	75	0.004660	0.004890	0.004646	0.005202	0.004712	0.004906
37.8	100	0.005900	0.006285	0.006072	0.006069	0.006414	0.006640

Test method: ASTM F1249.

Table 10.53 Permeation of Oxygen, Carbon Dioxide, and Nitrogen at 25 °C through Honeywell Aclar® PCTFE Film[28]

Aclar® Grade	33C		22C			22A		
Permeant gas	Oxygen	Carbon dioxide	Oxygen	Nitrogen	Carbon dioxide	Oxygen	Nitrogen	Carbon dioxide
Source document units, permeability coefficient (cm² mil/100 in.² day)	7	16	15	2.5	40	12	2.5	30
Normalized units, permeability coefficient (cm² mm/m² day atm)	2.8	6.3	5.9	1	15.7	4.7	1	11.8

Table 10.54 Permeation of Gases through Daikin Industries Neoflon® CTFE Film[29]

	Permeability Coefficient	
Permeant Gas	Source Document Units (cm³ cm/cm² s atm) × 10⁻¹⁰	Normalized Units (cm³ mm/m² day atm)
Nitrogen	0.18	0.16
Oxygen	1.5	1.3
Hydrogen	56.4	48.7
Carbon dioxide	2.9	2.5

Table 10.55 Permeation of Gases through Arkema Voltalef® PCTFE Film[30]

Temperature (°C)	Nitrogen	Oxygen	Carbon Dioxide	Hydrogen	Hydrogen Sulfide	Water Vapor
	Source Document Units Permeability Coefficient, $\times 10^{-10}$ (cm^3 mm/cm^2 s cm Hg)					
0		0.07	0.35	3.2		
25	0.05	0.4	1.4	9.8		1
50	0.3	1.4	2.4	24	0.35	10
75	0.91	5.7	15		2	28
100						100
	Normalized Units Permeability Coefficient (cm^3 mm/m^2 day atm)					
0		0.46	2.3	21		
25	0.33	2.6	9.2	64		6.57
50	2.0	9.2	16	158	2.3	66
75	6.0	37	98		13	184
100						657

10.9 Polyvinylidene Fluoride (PVDF)

The polymers made from 1,1-di-fluoro-ethene (or vinylidene fluoride) are known as PVDF. They are resistant to oils and fats, water and steam, and gas and odors, making them of particular value for the food industry. PVDF is known for its exceptional chemical stability and excellent resistance to UV radiation. It is used chiefly in the production and coating of equipment used in aggressive environments, and where high levels of mechanical and thermal resistance are required. It has also been used in architectural applications as a coating on metal siding where it provides exceptional resistance to environmental exposure. The chemical structure of PVDF is shown in Fig. 10.19. Its CAS number is 24937799. Some products are comonomers.

The alternating CH_2 and CF_2 groups along the polymer chain provide a unique polarity that influences its solubility and electric properties. At elevated temperatures PVDF can be dissolved in polar solvents such as organic esters and amines. This selective solubility offers a way to prepare corrosion resistant coatings for chemical process equipment and long-life architectural finishes on building panels.

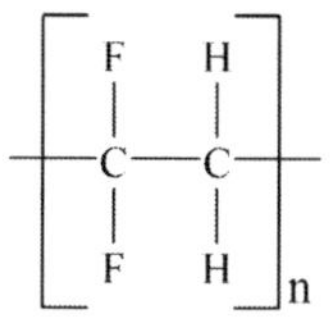

Key attributes of PVDF include:

Mechanical strength and toughness
High abrasion resistance
High thermal stability
High dielectric strength
High purity
Readily melt processable
Resistant to most chemicals and solvents
Resistant to UV and nuclear radiation
Resistant to weathering
Resistant to fungi
Low permeability to most gases and liquids
Low flame and smoke characteristics.

Manufacturers and trade names: Arkema Kynar®, Solvay Solexis Solef® and Hylar®.

Applications and uses: Coatings, piping for ultra-high purity water and hot concentrated acids, high-purity pharmaceutical grade chemicals, pumps, tubing, and automotive fuel systems (Tables 10.56 and 10.57).

See also Figs. 10.20–10.26.

Figure 10.19 Chemical structure of PVDF.

Table 10.56 Permeation of Gases at 23 °C through Solvent Cast Arkema Kynar® PVDF[31]

Penetrant	Permeability Coefficient	
	Source Document Units (cm³/m² day bar)	Normalized Units (cm³ mm/m² day atm)
Oxygen	20	1.96
Nitrogen	30	2.94
Helium	600	58.8
Carbon dioxide	100	9.8
Ammonia	65	6.6
Helium	850	86
Chlorine	12	1.2
Hydrogen	210	21.3
Air	7	0.69

Test methods: ASTM D1434, DIN 53122. Thickness: 0.1 mm.

Table 10.57 Permeation of Gases at 23 °C through Solvay Solexis Solef® 1008 PVDF[31]

Penetrant	Film Thickness (mm)	Permeability Coefficient	
		Source Document Units (cm³ N/m² day bar)	Normalized Units (cm³ mm/m² day atm)
Oxygen	0.1	21	2.13
Nitrogen	0.1	30	3.04
Carbon dioxide	0.1	70	7.09
Freon® 12	0.025	6.3	0.16
Freon® 114	0.025	10	0.25
Freon® 115	0.025	4	0.1
Freon® 318	0.025	7	0.18
Nitrous oxide	0.025	900	22.8
Hydrogen sulfide	0.025	60	1.52
Sulfur dioxide	0.025	60	1.52

Test methods: ASTM D1434.

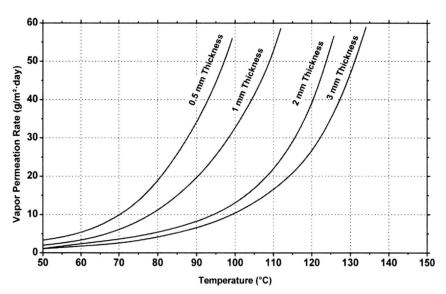

Figure 10.20 Permeation of water vapor vs. temperature through PVDF.[32]

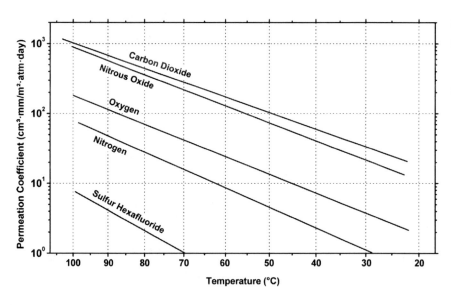

Figure 10.21 Permeability of gases vs. temperature through Extruded Solvay Solexis Solef® PVDF film.[26] Film thickness (mm): 0.1. Method: gas chromatography.

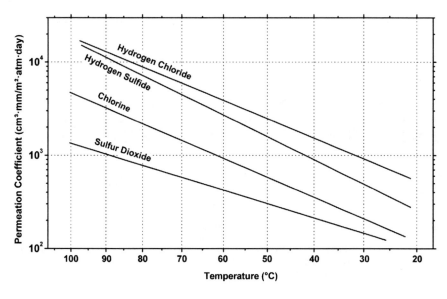

Figure 10.22 Permeability of gases vs. temperature through Extruded Solvay Solexis Solef® PVDF films. Film thickness (mm): 0.1. Method: ASTM D1434.

Figure 10.23 Water permeation rate vs. temperature through Solvay Solexis Solef® PVDF. Method: ASTM D1249-89.

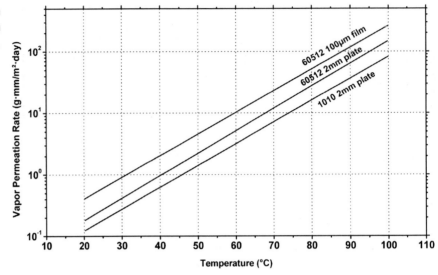

Figure 10.24 Vapor permeation rate of solvents vs. temperature through Extruded Solvay Solexis Solef® 1008 or 1010 PVDF.

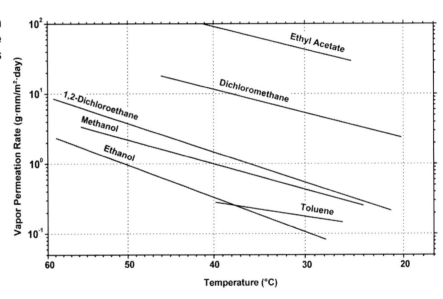

Figure 10.25 Vapor permeation rate of more solvents vs. temperature through Extruded Solvay Solexis Solef® 1008 or 1010 PVDF.

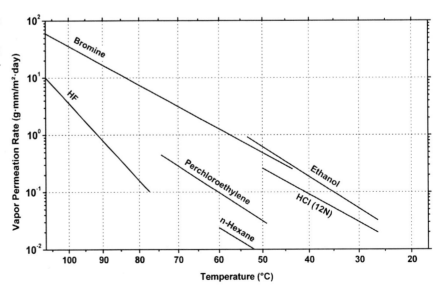

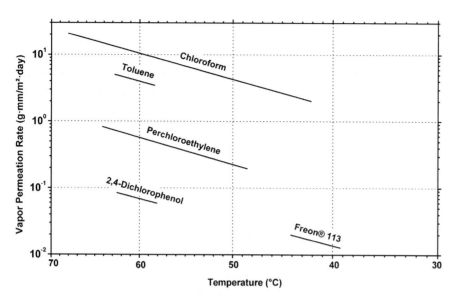

Figure 10.26 Vapor permeation rate of solvents vs. temperature through Extruded Solvay Solexis Solef® 11010 PVDF copolymer.

10.10 Ethylene–Tetrafluoroethylene Copolymer (ETFE)

ETFE is a copolymer of ethylene and TFE. The basic molecular structure of ETFE is shown in Fig. 10.27.

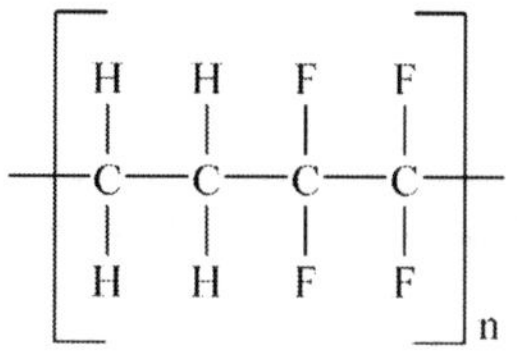

Figure 10.27 Chemical structure of polyethylene tetrafluoroethylene.

It is sometimes called polyethylene tetrafluoroethylene. The depicted structure in Fig. 10.28 shows alternating units of TFE and ethylene. While this can be readily made, many grades of ETFE vary the ratio of the two monomers slightly to optimize properties for specific end uses. Its CAS number is 25038-71-5.

ETFE is a fluoroplastic with excellent electrical and chemical properties. It also has excellent mechanical properties. ETFE is especially suited for uses requiring high mechanical strength, chemical, thermal, and/or electrical properties. The mechanical properties of ETFE are superior to those of PTFE and FEP. ETFE has:

Excellent resistance to extremes of temperature, that is, ETFE has a working temperature range of −200 °C to 150 °C.

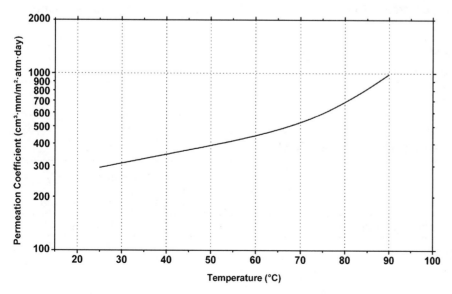

Figure 10.28 Permeability of chlorine gas vs. temperature through ETFE.[13]

Excellent Chemical Resistance

Mechanical strength of ETFE is good with excellent tensile strength and elongation and has superior physical properties compared to most fluoropolymers.

With low smoke and flame characteristics, ETFE is rated 94V-0 by the Underwriters Laboratories Inc. It is odorless and nontoxic.

Outstanding resistance to weather and aging

Excellent dielectric properties

Nonstick characteristics

Manufacturers and trade names: DuPont™ Tefzel®, Asahi Glass Fluon®, 3M Dyneon™.

Applications and uses: Pressure-sensitive tapes, flexible printed circuits, liquid pouches, and other applications demanding high-flex life/crack resistance, exposure to high temperatures, and wear (Tables 10.58–10.65).

See also Fig. 10.29.

Table 10.58 Permeation of Carbon Dioxide, Nitrogen, Oxygen, Helium, and Water Vapor through DuPont™ Tefzel®[33]

Gas Penetrant	Permeability Coefficient	
	Source Document Units (cm^3 mil/100 $in.^2$ day)	Normalized Units (cm^3 mm/m^2 day atm)
Carbon dioxide	250	98.4
Nitrogen	30	11.8
Oxygen	100	39.4
Helium	900	354
Vapor Permeant	**Vapor Transmission Rate**	
	Source Document Units (g mil/100 in^2 day)	Normalized Units (g mm/m^2 day)
Water vapor	1.65	0.65

Sample thickness (mm): 0.102. Temperature: 25 °C. Test methods: Gas permeability by ASTM D1434 and Vapor Transmission by ASTM E96.

Table 10.59 Permeation of Water Vapor through DuPont™ Tefzel® T^2 ETFE Film (Machine Oriented—One Direction)[34]

Source document units, vapor transmission rate (g mil/100 $in.^2$ day)	0.8
Normalized units, vapor transmission rate (g mm/m^2 day)	0.3

Table 10.60 Permeation of Oxygen through Dyneon™ 6235G ETFE[3]

Permeability Coefficient	Temperature (°C)		
	20	40	80
Source document units (cm^3 100 μm/m^2 day bar)	666	1550	6020
Normalized units (cm^3 mm/m^2 day atm)	67	157	610

Sample thickness (mm): 0.1. Test method: DIN 53380 Part 4.1.2.

Table 10.61 Permeation of Carbon Dioxide through Dyneon™ 6235G ETFE[34]

Permeability Coefficient	Temperature (°C)		
	20	40	80
Source document units (cm^3 100 μm/m^2 day bar)	3790	5870	16,100
Normalized units (cm^3 mm/m^2 day atm)	384	595	1631

Sample thickness (mm): 0.1. Test method: DIN 53380 Part 4.1.2.

Table 10.62 Permeation of Nitrogen through Dyneon™ 6235 G ETFE[34]

Permeability Coefficient	Temperature (°C)		
	20	40	80
Source document units (cm^3 100 μm/m^2 day bar)	217	580	1540
Normalized units (cm^3 mm/m^2 day atm)	22	59	156

Sample thickness (mm): 0.1. Test method: DIN 53380 Part 4.1.2.

Table 10.63 Permeation of Water Vapor through Dyneon™ 6235 G ETFE[3,34]

Permeability Coefficient	Temperature (°C)				
	20	23	40	80	100
Source document units (g 100 μm/m^2 day)	1.03	1.1	3.13	26.9	70
Normalized units (g mm/m^2 day)	0.1	0.11	0.31	2.69	7.0

Sample thickness (mm): 0.2. Test method: DIN 53122 Part 2.

Table 10.64 Permeation of Hydrogen Chloride through Dyneon™ 6235 G ETFE[35]

Permeability Coefficient	Temperature (°C)	
	25	100
Source document units (cm^3 1000 μm/m^2 day bar)	190	2700
Normalized units (cm^3 mm/m^2 day atm)	193	2736

Test method: DIN 53380 Part 4.1.2.

Table 10.65 Fuel Vapor Permeation at 60 °C through Dyneon™ ETFE 6235 G[6]

Fuel	Description	Vapor Transmission Rate (g mm/m^2 day)
CM15	15% Methanol	7.2
CE10	10% Ethanol	3.0
RFC	Reference fuel C	1.6

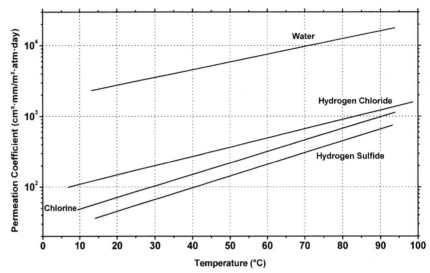

Figure 10.29 Permeability of various gases vs. temperature through ETFE.[32]

10.11 Ethylene–Chlorotrifluoroethylene Copolymer (ECTFE)

Ethylene–chlorotrifluoroethylene copolymer, also called polyethylene chlorotrifluoroethylene or ECTFE, is a copolymer of ethylene and chlorotrifluoroethylene. Its CAS number is 25101-45-5. Figure 10.30 shows the molecular structure of ECTFE.

This simplified structure shows the ratio of the monomers being 1:1 and strictly alternating, which is the desirable proportion. Commonly known by the trade name, Halar®, ECTFE is an expensive, melt-processible, semicrystalline, whitish semiopaque thermoplastic with good chemical resistance and barrier properties. It also has good tensile and creep properties and good high-frequency electrical characteristics.

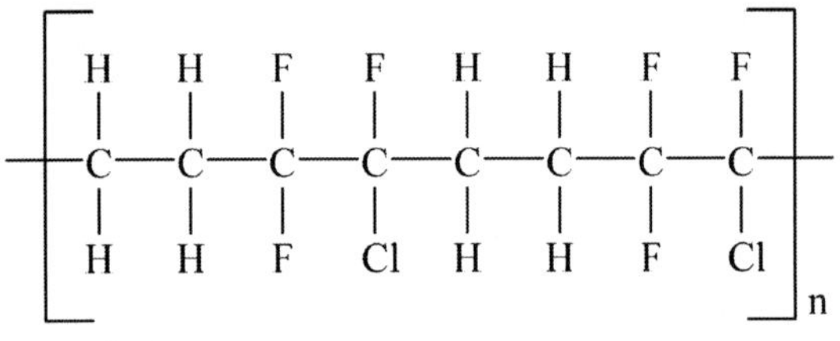

The primary producer is Solvay Solexis under the trade name Halar®.

Processing methods include extrusion, compression molding, rotomolding, blow molding, and liquid and powder coating.

Manufacturers and trade names: Solvay Solexis Halar®.

Applications and uses: Chemically resistant linings and coatings, valve and pump components, hoods, tank and filter house linings, nonwoven filtration fibers, barrier films and release/vacuum bagging films. It is used in food processing particularly involving acidic food and fruit juice processing. See also Figs. 10.31–10.38.

Figure 10.30 Chemical structure of ECTFE.

Table 10.66 Permeation of Hydrogen vs. Temperature and Pressure through Solvay Solexis Halar® ECTFE[13]

Pressure (kPa)	1724			3447			6895		
Temperature (°C)	−22	25	66	−22	25	66	−22	25	66
Source document units, permeation coefficient (cm³ mm/ cm² kPa s)	1.19×10^{-10}	1.21×10^{-9}	6.58×10^{-9}	1.18×10^{-10}	1.25×10^{-9}	6.65×10^{-9}	1.18×10^{-10}	1.23×10^{-9}	6.74×10^{-9}
Normalized units, permeation coefficient (cm³ mm/ m² day atm)	10.4	106	576	10.3	109	582	10.3	108	590

Sample thickness (mm): 0.02. Test method: mass spectrometry and calibrated standard gas leaks; developed by McDonnell Douglas Space Systems Company Chemistry Laboratory.

10: Fluoropolymers

Table 10.67 Permeation of Nitrogen versus Temperature and Pressure through Solvay Solexis Halar® ECTFE[13]

Pressure (kPa)	1724			3447			6895		
Temperature (°C)	10	25	70	10	25	70	10	25	70
Source document units, permeation coefficient (cm³ mm/cm² kPa s)	5.53×10^{-12}	1.29×10^{-11}	2.43×10^{-10}	5.53×10^{-12}	1.49×10^{-11}	4.27×10^{-10}	6.09×10^{-12}	1.43×10^{-11}	2.48×10^{-10}
Normalized units, permeation coefficient (cm³ mm/m² day atm)	0.48	1.13	21.3	0.48	1.30	37.4	0.53	1.25	21.7

Sample Thickness (mm): 0.02. Test Method: mass spectrometry and calibrated standard gas leaks; developed by McDonnell Douglas Space Systems Company Chemistry Laboratory.

Table 10.68 Permeation of Ammonia through Solvay Solexis Halar® ECTFE[13]

Pressure (kPa)	965		
Temperature (°C)	−1	25	65
Source document units, permeation coefficient (cm³ mm/cm² kPa s)	3.73×10^{-10}	1.29×10^{-9}	7.05×10^{-9}
Normalized units, permeation coefficient (cm³ mm/m² day atm)	32.6	113	617

Sample thickness (mm): 0.02. Test method: mass spectrometry and calibrated standard gas leaks; developed by McDonnell Douglas Space Systems Company Chemistry Laboratory.

Table 10.69 Permeation of Oxygen through Solvay Solexis Halar® ECTFE[13]

Pressure (kPa)	1724			3447		
Temperature (°C)	−18	25	55	−15	25	556
Source document units, permeation coefficient (cm³ mm/cm² kPa s)	5.52×10^{-12}	1.16×10^{-10}	5.16×10^{-10}	5.73×10^{-12}	1.1×10^{-10}	5.26×10^{-10}
Normalized units, permeation coefficient (cm³ mm/m² day atm)	0.48	10.2	45.2	0.5	9.6	46.0

Sample thickness (mm): 0.02. Test method: mass spectrometry and calibrated standard gas leaks; developed by McDonnell Douglas Space Systems Company Chemistry Laboratory.

Table 10.70 Permeabilities of Hydrochloric and Nitric Acids through Solvay Solexis Halar® ECTFE[13]

Permeant	Permeability (g mm/m² day)
HCl from 37% solution	0.01958
HNO₃ from 65% solution	0.00102

Table 10.71 Liquid Permeabilities of Common Solvents in Solvay Solexis Halar® ECTFE[7]

Permeant	Permeability (g mm/m² day)
Hexane	4.4
Methylene chloride	769
Dimethylacetamide	5.2
Methanol	6.1

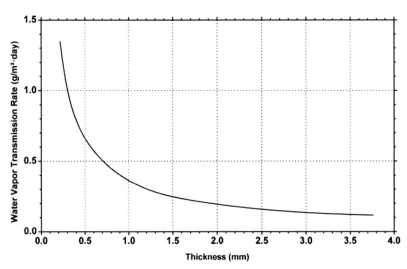

Figure 10.31 Water vapor permeation vs. thickness through Solvay Solexis Halar® ECTFE.[36]

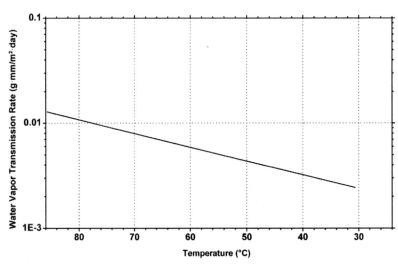

Figure 10.32 Water vapor permeation vs. temperature through Solvay Solexis Halar® ECTFE.[7]

10: Fluoropolymers

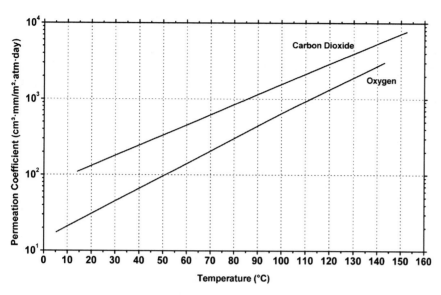

Figure 10.33 Permeation of carbon dioxide and oxygen through Solvay Solexis Halar® ECTFE.[7]

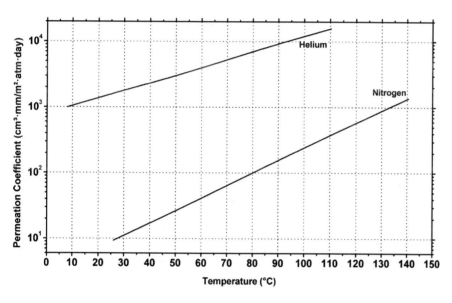

Figure 10.34 Permeation of nitrogen and helium vs. temperature through Solvay Solexis Halar® ECTFE.[7]

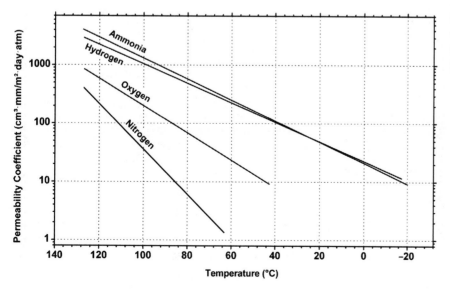

Figure 10.35 Permeation of various gases vs. temperature through Solvay Solexis Halar® ECTFE.[12]

Figure 10.36 Permeation of chlorine vs. temperature through Solvay Solexis Halar® ECTFE.[7]

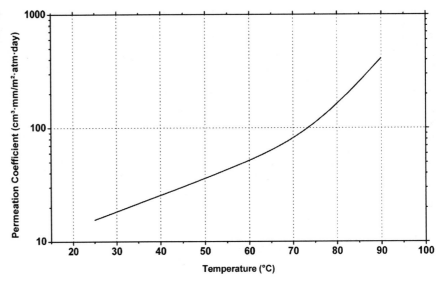

Figure 10.37 Permeation of hydrogen, nitrogen, oxygen, and ammonia vs. temperature through Solvay Solexis Halar® ECTFE.[7]

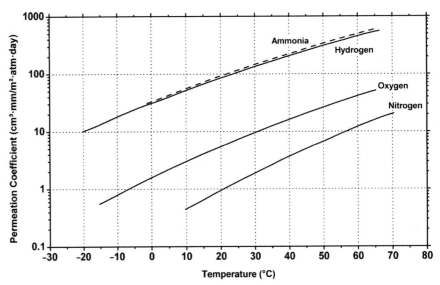

Figure 10.38 Permeation of hydrogen sulfide and hydrogen chloride vs. temperature through Solvay Solexis Halar® ECTFE.[33]

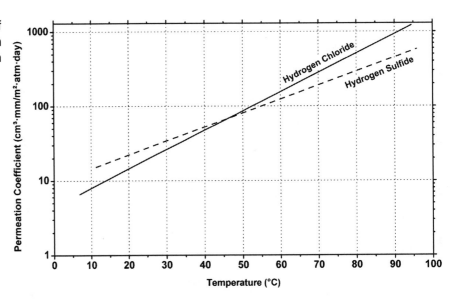

References

1. Adam SJ, David CE. Permeation measurement of fluoropolymers using mass spectrometry and calibrated standard gas leaks. *23rd International SAMPE Technical Conference, conference proceedings, SAMPE*; 1991.
2. Technical service supplied data. Dyneon, a 3M Company; 2001.
3. Teflon fluoropolymer resins. E.I. DuPont de Nemours and Company; 2000.
4. Dyneon™ TFM™ PTFE Product Brochure; 2002.
5. Dyneon™ fluoropolymers for the toughest jobs in the chemical industry. Dyneon™; 2001.
6. Permeation of Dyneon™ fluoropolymers. Dyneon; 2002.
7. Felder RM, Spence RD, Ferrell JK. Permeation of sulfur dioxide through polymers. *J Chem Eng Data* 1975;**20**:235–42.
8. http://www2.dupont.com/Teflon_Industrial/en_US/products/product_by_name/teflon_nxt/benefits.html
9. Dyneon™ TFM™ PTFE brochure. Dyneon; 2002.
10. DuPont FEP, Information Bulletin, H-50102; 1996.
11. Journal of Teflon®, vol. 22, December 1964.
12. *Neoflon™ FEP Pellets, Product Information EG-61K*. Daikin America, Inc.; 2001.
13. *Hyflon® PFA design and processing guide*. Solvay Solexis; 2008.
14. High Performance Films, DuPont PFA.E.I. DuPont de Nemours and Company; 2001.
15. *Halar® ECTFE design and processing guide*. Solvay Solexis; 2006.
16. *Hyflon® MFA design and processing guide, document no. Y42E001*. Solvay Solexis; 2008.
17. Stannett V. The transport of gases in synthetic polymeric membranes—an historic perspective. *J Membr Sci* 1978;**3**:97–115.
18. *Permeation resistance*. Ausimont, USA: Hyflon PFA & MFA; 2002.
19. Pinnau I, Toy L. Gas and vapor transport properties of amorphous perfluorinated copolymer membranes based on 2,2-bistrifluoromethyl-4,5-difluoro-1,3-dioxole/tetrafluoroethylene. Available at: *J Membr Sci*(1):125–33 http://linkinghub.elsevier.com/retrieve/pii/037673889500193X, 1996;**109**.
20. Nemser SM, Roman IC. Perfluorinated membranes. US Pat 5.051 1991:114.
21. DuPont™ Teflon® AF amorphous fluoropolymers, product information, H44587–4; 2010.
22. Lin FMC. *Chain microstructure studies of poly(vinyl fluoride) by high resolution NMR spectroscopy*. Ph.D. dissertation. University of Akron; 1981.
23. Technical information, Tedlar® PVF. DuPont™: 234427B; 1995.
24. Tedlar® SP product and properties guide. DuPont, 234452C; 1996.
25. *Aclar® data sheets*. Honeywell; 2001.
26. Solef® & Hylar® PVDF, Design and Processing Guide, BR2001C-B-2–1106; 2006.
27. Aclar® data sheets. Honeywell; 2009.
28. Aclar® performance films, supplier technical report (SFI-14 Rev. 9–89), Allied-Signal Engineered Plastics; 1989.
29. *Product information, Neoflon® CTFE molding powders, EG-71h*. Daikin Industries; 2001.
30. Voltalef® PCTFE technical brochure. Arkema; 2004.
31. Technical information, gas permeability of fluoropolymers. Atofina Chemicals; 2000.
32. *Tefzel® properties handbook (H96518–1)*. DuPont Company; 2003.
33. *Halar® ECTFE, fluoropolymer coatings for clean room exhaust duct systems*. Solvay Solexis; 2003.
34. *DuPont teflon and tefzel films, high performance films*. In: *DuPont*. E.I. du Pont de Nemours and Company; 2000.
35. *Dyneon™ fluoropolymers for the toughest jobs in chemical industry*. Dyneon™: a 3M Company; 2001.
36. Chemical resistance of Halar fluoropolymer, Supplier technical report (AHH), Ausimont.

11 High-Temperature and High-Performance Polymers

This chapter covers several high-temperature and high-performance plastics. They might be classified or been appropriate to include in another chapter, but they are grouped in this chapter because of their performance levels.

11.1 Polyether Ether Ketone

Polyetheretherketones (PEEKs) are also referred to as polyketones. The most common structure is given in Fig. 11.1. The CAS number is 31694-16-3.

PEEK is a thermoplastic with extraordinary mechanical properties. The Young's modulus of elasticity is 3.6 GPa and its tensile strength is 170 MPa. PEEK is partially crystalline, melts at around 350 °C, and is highly resistant to thermal degradation. The material is also resistant to both organic and aqueous environments and is used in bearings, piston parts, pumps, compressor plate valves, and cable insulation applications. It is one of the few plastics compatible with ultra-high vacuum applications. In summary, the properties of PEEK include:

- Outstanding chemical resistance
- Outstanding wear resistance
- Outstanding resistance to hydrolysis
- Excellent mechanical properties
- Outstanding thermal properties

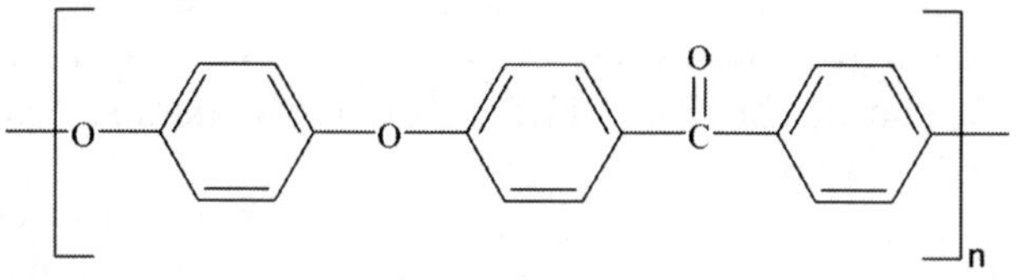

Figure 11.1 The structure of polyetheretherketone (PEEK).

- Very good dielectric strength, volume resistivity, and tracking resistance
- Excellent radiation resistance

Manufacturers and trade names: Victrex PLC VICTREX® and APTIV®, Greene, Tweed & Co. Arlon®, Solvay Advanced Polymers GATONE™ and KetaSpire®.

Applications and uses: Metal and glass replacement in aerospace, automotive, oil and gas, medical, semiconductor industries (Tables 11.1 and 11.2).

11.2 Polysiloxane

Silicone rubber is a semiorganic synthetic. Its structure consists of a chain of silicon and oxygen atoms rather than carbon and hydrogen atoms, as in the case with other types of rubber. The molecular structure of silicone rubber results in a very flexible—but weak—chain. Silicones are very stable at low and high temperatures. Although fillers may improve properties somewhat, tear and tensile strengths remain relatively low. Figure 11.2 shows four of the primary groups that make up a typical polysiloxane. To simply discussion that are identified by letters.

"M" stands for Me_3SiO
"D" for Me_2SiO_2
"T" for $MeSiO_3$
"Q" for SiO_4
"P" for replace Me with phenyl side groups
"V" for replace Me with vinyl side groups (typically <1%)
"F" for replace Me with fluorine

Some common abbreviations for the polymers include: MQ, VMQ, PMQ, PVMQ, PDMS poly(1-trimethylsilyl-1-propyne) or PTMSP.

Manufacturers and trade names: Bayer Corporation Baysilone®; Shincor Silicones KE®; Dow

Permeability Properties of Plastics and Elastomers. DOI: 10.1016/B978-1-4377-3469-0.10011-6

Table 11.1 Permeability of Amorphous and Crystalline VICTREX Morphology PEEK-Based APTIV™ Film

Gas	Morphology	Permeability Coefficient	
		Source Document (cm³ 100 μm/m² day bar)	Normalized (cm³ mm/m² day atm)
Carbon dioxide	Crystalline	424	10.8
	Amorphous	952	24.2
Helium	Crystalline	1572	39.9
	Amorphous	1336	33.9
Hydrogen	Crystalline	1431	36.3
	Amorphous	3178	80.7
Methane	Crystalline	8	0.2
	Amorphous	24	0.6
Nitrogen	Crystalline	15	0.4
	Amorphous	27	0.7
Oxygen	Crystalline	76	1.9
	Amorphous	171	4.3

100 μm Thickness at 1 bar.[1]

Table 11.2 Water Vapor Transmission Rate of Amorphous Gas and Crystalline VICTREX Morphology PEEK-Based APTIV™ Film[1]

Morphology	Vapor Transmission Rate	
	Source Document (g 100 μm/m² day)	Normalized (g mm/m² day)
Crystalline	4	0.4
Amorphous	9	0.9

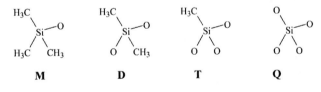

Figure 11.2 Structure of groups that make up polysiloxanes.

Corning Corp. Silastic®; General Electric Silplus®, Tufel®, SE; Rhone-Poulenc Inc. Rhodorsil.

Applications and uses: Caulking and nonstick coatings (Table 11.3).

See also Figs. 11.3–11.7.

11.3 Polyphenylene Sulfide

Polyphenylene sulfide (PPS) is a semicrystalline material. It offers an excellent balance of properties, including high-temperature resistance, chemical resistance, flowability, dimensional stability, and electrical characteristics. PPS must be filled with fibers and fillers to overcome its inherent brittleness. Because of its low viscosity, PPS can be molded with high loadings of fillers and reinforcements. Because of its outstanding flame resistance, PPS is ideal for high-temperature electrical applications. It is unaffected by all industrial solvents. The structure of PPS is shown in Fig. 11.8. The CAS number is 26125-40-6.

There are several variants to regular PPS that may be talked about by suppliers or may be seen in the literature. They are:

- Regular PPS is of "modest" molecular weight. Materials of this type are often used in coating products.

Table 11.3 Permeability of Gases Through Dimethylsilicone Rubber[2]

Gas	Permeability Coefficient	
	Source Document Units, ×10⁹ (cm^3 cm/s cm^2 cm Hg)	Normalized Units (cm^3 mm/m^2 day atm)
Hydrogen	65	42,700
Helium	35	23,000
Carbon dioxide	323	212,000
Nitrogen	28	18,400
Oxygen	62	40,700
Methane	95	62,400

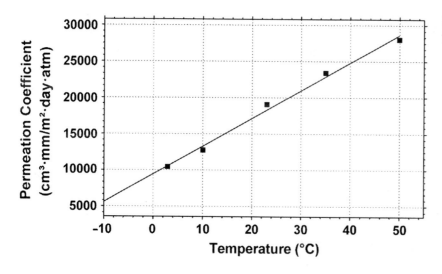

Figure 11.3 Permeability of helium through a PVMQ polysilxane.[3]

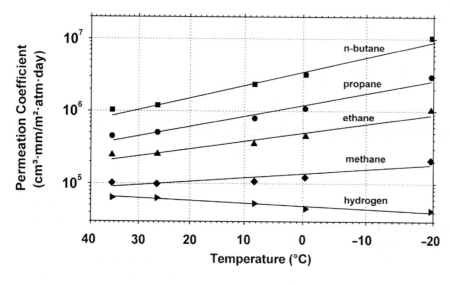

Figure 11.4 Permeability of gases vs. temperature through a polydimethyl siloxane (PDMS) membrane.[4]

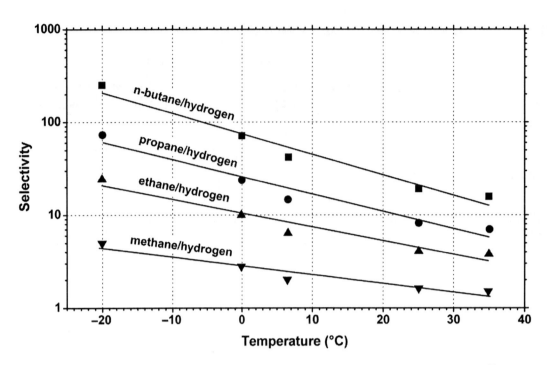

Figure 11.5 Selectivity of separation of gases vs. temperature through a PDMS membrane.[5]

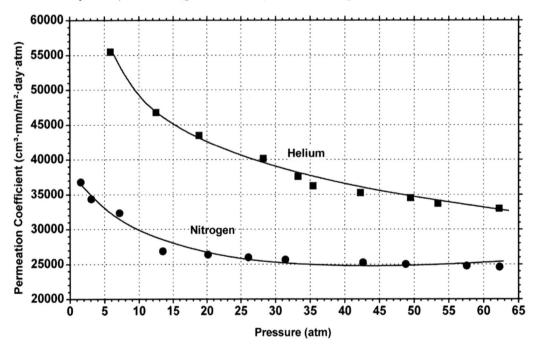

Figure 11.6 Permeation of helium and nitrogen at 35 °C through a PDMS membrane.[6]

- Cured PPS is PPS that has been heated to high temperature, above 300 °C, in the presence of air or oxygen. The oxygen causes some crosslinking and chain extension called oxidative crosslinking. This results in some thermoset-like properties such as improved thermal stability, dimensional stability, and improved chemical resistance.

- High-Molecular Weight (HMW) Linear PPS has a molecular weight about double that of regular PPS. The higher molecular weight improves elongation and impact strength.

- High-Molecular Weight (HMW) Branched PPS has not only higher molecular weight than regular

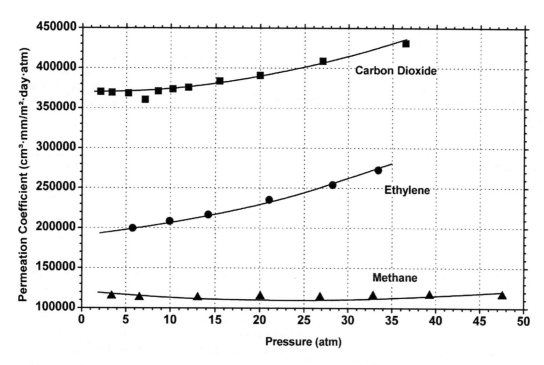

Figure 11.7 Permeation of gases at 35 °C through a PDMS membrane.[7]

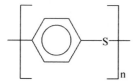

Figure 11.8 Structure of PPS.

PPS, but it also has polymer chain branches along the main molecule backbone. This provides improved mechanical properties.

PPS properties are summarized:

- Continuous use temperature of 220 °C
- Excellent dimensional properties
- Transparent
- Improved impact strength and toughness as compared to PES
- Excellent hydrolytic stability
- High stress cracking resistance
- Good chemical resistance
- Good surface release properties
- Expected continuous temperature of 180 °C

Manufacturers and trade names: Dinippon Ink, Chevron Phillips Ryton®, Ticona Fortran®, Toray Torelina®.

Applications and uses: Chemical process pipe, pump housings, shafts and impellers, oil field equipment, valves, corrosion resistant industrial parts, nonstick cookware, capacitor housings electrical packaging, and connectors and sockets (Tables 11.4–11.6).

Permeability of Fuel CE10 at 40 °C through Ticona Fortron® SKX-382 PPS film is 0.12 g mm/m^2 day.[7]

11.4 Polysulfone

Polysulfone (PSU) is a rigid, strong, tough, high-temperature amorphous thermoplastic. The structure of PSU is shown in Fig. 11.9. Its CAS number is 25135-51-7.

Its properties summarized:

- High thermal stability
- High toughness and strength
- Good environmental stress crack resistance
- Inherent fire resistance
- Transparent

Manufacturers and trade names: Solvay Advanced Polymers Udel®.

Applications and uses: Solvay Advanced Polymers Udel® polysulfone membranes can be used

Table 11.4 Liquid Vapor Transmission at 23 °C through Chevron Phillips Ryton® PPS Films[5]

Permeant Vapor	Vapor Transmission Rate	
	Source Document Units (g mil/100 in.^{2} day)	Normalized Units (g mm/m^2 day)
Water	0.8	0.3
Hydrochloric acid (37%)	0.1	0.04
Acetic acid	2.0	0.79
Benzene	6.3	2.5
Methyl alcohol	0.3	0.12

Table 11.5 Permeability of Gases at 23 °C through Chevron Phillips Ryton® PPS Films[8]

Permeant Gas	Permeability Coefficient	
	Source Document Units (cm^3 mil/100 in.^{2} day atm)	Normalized Units (cm^3 mm/m^2 day atm)
Oxygen	30	11.8
Carbon dioxide	75	29.6
Hydrogen	420	165
Ammonia	15	5.9
Hydrogen sulfide	3	1.2

Table 11.6 Permeability of Carbon dioxide through Ticona Fortron® PPS Films[9]

Temperature (°C)	Permeability Coefficient	
	Source Document Units (cm^3 100 μm/100 in.^{2} day atm)	Normalized Units (cm^3 mm/m^2 day atm)
60	708	1097
69	822	1274

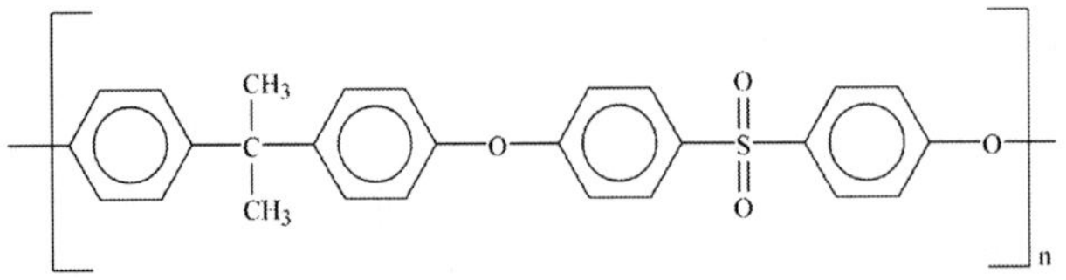

Figure 11.9 Structure of PSU.

for production of cheese, whey, orange juice, and apple juice, as well as for recovery of protein and lactose and the sterilization and clarification of beer, wine, and vinegar. Udel® resin offers unique properties, such as the ability to be put into solution for creating porous filaments or casting into flat sheet that allow it to be used in micro, ultra, and reverse osmosis membranes (Tables 11.7 and 11.8).

See also Figs. 11.10–11.13.

Table 11.7 Permeation of Gases through BASF Udel® PSU[8]

Permeant Gas	Permeability Coefficient	
	Source Document Units (cm^3 mil/100 $in.^2$ day)	Normalized Units (cm^3 mm/m^2 day atm)
Ammonia	1070	421
Carbon dioxide	950	374
Helium	1960	772
Hydrogen	1800	709
Methane	38	14.8
Nitrogen	30	15.7
Oxygen	230	91
Sulfur Hexafluoride	2	0.71
Dichlorodifluoromethane	0.59	0.23
Dichlorotetrafluoroethane	0.25	0.096

Test method: ASTM 1434.

Table 11.8 Permeation of Water Vapor through BASF Udel® PSU[8]

Temperature and Relative Humidity	Vapor Transmission Rate	
	Source Document Units (g mil/100 $in.^2$ day)	Normalized Units (g mm/m^2 day)
38 °C/90% RH	18	7.1
71 °C/100% RH	69	27.2

Test method: ASTM E96.

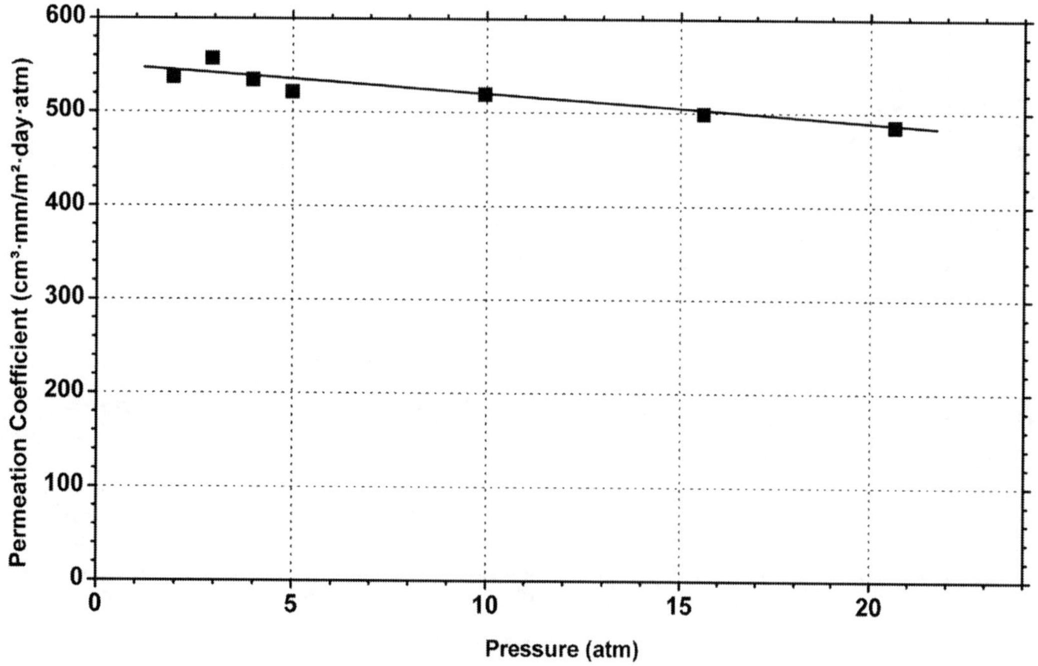

Figure 11.10 Permeability of helium vs. driving pressure at 35 °C through polysulfone.[10]

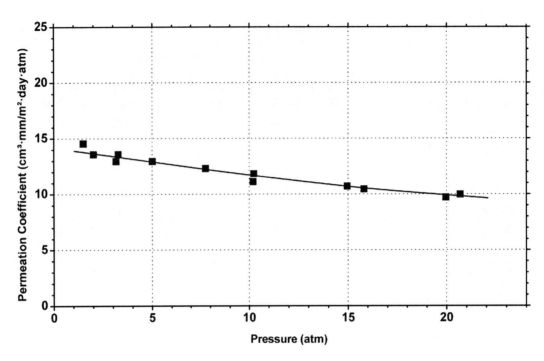

Figure 11.11 Permeability of methane vs. driving pressure at 35 °C through polysulfone.[11]

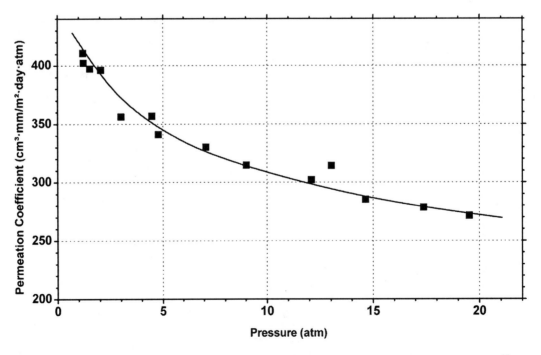

Figure 11.12 Permeability of carbon dioxide vs. driving pressure at 35 °C through polysulfone.[11]

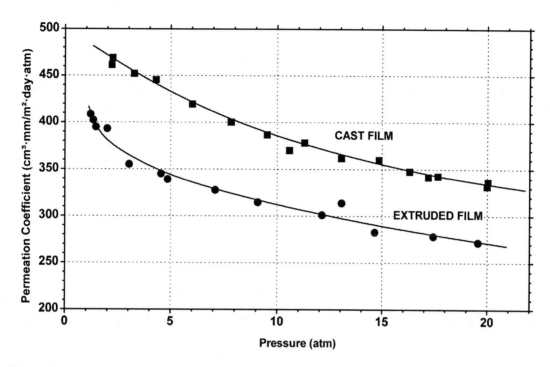

Figure 11.13 Permeability of carbon dioxide vs. driving pressure at 35 °C through cast and extruded polysulfone.[11]

11.5 Polyethersulfone

Polyethersulfone (PES) is an amorphous polymer and a high-temperature engineering thermoplastic. Even though PES has high-temperature performance, it can be processed on conventional plastics processing equipment. Its chemical structure is shown in Fig. 11.14. Its CAS number is 25608-63-3. PES has an outstanding ability to withstand exposure to elevated temperatures in air and water for prolonged periods.

Because PES is amorphous, mold shrinkage is low and is suitable for applications requiring close tolerances, and little dimensional change over a wide temperature range. Its properties include:

- Excellent thermal resistance—T_g 224 °C
- Outstanding mechanical, electrical, flame, and chemical resistance
- Very good hydrolytic and sterilization resistance
- Good optical clarity

Manufacturers and trade names: BASF Ultrason® E, Sumitomo Chemical Co., Ltd. SUMIKAEXCEL® PES, Solvay Advanced Polymers Veradel®.

Applications and uses: Automotive lighting, electrical and electronic components, baby bottles, coatings, cookware, food service, membranes for low-pressure water filtration, food and beverage processing and industrial filtration (Tables 11.9 and 11.10).

See also Figs. 11.15–11.17.

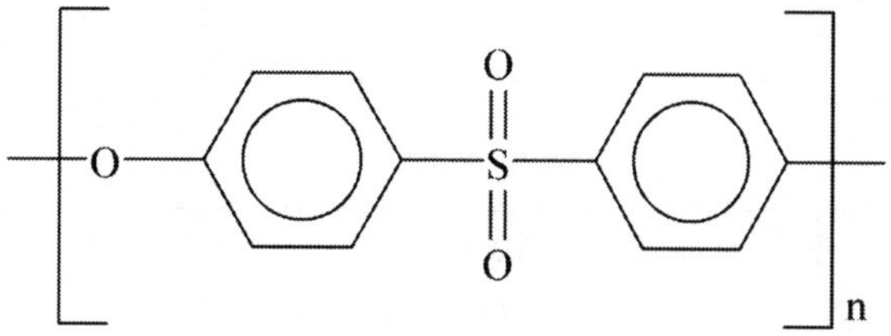

Figure 11.14 Structure of PES.

Table 11.9 Permeation of Gases at 35 °C through PES[12]

Permeant Gas	Permeability Coefficient	
	Source Document Units (Barrers)	Normalized Units (cm^3 mm/m^2 day atm)
Helium	8.0	525
Carbon dioxide	2.8	184
Methane	0.10	6.6

Table 11.10 Permeation through Sabic Innovative Plastics Ultem® 1000 PES[13]

Permeant	Vapor Transmission Rate	
	Source Document Units (g mil/100 in.^{2} day)	Normalized Units (g mm/m^2 day)
Water vapor	7.9	3.1
Permeant	**Permeability Coefficient**	
	Source Document Units (cm^3 mil/100 in.^{2} day)	Normalized Units (cm^3 mm/m^2 day atm)
Oxygen	37	14.6

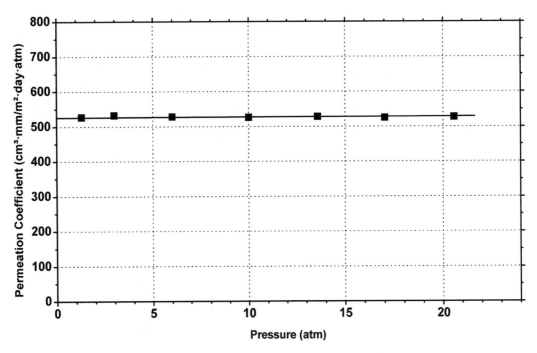

Figure 11.15 Permeation of helium at 35 °C vs. pressure differential of PES.[10]

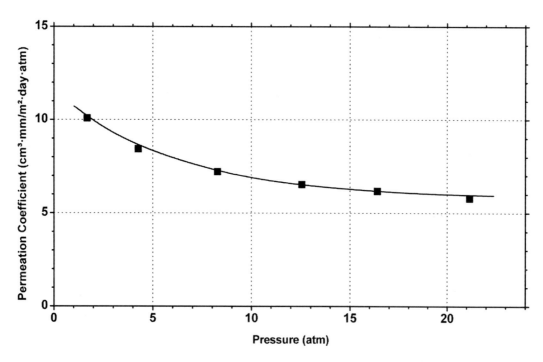

Figure 11.16 Permeation of methane at 35 °C vs. pressure differential of PES.[12]

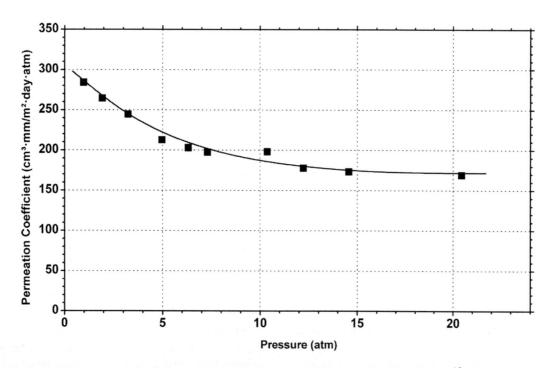

Figure 11.17 Permeation of carbon dioxide at 35 °C vs. pressure differential of PES.[12]

11.6 Polybenzimidazole

Polybenzimidazole (PBI) is a unique and highly stable linear heterocyclic polymer. The chemical structure is shown in Fig. 11.18. PBI exhibits excellent thermal stability, resistance to chemicals, acid and base hydrolysis, and temperature resistance. PBI can withstand temperatures as high as 430 °C, and in short bursts, to 760 °C. PBI does not burn and maintains its properties as low as −196 °C.

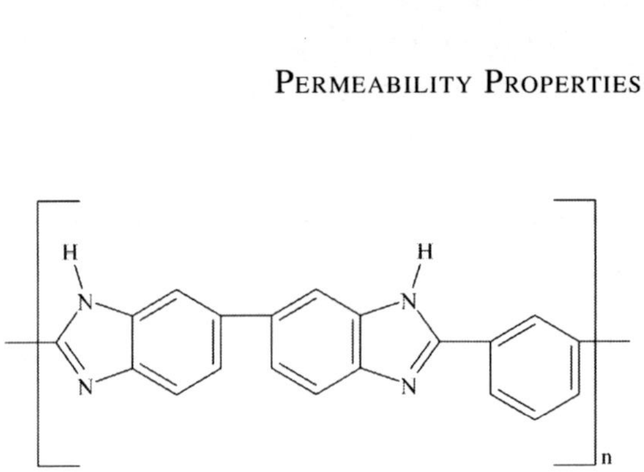

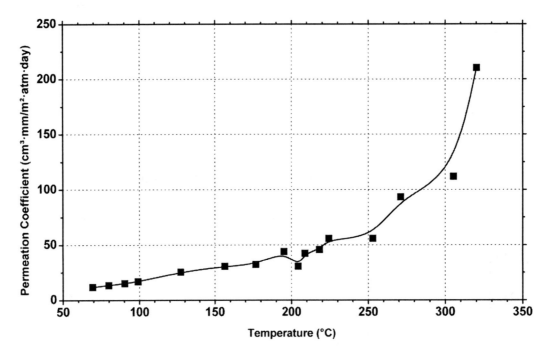

Figure 11.18 Structure of PBI.

Figure 11.19 Permeation of hydrogen vs. temperature through PBI Performance Products Inc. Celazole® PBI membrane.[14]

Ideally suited for its application in extreme environments, PBI can be formed into stock shapes and subsequently machined into high precision finished parts. Since PBI does not have a melt point, moldings from virgin PBI polymer can only be formed in a high temperature, high pressure compression molding process.

PBI is highly resistant to deformation and has low hysteresis loss and high elastic recovery. PBI exhibits ductile failure and may be compressed to over 50% strain without fracture. Celazole® PBI has the highest compressive strength of any thermoplastic or thermosetting resin at 400 MPa. There is no weight loss or change in compressive strength of Celazole® PBI exposed to 260 °C in air for 500 h. At 371 °C, no weight or strength change takes place for 100 h. In spite of these unusual properties, PBI is usually blended with other plastics, particularly polyesters and PEEK.

Manufacturers and trade names: PBI Performance Products Inc. Celazole®.

Applications and uses: Semiconductor: PBI is especially suited for oxide etching, sputtering, and spincoating; Petro/Chemical: seals, bearings, bushings, thrust washers, and other parts in which wear resistance, strength, heat, and chemical resistance are essential to performance and long process run time; Industrial: soldering, glass making, metal cutting, and even business machines; Aerospace: for nose cones, leading edges, ablative heat shields; fire resistant fibers.

See also Figs. 11.19–11.21.

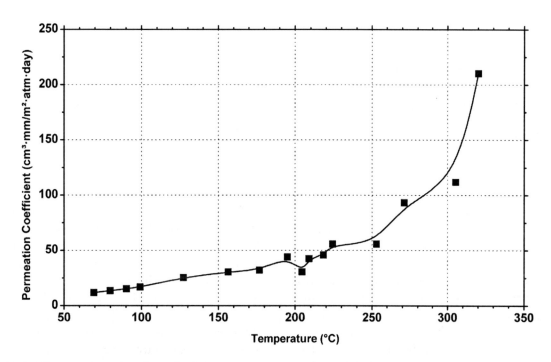

Figure 11.20 Permeation of carbon dioxide vs. temperature through PBI Performance Products Inc. Celazole® PBI membrane.[3]

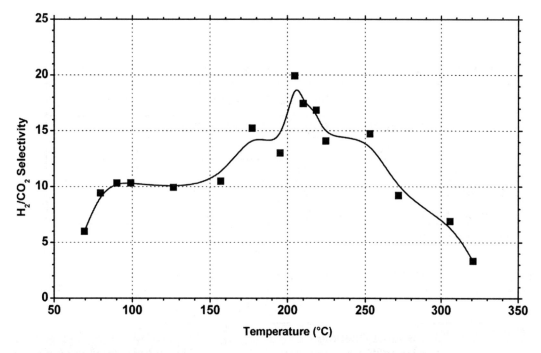

Figure 11.21 Selectivity of hydrogen/carbon dioxide vs. temperature through PBI Performance Products Inc. Celazole® PBI membrane.[3]

11.7 Parylene (poly(p-xylylene))

Parylene is the generic name for members of a series of polymers. The basic member of the series, called Parylene N, is poly-*para*-xylylene, a completely linear, highly crystalline material. The structures of four Parylene types are shown in Fig. 11.22.

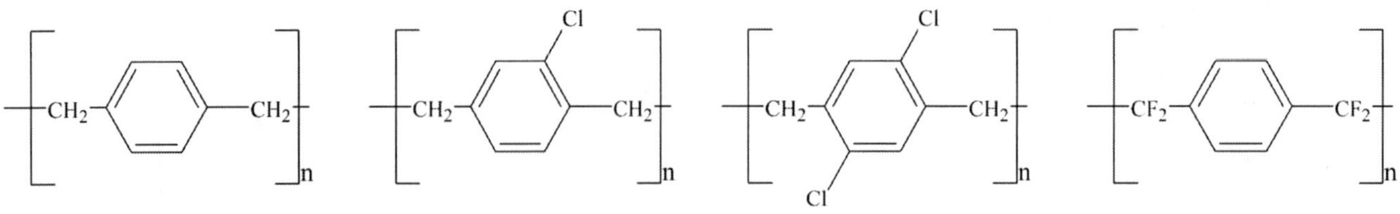

Parylene N   Parylene C   Parylene D   Parylene HT

Figure 11.22 Structures of the parylene polymer molecules.

Parylene polymers are not manufactured and sold directly. They are deposited from the vapor phase by a process that in some respects resembles vacuum metalizing. The Parylenes are formed at a pressure of about 0.1 torr from a reactive dimmer in the gaseous or vapor state. Unlike vacuum metalizing, the deposition is not line of sight and all sides of an object to be encapsulated are uniformly impinged by the gaseous monomer. Due to the uniqueness of the vapor phase deposition, the Parylene polymers can be formed as structurally continuous films from as thin as a fraction of a micrometer to as thick as several mils.

The first step is the vaporization of the solid dimer at approximately 150 °C. The second step is the quantitative cleavage (pyrolysis) of the dimer vapor at the two methylene–methylene bonds at about 680 °C to yield the stable monomeric diradical, *para*-xylylene. Finally, the monomeric vapor enters the room temperature deposition chamber where it spontaneously polymerizes on the substrate. The substrate temperature never rises more than a few degrees above ambient.

Parylene is used as a coating on electronics ranging from advanced military and aerospace electronics to general-purpose industrial products,

Table 11.11 Permeability of Various Gases at 25 °C through Parylene Polymers[15]

Polymer	Permeability Coefficient (cm^3 mm/m^2 day atm)			
	Nitrogen	Oxygen	Carbon Dioxide	Hydrogen
Parylene N	3.0	15.4	84.3	212.6
Parylene C	0.4	2.8	3.0	43.3
Parylene D	1.8	12.6	5.1	94.5
Parylene HT	4.8	23.5	95.4	—

Test standard: ASTM D 1434.

Table 11.12 Permeability of Various Gases at 23 °C through Parylene Polymers[16]

Polymer	Gas Permeability					
	Source Document (cm^3 mil)/(100 $in.^2$ day atm)			Normalized (cm^3 mm/m^2 day atm)		
	Hydrogen Sulfide	Sulfur Dioxide	Chlorine	Hydrogen Sulfide	Sulfur Dioxide	Chlorine
Parylene N	795	1890	74	313	744	29
Parylene C	13	11	0.35	5.1	4.3	0.1
Parylene D	1.45	4.75	0.55	0.6	1.9	0.2

Test standard: ASTM D 1434.

Table 11.13 Water Vapor Permeability through Parylene polymers[4]

Polymer	Water Vapor Transmission Rate (g mm/m² day)	Temperature (°C)	Relative Humidity (%)	Test Standard
Parylene N	0.59	37	90	ASTM E96
Parylene C	0.08	37	90	ASTM F1249
Parylene D	0.09	37	90	ASTM E96
Parylene HT	0.22	38	100	ASTM F1249

medical devices ranging from silicone tubes to advanced coronary stents, synthetic rubber products ranging from medical grade silicone rubber to EPDM.

The manufacturer of coating equipment and starting materials is Para Tech Coating, Inc. They also offer coating services.

Manufacturers and trade names: Para Tech Coating, Inc. Parylene.

Applications and uses: Electronics: circuit boards, sensors, integrated circuits/hybrids, MEMs devices, motor assemblies, coil forms, silicon wafers; Medical: needles, prosthetic devices, implantable components, catheter, electrodes, stents, epidural probes, cannulae assemblies; Aerospace: deep space vision systems, navigation and controls, optical devices, satellite and spacecraft devices, flight deck controls (Tables 11.11–11.13).

11.8 Polyoxymethylene (POM or Acetal Homopolymer)/ Polyoxymethylene Copolymer (POM-Co or Acetal Copolymer)

Acetal polymers, also known as polyoxymethylene (POM) or polyacetal, are formaldehyde-based thermoplastics that have been commercially available since the 1960s. Polyformaldehyde is thermally unstable. It decomposes on heating to yield formaldehyde gas. Two methods of stabilizing polyformaldehyde for use as an engineering polymer were developed and introduced by DuPont, in 1959, and Celanese in 1962 (now Ticona).

DuPont's method for making polyacetal yields a homopolymer through the condensation reaction of polyformaldehyde and acetic acid (or acetic anhydride). The acetic acid puts acetate groups (CH_3COO-) on the ends of the polymer as shown in Fig. 11.23, which provide thermal protection against decomposition to formaldehyde.

Further stabilization of acetal polymers also includes the addition of antioxidants and acid scavengers. Polyacetals are subject to oxidative and acidic degradation, which leads to molecular weight decline. Once the chain of the homopolymer is ruptured by such an attack, the exposed polyformaldehyde ends may decompose to formaldehyde and acetic acid.

The Celanese route for the production of polyacetal yields a more stable copolymer product via the reaction of trioxane, a cyclic trimer of formaldehyde, and a cyclic ether, such as ethylene oxide or 1,3 dioxolane. The structures of these monomers are shown in Fig. 11.24. The polymer structure is given in Fig. 11.25.

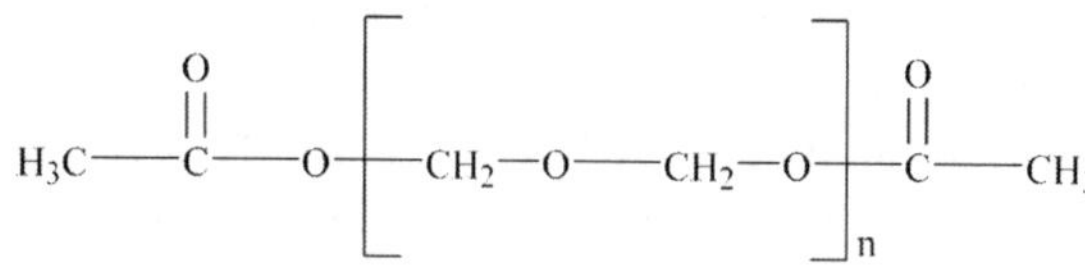

Figure 11.23 Chemical structure of acetal homopolymer.

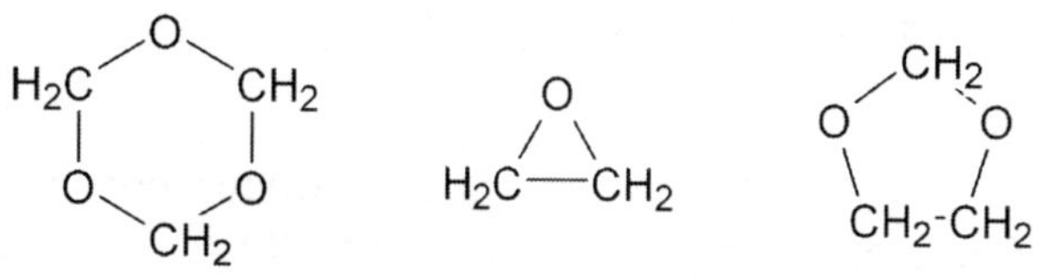

Figure 11.24 Chemical structure of polyoxymethylene copolymer monomers.

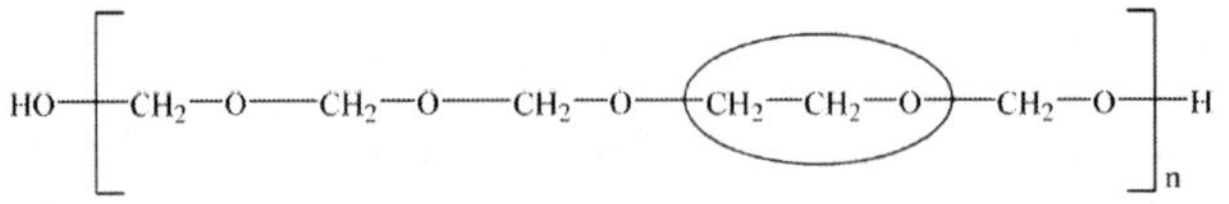

Figure 11.25 Chemical structure of acetal copolymer.

The improved thermal and chemical stability of the copolymer versus the homopolymer is a result of randomly distributed oxyethylene groups, which is circled in Fig. 5.5. All polyacetals are subject to oxidative and acidic degradation, which leads to molecular weight reduction. Degradation of the copolymer ceases, however, when one of the randomly distributed oxyethylene linkages is reached. These groups offer stability to oxidative, thermal, acidic, and alkaline attack. The raw copolymer is hydrolyzed to an oxyethylene end cap to provide thermally stable polyacetal copolymer.

The copolymer is also more stable than the homopolymer in an alkaline environment. Its oxyethylene end cap is stable in the presence of strong bases. The acetate end cap of the homopolymer, however, is readily hydrolyzed in the presence of alkalis, causing significant polymer degradation.

The homopolymer is more crystalline than the copolymer. The homopolymer provides better mechanical properties, except for elongation. The oxyethylene groups of the copolymer provide improved long-term chemical and environmental stability. The copolymer's chemical stability results in better retention of mechanical properties over an extended product life.

Acetal polymers have been particularly successful in replacing cast and stamped metal parts due to their toughness, abrasion resistance, and ability to withstand prolonged stresses with minimal creep. Polyacetals are inherently self-lubricating. Their lubricity allows the incorporation of polyacetal in a variety of metal-to-polymer and polymer-to-polymer interface applications such as bearings, gears, and switch plungers. These properties have permitted the material to meet a wide range of market requirements.

The properties of polyacetals can be summarized as follows:

- Excellent wear resistance
- Very good strength, stiffness
- Good heat resistance
- Excellent chemical resistance
- Opaque
- Moderate to high price
- Somewhat restricted processing

Manufacturers and trade names: DuPont™ Delrin®, Ticona Celcon®.

Applications and uses: Metal and glass replacement in aerospace, automotive, oil and gas, medical, semiconductor industries (Tables 11.14–11.17).

See also Fig. 11.26.

Table 11.14 Gas Permeability at 23 °C of Ticona Celcon® M25™, M90™ and M270™ POM Copolymer[11]

Permeant Gas	Permeability Coefficient	
	Source Document Units (cm^3 mil/100 in.^{2} day atm)	Normalized Units (cm^3 mm/m^2 day atm)
Air	2.2–3.2	0.9–1.3
Nitrogen	2.2–3.2	0.9–1.3
Oxygen	5.0–7.4	2.0–2.9

Thickness: 0.15 mm.

Table 11.15 Permeability of DuPont™ Delrin® POM Homopolymer[12]

	Vapor Transmission Rate			
	Source Document Units (g mil/100 in.² day)		Normalized Units (g mm/m² day)	
Permeate Vapor	23 °C, 50% RH	38 °C	23 °C, 50% RH	38 °C
Cologne formulations	0.6	4.5	0.23	1.8
Ethyl alcohol/water (90%/10% by wt.)	0.25	—	0.1	
Ethyl alcohol/water (70%/30% by wt.)	1.5	7.8	0.6	3.1
Freon® 12/11 propellant (30/70)	0.2	0.54	0.08	.21
Freon® 12/114 propellant (20/80)	0.2	0.42	0.08	0.17
Gasoline	0.1	—	0.04	
Hair spray formulations	0.8	6.0	0.3	2.4
Methylsalicylate	0.3	—	0.12	
Nitrogen at 6.1 atm	0.05	—	0.02	
Oils (motor, mineral, vegetable)	0	0	0	0
Perchloroethylene	0.2	—	0.08	
Shampoo formulations	2.4	8.5	0.95	3.3
Tar and road-oil remover	0.03	0.19	0.01	0.07
Trichloroethylene	25	56	9.9	22.1
Toluene	0.6	—	0.24	

Table 11.16 Permeability of DuPont™ Delrin® POM Homopolymer[10]

	Permeability Coefficient	
Permeant Gas	Source Document Units (cm³ mil/100 in.² day atm)	Normalized Units (cm³ mm/m² day atm)
Carbon dioxide	37–50	14.6–19.7
Oxygen	12–17	4.7–6.7

Table 11.17 Permeability at 23 °C of Ticona Hostaform® POM Copolymer[17]

Permeant Gas	Source Document Units, Permeability (cm³ (STP)/m² day bar)	Normalized Units, Permeability Coefficient (cm³ mm/m² day atm)
Oxygen	49	4.0
Carbon dioxide	1110	90
Helium	7.0	0.57
Permeant Vapor	Vapor Transmission Rate	
	Source Document Units (g/m² day)	Normalized Units (g mm/m² day)
Water (3 mm thickness)	32	96

Thickness: 0.08 mm.

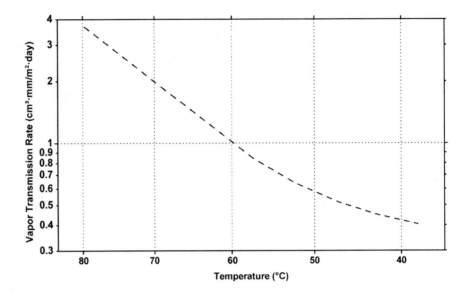

Figure 11.26 Permeability of super unleaded gasoline vs. temperature through Ticona Hostaform® C 27021 POM Copolymer.[14]

References

1. *Victrex materials guide.* Victrex PLC; 2010.
2. Robb WL. Thin silicone membranes—their permeation properties and some applications. *Ann N Y Acad Sci* 1968;**146**:119—37.
3. *Ryton® PPS chemical properties.* Chevron Phillips; 1995.
4. *Fortron® Brochure.* Ticona; 2007.
5. *Fuel permeation chart test method.* Ticona; 2005.
6. *Udel® design guide, version 2.1.* Solvay Advanced Polymers; 2002.
7. *Oxygen and water permeation data.* Sabic Innovative Plastics; 2008.
8. *SCS parylene properties.* Specialty Coating Systems; 2008.
9. *Parylene properties and characteristics.* V&P Scientific, Inc.; 2010.
10. *Designing with Celcon®.* Ticona; 2002.
11. *Delrin design guide—Module III, H-57472.* DuPont™; 1997.
12. *Hostaform®.* Ticona; 2006.
13. Zhang H, Cloud A. *The permeability characteristics of silicone rubber.* Haibing Zhang et al.pdf. Conference Proceedings, Society. Available from:, http://www.arlonstd.com/Library/Guides/D116; 2006.
14. Li NN. Advanced membrane technology and applications. 2008. p. 648.
15. Jordan SM, Koros WJ. Permeability of pure and mixed gases in silicone rubber at elevated pressures. *J Polymer Sci B Polymer Phys* 1990;**28**(6):795—809. Available from: http://doi.wiley.com/10.1002/polb.1990.090280602.
16. Chiou J, Maeda Y, Paul D. Gas permeation in polyethersulfone. *J Appl Polymer Sci* 1987;**33**(5):1823—8. Available from: http://www3.interscience.wiley.com/journal/104028508/abstract.
17. Li NN, Fane AG, Ho, Winston WS, Matsuura T. Advanced membrane technology and applications. John Wiley & Sons. p. 652—4.

12 Elastomers and Rubbers

An elastomer is a polymer with the property of "elasticity," generally having notably low Young's modulus and high yield strain compared with other materials.[1] The term is often used interchangeably with the term rubber. Elastomers are amorphous polymers existing above their glass transition temperature, so that considerable segmental motion is possible, so it is expected that they would also be very permeable. At ambient temperatures rubbers are thus relatively soft and deformable. Their primary uses are for seals, adhesives, and molded flexible parts. Elastomers may be thermosets (requiring vulcanization, a form of crosslinking) or thermoplastic, called thermoplastic elastomer or TPE.

TPEs have two big advantages over the conventional thermoset (vulcanized) elastomers. Those are ease and speed of processing. Other advantages of TPEs are recyclability of scrap, lower energy costs for processing, and the availability of standard, uniform grades (not generally available in thermosets).

TPEs are molded or extruded on standard plastics-processing equipment in considerably shorter cycle times than those required for compression or transfer molding of conventional rubbers. They are made by copolymerizing two or more monomers, using either block or graft polymerization techniques. One of the monomers provides the hard, or crystalline, polymer segment that functions as a thermally stable component; the other monomer develops the soft or amorphous segment, which contributes the elastomeric or rubbery characteristic.

Physical and chemical properties can be controlled by varying the ratio of the monomers and the length of the hard and soft segments. Block techniques create long-chain molecules that have various or alternating hard and soft segments. Graft polymerization methods involve attaching one polymer chain to another as a branch. The properties that are affected by each phase can be generalized:

"Hard phase"—Plastic properties:

1. Processing temperatures
2. Continuous use temperature
3. Tensile strength
4. Tear strength
5. Chemical and fluid resistance
6. Adhesion to inks, adhesives, and over-molding substrates.

"Soft phase"—Elastomeric properties:

1. Lower service temperature limits
2. Hardness
3. Flexibility
4. Compression set and tensile set

This chapter has data on many thermoset and TPEs. TPEs will be discussed first.

12.1 Thermoplastic Polyurethane Elastomers (TPU)

Urethanes are a reaction product of a diisocyanate and long- and short-chain polyether, polyester, or caprolactone glycols.[2] The polyols and the short-chain diols react with the diisocyanates to form linear polyurethane molecules. This combination of diisocyanate and short-chain diol produces the rigid or hard segment. The polyols form the flexible or soft segment of the final molecule. Figure 12.1 shows the molecular structure in schematic form.

The properties of the resin depend on the nature of the raw materials, the reaction conditions, and the ratio of the starting raw materials. The polyols used have a significant influence on certain properties of

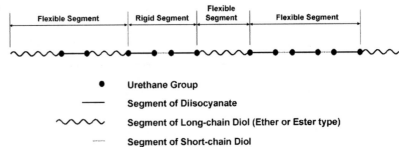

Figure 12.1 Molecular structure of a thermoplastic polyurethane elastomer.

the thermoplastic polyurethane. Polyether and polyester polyols are both used to produce many products.

The polyester-based TPUs have the following characteristic features:

- Good oil/solvent resistance
- Good UV resistance
- Abrasion resistance
- Good heat resistance
- Mechanical properties

The polyether-based TPUs have the following characteristic features:

- Fungus resistance
- Low-temperature flexibility
- Excellent hydrolytic stability
- Acid/base resistance

In addition to the basic components described above, most resin formulations contain additives to facilitate production and processability. Other additives can also be included such as:

- Demolding agents
- Flame retardants
- Heat/UV stabilizers
- Plasticizers

The polyether types are slightly more expensive and have better hydrolytic stability and low-temperature flexibility than the polyester types.

Manufacturers and trade names: Lubrizol Estane® TPU, Bayer MaterialScience Texin and Desmopan, BASF Elastollan® (Tables 12.1–12.5). See also Figs. 12.2–12.6.

Table 12.1 Permeation of Gases and Vapors through Lubrizol Estane® TPU[3]

Permeant Gas	Permeability Coefficient	
	Source Document Units (m^3 mm/ m^2 day)	Normalized Units (cm^3 mm/ m^2 day atm)
Air	0.00052	5200
Oxygen	0.00162	16,200
Nitrogen	0.0004	400
Carbon dioxide	0.01023	102,300
Helium	0.00291	29,100
Argon	0.00112	11,200
Freon® 12	0.00122	12,200
Freon® 22	0.00106	10,600
	Vapor Permeation Rate	
Permeant Vapor	(g/m^2 day)	(g mm/m^2 day)
Water	2.0	0.1

Thickness: 1.14 mm.

Table 12.2 Permeation of Water Vapor at 23 °C and 50% RH through Lubrizol Estane® Breathable TPU[4]

Estane® Code	Water Vapor Transmission Rate (g/m^2 day)
75AT3	380
80AF3	650
90AF3	500
58245	650
58315	250
58237	550

Test method: ASTM E96 B.

Table 12.3 Permeation of Solvent Vapors through Lubrizol Estane® TPU[2]

Permeant Vapor	Temperature (°C)	Thickness (mm)	Vapor Permeation Rate	
			Source Document Units (cm^3/m^2 day)	Normalized Units (g mm/m^2 day)
Gasoline fuel B	23	0.18	117.2	15.4
Gasoline fuel B	23	0.26	16.9	3.2
Gasoline fuel B	23	0.18	19.1	4.6
JP-4 fuel	27	0.47	3.8	0.13
JP-4 fuel	48	0.47	40.1	13.8
JP-4 fuel	66	0.47	112	38.4
JP-4 fuel	70	0.47	216	74.1
JP-4 fuel	83	0.47	385	132
JP-4 fuel	121	0.47	917	315

Test method: ASTM E96 B.

Table 12.4 Permeation of Gases at 20 °C through Various BASF Elastollan® TPUs[3]

Elastollan® Type	Source Document Units, Permeability Coefficient (m^2/s Pa) × 10^{-18}						
	Argon	Methane	Carbon Dioxide	Hydrogen	Helium	Nitrogen	Oxygen
C 80 A	12	11	200	45	35	4	14
C 85 A	9	6	150	40	30	3	10
C 90 A	5	4	40	30	25	2	7
C 95 A	3	2	20	20	20	1	4
1180 A	14	18	230	70	50	6	21
1185 A	9	14	180	60	40	5	16
1190 A	7	9	130	50	30	4	12
1195 A	6	5	90	40	20	3	8
Normalized Units, Permeability Coefficient (cm^3 mm/m^2 day atm)							
C 80 A	105	96	1751	394	306	35	123
C 85 A	79	53	1313	350	263	26	88
C 90 A	44	35	350	263	219	18	61
C 95 A	26	18	175	175	175	9	35
1180 A	123	158	2014	613	438	53	184
1185 A	79	123	1576	525	350	44	140
1190 A	61	79	1138	438	263	35	105
1195 A	53	44	788	350	175	26	70

Test method: DIN 53536.

Table 12.5 Water Vapor Permeation at 23 °C through Various BASF Elastollan® TPUs[3]

Elastollan® Type	Water Vapor Permeation Rate (g mm/m² day)
C 80 A	18
C 85 A	15
C 90 A	20
C95 A	8
1180 A	21
1185 A	17
1190 A	15
1195 A	12

Test method: DIN 53122 Part 1; RH differential of 93%.

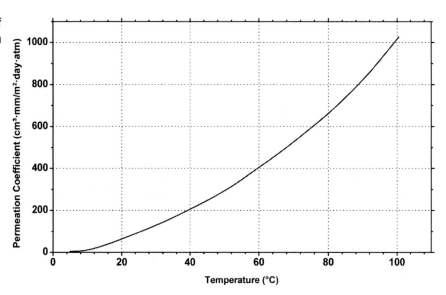

Figure 12.2 Permeability of nitrogen vs. temperature through BASF Elastollan® 1185A TPU.[5]

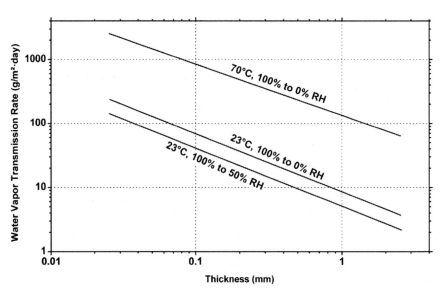

Figure 12.3 Water vapor transmission vs. thickness, temperature, and relative humidity (RH) of Bayer MaterialScience Texin® 285 TPU.[6]

12: Elastomers and Rubbers

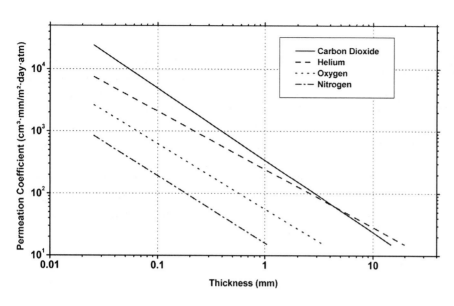

Figure 12.4 Permeability of gases vs. thickness, temperature, and RH of Bayer MaterialScience Texin® 390 TPU.[4]

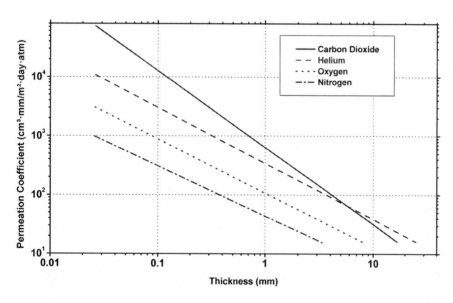

Figure 12.5 Permeability of gases vs. thickness, temperature, and RH of Bayer MaterialScience Texin® 285 TPU.[4]

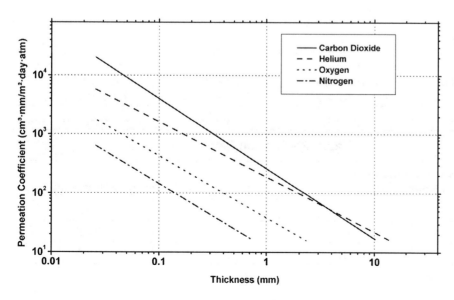

Figure 12.6 Permeability of gases vs. thickness, temperature, and RH of Bayer MaterialScience Texin® 453 TPU.[4]

12.2 Olefinic TPEs (TPO)

Polyolefin TPE (TPO) materials are defined as compounds (mixtures) of various polyolefin polymers, semicrystalline thermoplastics, and amorphous elastomers. Most TPOs are composed of polypropylene and a copolymer of ethylene and propylene called ethylene–propylene rubber (EPR).[2] A common rubber of this type is called EPDM rubber, which has a small amount of a third monomer, a diene (two carbon–carbon double bonds in it). The diene monomer leaves a small amount of unsaturation in the polymer chain that can be used for sulfur crosslinking. Like most TPEs, TPO products are composed of hard and soft segments. TPO compounds include fillers, reinforcements, lubricants, heat stabilizers, antioxidants, UV stabilizers, colorants, and processing aids. They are characterized by high-impact strength, low density, and good chemical resistance; they are used when durability and reliability are primary concerns.

Manufacturers and trade names: Advanced Elastomer Systems Santoprene®, LyondelBasell Advanced Polyolefins Dexflex®.

Applications and uses: Roofing and automotive exterior parts; capping distilled water, dairy products, fruit juices, sports drinks, beer, wine; and food, cosmetics, toiletries; and pharmaceutical packaging, sterilized closures, seals, and liners (Tables 12.6 and 12.7).

Table 12.6 Permeation of Gases at 23 °C through Advanced Elastomer Systems Santoprene® TPO[7]

Permeant Gas	Santoprene® Grade					
	201–73	201–87	203–50	201–73	201–87	203–50
	Permeability Coefficient					
	Source Document Units (cm³ 0.5mm/100 in.² day atm)			Normalized Units (cm³ mm/m² day atm)		
Air	31	39	18	240	302	140
Nitrogen	25	34	12	194	264	93
Oxygen	65	56	36	504	434	279
Carbon dioxide	390	260	170	3023	2015	1318
Argon	67	77	51	519	597	395
Propane	150	430	250	1163	3333	1938

Thickness: 0.5 mm. Test method: ASTM D1434.

Table 12.7 Water Vapor Transmission at 25 °C through Advanced Elastomer Systems Santoprene® TPO[7]

Santoprene® Grade	ASTM E96, Procedure A	ASTM E96, Procedure BW	ASTM E96, Procedure A	ASTM E96, Procedure BW
	25% RH	75% RH	25% RH	75% RH
	Vapor Permeation Rate			
	Source Document Units (g 0.5 mm/m² day)		Normalized Units (g mm/m² day)	
201–73	0.97	0.45	0.49	0.23
201–87	0.32	0.45	0.16	0.23
203–50	0.45	1.61	0.23	0.81

Thickness: 0.5 mm. Test method: ASTM E96.

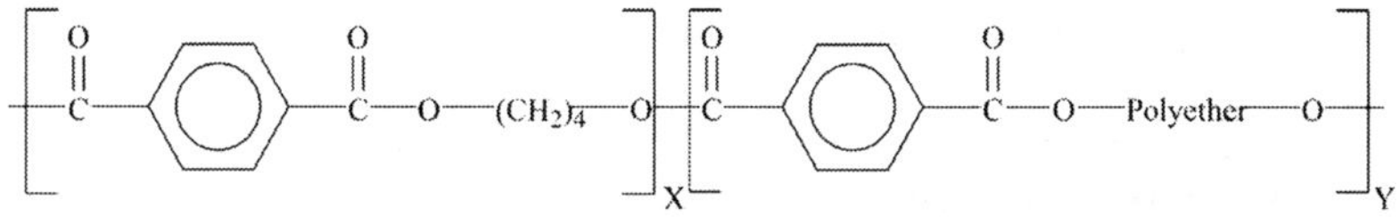

Figure 12.7 Molecular structure of Ticona Riteflex® thermoplastic copolyester elastomers.

12.3 Thermoplastic Copolyester Elastomers (TPE-E or COPE)

Thermoplastic copolyester elastomers (TPE-E or COPE) are block copolymers.[2] The chemical structure of one such elastomer is shown in Fig. 12.7.

These TPEs are generally tougher over a broader temperature range than the urethanes described in Section 10.1.1. Also, they are easier and more forgiving in processing.

- Excellent abrasion resistance
- High tensile, compressive, and tear strength
- Good flexibility over a wide range of temperatures
- Good hydrolytic stability
- Resistance to solvents and fungus attack
- Selection of a wide range of hardness

In these polyester TPEs, the hard polyester segments can crystallize, giving the polymer some of the attributes of semicrystalline thermoplastics, most particularly better solvent resistance than ordinary rubbers, but also better heat resistance. Above the melting temperature of the crystalline regions, these TPEs can have low viscosity and can be molded easily in thin sections and complex structures. Properties of thermoplastic polyester elastomers can be fine-tuned over a range by altering the ratio of hard to soft segments.

In DuPont Hytrel® polyester TPEs, the resin is a block copolymer. The hard phase is polybutylene terephthalate (PBT). The soft segments are long-chain polyether glycols.

Manufacturers and trade names: Ticona Riteflex®, DuPont™ Hytrel®, Eastman Ecdel®, DSM engineering plastics Arnitel® (Tables 12.8—12.13).

Table 12.8 Permeability of Various Gases at 21.5 °C through DuPont™ Hytrel® 4056 Thermoplastic Copolyester Elastomer[8]

Gas	Permeability Coefficient	
	Source Document Units [cm³(STP) mm/m² s Pa]	Normalized Units (cm³ mm/ m² day atm)
Air	2.4×10^{-8}	210
Nitrogen	1.7×10^{-8}	150
Carbon dioxide	3.5×10^{-7}	3100
Helium	1.57×10^{-7}	1370
Propane	$<2.0 \times 10^{-9}$	<18
Water	3.1×10^{-5}	270,000
Freon® 12	1.4×10^{-8}	120
Freon® 22	4.7×10^{-9}	41
Freon® 114	4.1×10^{-7}	3600

Thickness: 0.5 mm. Test method: ASTM E96.

Table 12.9 Permeability of Various Gases at 21.5 °C through DuPont™ Hytrel® 5556 Thermoplastic Copolyester Elastomer[2]

Gas	Permeability Coefficient	
	Source Document Units [cm³(STP) mm/m² s Pa]	Normalized Units (cm³ mm/ m² day atm)
Air	1.8×10^{-8}	160
Nitrogen	1.4×10^{-8}	120
Carbon dioxide	1.8×10^{-7}	1600
Helium	9.9×10^{-8}	870
Propane	$<2.0 \times 10^{-9}$	<18
Water	2.4×10^{-5}	210,000
Freon® 12	1.2×10^{-8}	105
Freon® 22	5.9×10^{-9}	52
Freon® 114	2.8×10^{-7}	2500

Table 12.10 Permeability of Various Gases at 21.5 °C through DuPont™ Hytrel® 6346 Thermoplastic Copolyester Elastomer[2]

Gas	Permeability Coefficient	
	Source Document Units [cm^3(STP) mm/m^2 s Pa]	Normalized Units (cm^3 mm/m^2 day atm)
Propane	<0.2 × 10^{-8}	<18
Freon® 12	1.20 × 10^{-8}	105
Freon® 22	<0.2 × 10^{-8}	<18
Freon® 114	4.60 × 10^{-8}	400

Table 12.11 Permeability of Various Gases at 21.5 °C through DuPont™ Hytrel® 7246 Thermoplastic Copolyester Elastomer[9]

Gas	Permeability Coefficient	
	Source Document Units [cm^3(STP) mm/m^2 s Pa]	Normalized Units (cm^3 mm/m^2 day atm)
Helium	3.20 × 10^{-8}	280
Freon® 12	8.20 × 10^{-9}	72
Freon® 114	2.70 × 10^{-8}	240

Table 12.12 Permeability of Oxygen at 23 °C through Ticona Riteflex® 663 Thermoplastic Copolyester Elastomer[9]

Test Conditions	Permeability Coefficient			
	Source Document Units (cm^3/100 $in.^2$ day atm)		Normalized Units (cm^3 mm/m^2 day atm)	
Film cast at temperature	38 °C	93 °C	38 °C	93 °C
0% RH	68.4	63.0	37.1	34.2
50% RH	68.4	64.0	37.1	34.7
100% RH	71.7	65.9	38.9	35.8

Cast film thickness: 0.035 mm.

Table 12.13 Permeability of Oxygen and Carbon Dioxide through Eastman Ecdel® 9966 Thermoplastic Copolyester Elastomer[10]

Gas	Temperature (°C)	Normalized Units, Permeability Coefficient (cm^3 mm/m^2 day atm)
Carbon dioxide	23	>1000
Oxygen	30	130

RH: 50%. Test method: ASTM D1434.

12.4 Thermoplastic Polyether Block Polyamide Elastomers (PEBA)

Polyether block amides are plasticizer-free TPEs.[2] The soft segment is the polyether and the hard segment is the polyamide (nylon). For example, Arkema PEBAX® 33 series products are based on nylon 12 (see Section 8.4) and polytetramethylene glycol segments (PTMG). They are easy to process by injection molding and profile or film extrusion. Often they can be easily melt-blended with other polymers, and many compounders will provide custom products by doing this. Their chemistry allows them to achieve a wide range of physical and mechanical properties by varying the monomeric block types and ratios.

- Light weight
- Great flexibility (extensive range)
- Resiliency
- Very good dynamic properties
- High strength
- Outstanding impact resistance properties at low temperature
- Easy processing
- Good resistance to most chemicals

Manufacturers and trade names: Arkema PEBAX®, EMS-Grivory Grilflex®.

Applications and uses: Medical. Surgical garments and sheeting; Textile. Sports, leisure, and workwear; Construction. Membranes, housewrap; and food and agriculture packaging (Tables 12.14–12.21).

See also Figs. 12.8 and 12.9.

Table 12.14 Water Vapor Permeation at 38 °C and 50% RH through Arkema PEBAX® Films[11]

PEBAX	Vapor Transmission Rate					
	Source Document Units (g/m² day)			Normalized Units (g mm/m² day)		
Thickness (mm)	0.012	0.025	0.050	0.012	0.025	0.050
MX1205	3000	1800	1400	36	45	70
MV1041	18,000	12,000	7000	216	300	350
MV3000	28,000	22,000	18,000	336	550	900
MV1074	30,000	25,000	21,000	360	625	1050
MV6100		6000			150	

Test method: ASTM E96 E.

Table 12.15 Water Vapor Permeation at 38 °C and 90% RH through Arkema PEBAX® Films[10]

PEBAX	Vapor Transmission Rate					
	Source Document Units (g/m² day)			Normalized Units (g mm/m² day)		
Thickness (mm)	0.012	0.025	0.050	0.012	0.025	0.050
MX1205	3000	1800	1400	36	45	70
MV1041	3500	2700	1800	42	68	90
MV3000	4500	3300	2200	54	82	110
MV1074	4800	4300	3600	57	107	180

Test method: ASTM E96 E.

Table 12.16 Permeability of Oxygen at 23 °C and 0% RH through Arkema PEBAX® Films[10,12]

PEBAX	Permeability Coefficient	
	Source Document Units (cm^3 mm/ cm^2 s cm Hg)	Normalized Units (cm^3 mm/ cm^2 day atm)
3533	131×10^{-10}	860
2533	150×10^{-10}	985
5533	35×10^{-10}	230
4033	59×10^{-10}	387
6333	31×10^{-10}	204
	(cm^3 25 μm/cm^2 day atm)	(cm^3 mm/cm^2 day atm)
MV3000	18,500	463
MV6100	6500	163
MV1205	24,000	600
PEBD	6000	150

Table 12.17 Permeability of Carbon Dioxide at 23 °C and 0% RH through Arkema PEBAX® Films[10]

PEBAX	Permeability Coefficient	
	Source Document Units (cm^3 mm/ cm^2 s cm Hg)	Normalized Units (cm^3 mm/ cm^2 day atm)
3533	1790×10^{-10}	11,800
2533	2600×10^{-10}	17,100
5533	500×10^{-10}	3280
4033	780×10^{-10}	5122
6333	420×10^{-10}	2760
	(cm^3 25 μm/ cm^2 day atm)	(cm^3 mm/ cm^2 day atm)
MV3000	177,000	4425
MV6100	72,000	1800
MV1205	175,000	4375
PEBD	27,000	675

Table 12.18 Permeability of Nitrogen at 23 °C and 0% RH through Arkema PEBAX® Films[10]

PEBAX	Permeability Coefficient	
	Source Document Units (cm^3 mm/ cm^2 s cm Hg)	Normalized Units (cm^3 mm/ cm^2 day atm)
3533	100×10^{-10}	657
2533	170×10^{-10}	1116
5533	11×10^{-10}	72
4033	39×10^{-10}	256
6333	5×10^{-10}	33
	(cm^3 25 μm/ cm^2 day atm)	(cm^3 mm/ cm^2 day atm)
MV3000	3000	75
MV6100	650	17
MV1205	3900	98
PEBD	2800	70

Table 12.19 Permeability of Helium at 23 °C and 0% RH through Arkema PEBAX® Films[10]

PEBAX	Permeability Coefficient	
	Source Document Units (cm^3 mm/ cm^2 s cm Hg)	Normalized Units (cm^3 mm/ cm^2 day atm)
3533	174×10^{-10}	1142
2533	235×10^{-10}	1543
5533	70×10^{-10}	460
4033	147×10^{-10}	965
6333	46×10^{-10}	302

12: Elastomers and Rubbers

Table 12.20 Permeability of Propane at 23 °C and 0% RH through Arkema PEBAX® Films[10]

PEBAX	Permeability Coefficient	
	Source Document Units (cm³ mm/cm² s cm Hg)	Normalized Units (cm³ mm/cm² day atm)
5533	120 × 10⁻¹⁰	789
6333	36 × 10⁻¹⁰	236

Table 12.21 Water Vapor Permeability at 38 C and 100% RH through Arkema PEBAX® Films[10]

PEBAX	Normalized Units, Vapor Transmission Rate (g mm/m² day)
6333	31
5533	34
4033	38
3533	67
2533	89

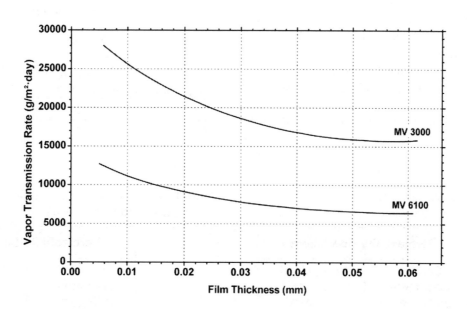

Figure 12.8 Water vapor permeation vs. film thickness at 38 °C and 50% RH through Arkema PEBAX® breathable PEBA films per ASTM E96.[13]

Figure 12.9 Water vapor permeation vs. film thickness at 38 °C and 90% RH through Arkema PEBAX® breathable PEBA films per ASTM E96.[11]

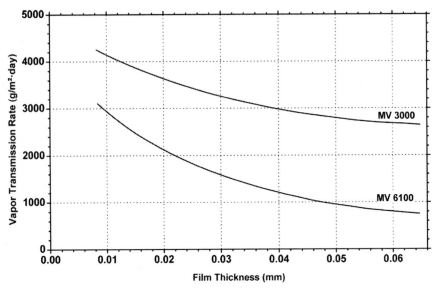

12.5 Styrenic Block Copolymer (SBC) TPEs

Styrenic block copolymer (SBC) TPEs are multiphase compositions in which the phases are chemically bonded by block copolymerization (see Section 2.2). At least one of the phases is a hard styrenic polymer. This styrenic phase may become fluid when the TPE composition is heated. Another phase is a softer elastomeric material that is rubber-like at room temperature. The polystyrene blocks act as cross-links, tying the elastomeric chains together in a three-dimensional network. SBC TPEs have no commercial applications when the product is just a pure polymer. They must be compounded with other polymers, oils, fillers, and additives to have any commercial value.

Manufacturers and trade names: BASF Styrolux®.

Applications and uses: Primarily food packaging, packed fruits and vegetables, fresh pasta and cheese, as thermoformed cups and lids, and also in applications including shrink film, must stay fresh as long as possible. Styrolux co-extruded with other thermoplastics, provides transparent barrier-layer composites (Tables 12.22–12.24).

Table 12.23 Permeability of Nitrogen at 23 °C through BASF Styrolux® Films[12]

Styrolux®	Permeability Coefficient	
	Source Document Units (cm³/m² day bar)	Normalized Units (cm³ mm/cm² day atm)
684 D	700	70.9
656 C	350	35.5

Thickness: 0.1 mm.

Table 12.22 Permeability of Oxygen at 23 °C through BASF Styrolux® Films[14]

Styrolux®	Permeability Coefficient	
	Source Document Units (cm³/m² day bar)	Normalized Units (cm³ mm/cm² day atm)
684 D	2600	263
656 C	1600	162

Thickness: 0.1 mm.

Table 12.24 Permeability of Carbon Dioxide at 23 °C through BASF Styrolux® Films[12]

Styrolux®	Permeability Coefficient	
	Source Document Units (cm³/m² day bar)	Normalized Units (cm³ mm/cm² day atm)
684 D	15,000	1520
656 C	8000	811

Thickness: 0.1 mm.

12.6 Ethylene Acrylic Elastomers (AEM)

AEM is mainly a copolymer of ethylene and methyl acrylate. These copolymers are generally cured with peroxides. Some AEM polymers add small amounts of a third carboxylic monomer to provide cure sites reactive to certain amines. AEM polymers do not have any crystallinity and the properties are related to the ratio of ethylene to methyl acrylate monomers used to make the polymer. The ethylene imparts good low-temperature properties, while the methyl acrylate content provides oil resistance (see Fig. 12.10).

Cured properties of AEM include:

- High-temperature durability
- Good oil resistance with service lubricants
- Excellent water resistance
- Good low-temperature flexibility
- Outstanding ozone/weather resistance

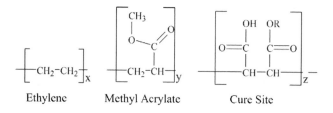

Figure 12.10 Structural units that are in DuPont Elastomers Vamac® AEM.

- Good mechanical strength
- Very good compression set resistance
- Good flex resistance
- Vibration-damping consistency
- Low permeability to many gases
- Colorability
- Nonhalogen, low-smoke emissions

Manufacturers and trade names: Ferro—Advanced Polymer Alloys, Alcryn®; DuPont Vamac® (Tables 12.25–12.28).

Table 12.25 Permeation of Various Gases at 0 °C through Ferro—Advanced Polymer Alloys Alcryn® 1170BK AEM[15]

	Permeability Coefficient	
Permeant Gas	**Source Document Units** [cm³ (STP) 0.5 mm/m² atm day]	**Normalized Units** (cm³(STP) mm/m² atm day)
Air	64	32
Nitrogen	48	24
Oxygen	598	299
Propane	192	96
Butane	174	87

Table 12.26 Permeation of Various Freon® Gases at 0 °C through Ferro—Advanced Polymer Alloys Alcryn® AEM[14]

	Permeability Coefficient			
	Source Document Units [cm³(STP) × 10¹⁰/s cm² cm Hg]		**Normalized Units** (cm³ mm/m² day atm)	
Product Code	**Freon® 12**	**Freon® 22**	**Freon® 12**	**Freon® 22**
Alcryn® 1060BK	5.3	32.3	348	2121
Alcryn® 1070BK	7.27	23.6	477	1550
Alcryn® 1080BK	7.07	39.9	464	2620

Table 12.27 Permeation of Various Liquids at 23 °C through Ferro—Advanced Polymer Alloys Alcryn® AEM[14]

Liquids	Vapor Transmission Rate	
	Source Document Units (g 30 mil/day m^2)	Normalized Units (g mm/m^2 day)
Distilled water (0–5 days)	2.4	1.8288
Distilled water (5–8 days)	6.8	5.1816
Reference fuel B	1820	1386.84
Exxon unleaded gasoline	1795	1367.79
Diesel fuel	70	53.34
ASTM #3 Oil	0	0

Test method: ASTM E96.

Table 12.28 Permeation of Various Gases at 23 °C through DuPont Elastomers Vamac® AEM[16]

Permeant Gas	Permeation Coefficient	
	Source Document Units (10^{-8} cm^3 cm/cm^2 atm s)	Normalized Units (cm^3 mm/m^2 atm day)
Air	0.3	25.9
Nitrogen	0.3	25.9
Methane	1	86
Freon® 12	1	86
Oxygen	1	86
Freon® 22	5	432
Carbon dioxide	7	605
Hydrogen	2.9	251
Helium	2.7	233

12.7 Bromobutyl Rubber

Bromobutyl rubber is a isobutylene–isoprene copolymer (halogenated butyl) containing reactive bromine replacing one of the hydrogens on some of the isoprene units, which are shown in Fig. 12.11. Bromobutyl rubber has a CAS number of 68441-14-5.

The bromine provides a crosslinking site. Crosslinking occurs by reaction of the bromines on different polymer chains or within the same polymer chain.[2] See Fig. 12.12 and the reference for chemistry details.

Manufacturers and trade names: Lanxess Bromobutyl.

Applications and uses: It is mainly used in tubeless tire inner liners. Nontire applications include conveyor belts for high-temperature resistance, tank linings for chemical resistance, and pharmaceutical closures and adhesives (Tables 12.29 and 12.30).

See also Figs. 12.13 and 12.14.

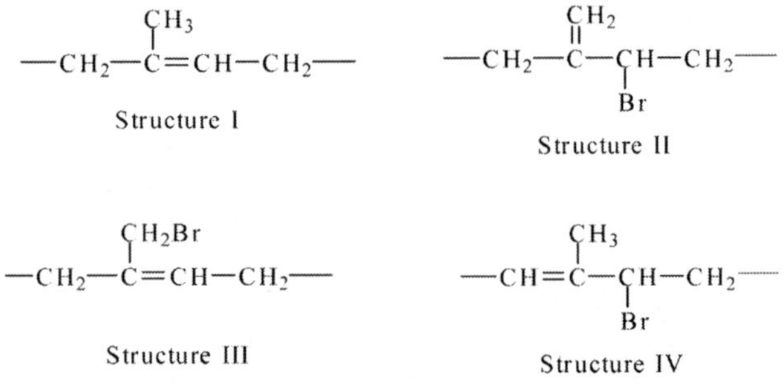

$$-CH_2-\underset{\text{Structure I}}{\overset{\overset{CH_3}{|}}{C}=CH-CH_2}- \quad -CH_2-\underset{\text{Structure II}}{\overset{\overset{CH_2}{\|}}{C}-\underset{\overset{|}{Br}}{C}H-CH_2}-$$

$$-CH_2-\underset{\text{Structure III}}{\overset{\overset{CH_2Br}{|}}{C}=CH-CH_2}- \quad -CH=\underset{\text{Structure IV}}{\overset{\overset{CH_3}{|}}{C}-\underset{\overset{|}{Br}}{C}H-CH_2}-$$

Figure 12.11 Bromobutyl rubber is a random mixture of the following isoprene-based structural units.

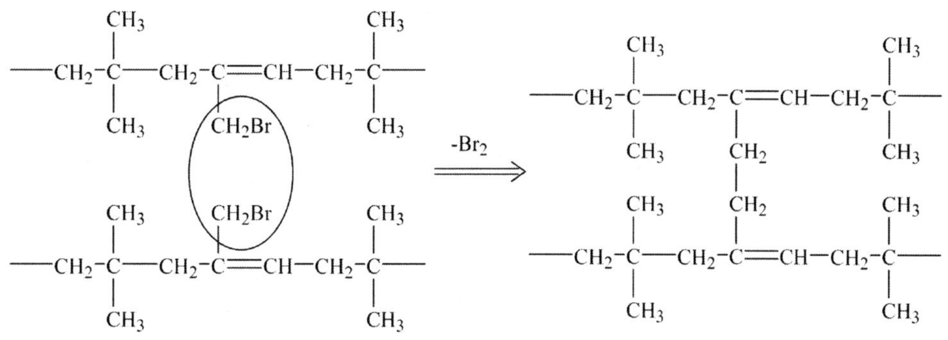

Figure 12.12 Bromobutyl rubber crosslinking (vulcanization) occurs through the bromine atoms on adjacent polymer chains.

Table 12.29 Permeation of Oxygen Gas through ExxonMobil Chemical Model Formula Compounds[17]

Grade	Source Document Units, Transmission Rate (cm³/m² day)	Normalized Units, Permeability Coefficient (cm³ mm/m² day atm)
At 40 °C		
Butyl Grade 065	120.0	135
Butyl Grade 068	126.0	141
Butyl Grade 165	121.0	136
Butyl Grade 0268	121.0	136
Butyl Grade 0269	118.0	129
At 60 °C		
Butyl Grade 2222		593
Butyl Grade 2235		570
Butyl Grade 2235		570
Butyl Grade 2255		540

Test method: Mocon® permeability test, ExxonMobil method, oxygen 21% concentration.

Table 12.30 Permeation of Oxygen Gas at 60 °C through ExxonMobil Chemical Model Formulations for Automobile Tire Innerliners[18]

Grade	Normalized Units, Permeability Coefficient (cm³ mm/m² day atm)
100% Butyl Grade 2222	530
90% Butyl Grade 2222 + 10% Natural Rubber TSR20	760
80% Butyl Grade 2222 + 20% Natural Rubber TSR20	910
70% Butyl Grade 2222 + 30% Natural Rubber TSR20	1060
60% Butyl Grade 2222 + 40% Natural Rubber TSR20	1290

Test method: Mocon permeability test, ExxonMobil method, oxygen 21% concentration.

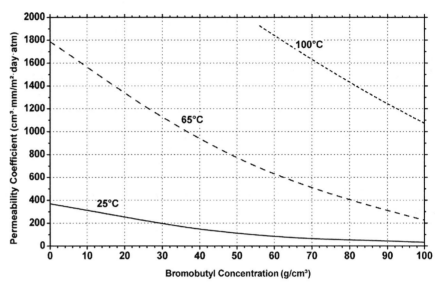

Figure 12.13 Permeation of air vs. bromobutyl concentration through bromoisobutylene–isoprene copolymer.[19]

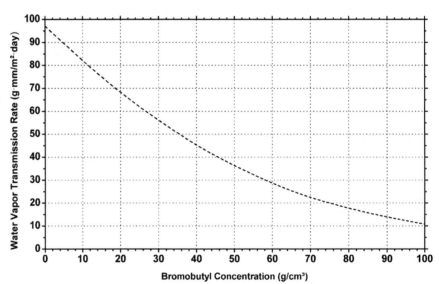

Figure 12.14 Permeation of moisture vapor vs. bromobutyl concentration through bromoisobutylene–isoprene copolymer.[18]

12.8 Butyl Rubber

Butyl is an elastomeric copolymer of isobutylene with small amounts of isoprene (1–2.5 mol%). Its structure is shown in Fig. 12.15 and its CAS Number is 9010-85-9.

Manufacturers and trade names: Lanxess Butyl, ExxonMobil Chemical (Polysar is obsolete).

Applications and uses: Major application area is the tire industry, mainly used for inner tubes and tire-curing bladders. Nontire applications include pharmaceutical closures, roof membranes, body mounts, and tank linings (Tables 12.31–12.34).

See also Fig. 12.16.

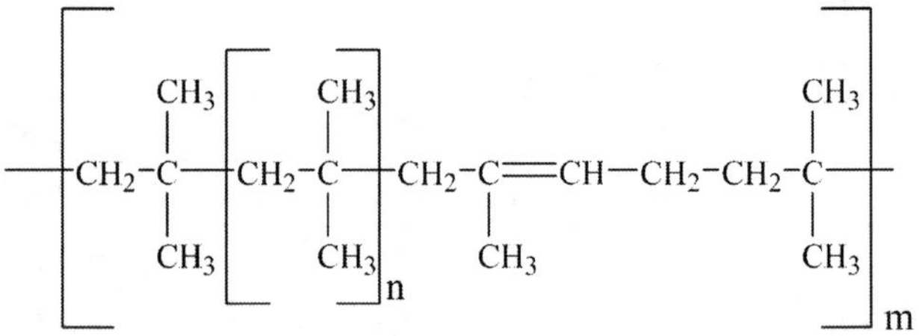

Figure 12.15 Chemical structure of butyl rubber.

12: Elastomers and Rubbers

Table 12.31 Permeation of Air vs. Temperature through Butyl Rubber[20,21]

Temperature (°C)	Permeability Coefficient (cm³ mm/m² day atm)	
	Polysar Butyl 301[a]	Exxon Butyl 268
24		36.4
40	51.8	
60	155	
66		375
80	3974	
93		1195

Table 12.32 Permeation of Various Gases at 0 °C through Butyl Rubber[15]

Permeant Gas	Permeability Coefficient		
	Source Document Units		Normalized Units (cm³ mm/ m² day atm)
	[cm³(STP) 0.5 mm/ m² day atm]	[10¹⁰ cm³ (STP) cm/ cm² s cm Hg]	
Air	45		23
Nitrogen	50		25
Oxygen	195		98
Propane	475		238
Butane	4360		2180
Freon® 12		22.8	1497
Freon® 22		33.6	2206

Table 12.33 Permeation of Oxygen at 40 °C through EXXON™ Butyl Rubber

Material Grade	Permeability Coefficient	
	Source Document Units (cm³ mm/m² day mm Hg)	Normalized Units (cm³ mm/m² day atm)
EXXON™ Butyl Grade 268	0.179	136
EXXON™ Butyl Grade 269	0.174	132
EXXON™ Butyl Grade 165	0.179	136
EXXON™ Butyl Grade 068	0.186	141
EXXON™ Butyl Grade 065	0.178	135

Oxygen concentration: 21%. Test method: Mocon Permeability Test, Exxon™ procedure.

Table 12.34 Permeability of Gas vs. Temperature through Bayer Lanxess Butyl® Butyl Rubber[22]

Permeant Gas	Temperature (°C)	Permeability Coefficient	
		Source Document Units (cm²/s bar) × 10⁻⁹	Normalized Units (cm³ mm/m² day atm)
Air	60	20	176
	80	50	438
Nitrogen	60	15	131
	80	35	306
Carbon dioxide	60	130	1138
	80	290	2539

Figure 12.16 Permeation of air vs. temperature through butyl rubber.[23]

12.9 Chlorobutyl Rubber (Polychloroprene)

Chlorobutyl rubber or polychloroprene is elastomeric isobutylene–isoprene copolymer (halogenated butyl) containing reactive chlorine. Polychloroprene was developed in 1930 by DuPont™ and is best known under the name neoprene. The polymer is made from chloroprene and its structure is given in Fig. 12.17. Its CAS number is 9010-98-4. The polymer is often modified to permit some degree of polymerization. Sulfur is a common modifier and the compounds are often called vulcanizates.

DuPont elastomer neoprene is available in many varieties including nonsulfur modified "W" and the more common sulfur modified "GN" types. Polychloroprene is known for its resistance to oil, gasoline, sunlight, ozone, and oxidation though there are other polymers that have better resistance to these same elements. Polychloroprene's advantage is its ability to combine these properties moderately into one all-purpose polymer.

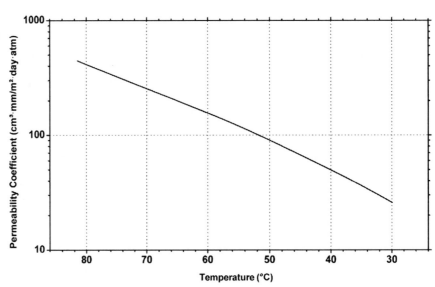

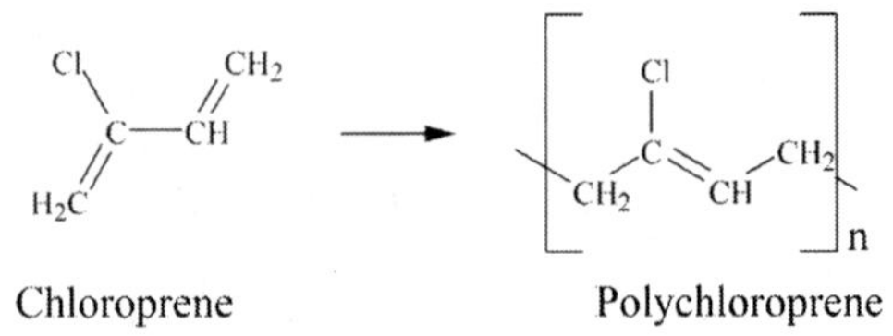

Chloroprene   Polychloroprene

Figure 12.17 Polymerization of chloroprene.

Table 12.35 Permeation of Gases through Lanxess Baypren® Polychloroprene Rubber Vulcanizate[24]

Permeant Gas	Temperature (°C)	Permeability Coefficient	
		Source Document Units (cm^2/s bar) × 10^{-9}	Normalized Units (cm^3 mm/m^2 day atm)
Air	60	70	613
	80	120	1051
Nitrogen	60	40	350
	80	80	700
Carbon dioxide	60	580	5078
	80	760	6653

Manufacturers and trade names: DuPont™ Performance Elastomers Neoprene (discontinued), Exxon™ Chlorobutyl, Lanxess Baypren®.

Applications and uses: Chlorobutyl's major application area is the tire industry. It is mainly used in tubeless tire inner liners, sidewalls, and inner tubes. Other applications include conveyor belts requiring high-temperature resistance, tank linings for chemical resistance, and pharmaceutical closures and adhesives, gloves, adhesives, binders, coatings, dipped goods, elasticized asphalt, and concrete (Tables 12.35–39).

Table 12.36 Permeation of gases at 23 °C through DuPont™ Performance Elastomers Neoprene Polychloroprene Rubber Vulcanizate[15]

Permeant Gas	Permeability Coefficient	
	Source Document Units [cm^3(STP)/m^2 day atm]	Normalized Units (cm^3 mm/m^2 day atm)
Air	210	105
Nitrogen	170	85
Oxygen	595	298
Methane	380	190
Propane	2100	445
Butane	6150	3075
Hydrogen	2185	1093
Helium	1800	900
Ammonia	26,350	13,175
Sulfur dioxide	40,120	20,060
Freon® 12	1155	578
Freon® 22	2415	1208

Thickness: 0.5 mm.

Table 12.37 Permeation of Vapors at 23 °C through DuPont™ Performance Elastomers Neoprene Polychloroprene Rubber Vulcanizate[22]

Permeant Vapor	Vapor Transmission Rate	
	Source Document Units (g/m² day)	Normalized Units (g mm/m² day)
Water (0–5 days)	1.2	0.7
Water (0–5 days)	3.4	2.6
Reference fuel B	2875	2191
Exxon unleaded gasoline	2960	2256
Diesel fuel	215	164
ASTM #3 oil	0	0

Thickness: 0.762 mm.

Table 12.38 Permeability of Various Gases at 23 °C and One Atmosphere Pressure Differential through Chlorobutyl Rubber[25]

Permeant Gas	Permeability Coefficient	
	Source Document Units, ×10^{-8} (cm³ cm/cm² s atm)	Normalized Units (cm³ mm/m² day atm)
Air	1.1	95
Nitrogen	1.0	86
Methane	2.9	248
Freon® 12	6.7	576
Oxygen	3.2	273
Freon® 22	14.2	1224
Carbon dioxide	22.5	1940
Hydrogen	12.6	1086
Helium	10.5	905

Table 12.39 Permeability of Air vs. Temperature through Exxon™ Chlorobutyl 1068 Chlorobutyl Rubber[21]

Temperature (°C)	23.9	65.6	93.3
Source document units, permeability coefficient (ft³ mil/ft² day psi)	0.00034	0.0032	0.0104
Normalized units, permeability coefficient (cm³ mm/m² day atm)	38.7	364	1183

Test apparatus: Aminco permeability apparatus.

12.10 Ethylene–Propylene Rubbers (EPM, EPDM)

There are two basic types of ethylene–propylene rubber available. ASTM classifies this synthetic elastomer as "EPM," meaning that it has a saturated polymer chain of the polymethylene type. Within this classification there are two basic kinds of ethylene–propylene rubber:

- EPM, the copolymer of ethylene and propylene.
- EPDM, the terpolymer of ethylene, propylene, and a nonconjugated diene with residual unsaturation in the side chain.

Manufacturers and trade names: Exxon™ Vistalon™ and Lanxess Buna® EP.

Applications and uses: Impact modification, hose, tubing, weather strips, insulation, jacketing, single-ply roofing sheet, window gaskets, sound deadening, solar pool panels, and face respirators (Tables 12.40 and 12.41).

See also Fig. 12.18.

Table 12.40 Permeation of Various Gases at 0 °C and One Atmosphere Pressure Differential through EPDM Rubber[15]

Permeant Gas	Permeability Coefficient	
	Source Document Units [cm^3(STP) 0.5 mm/m^2 day atm]	Normalized Units (cm^3 mm/m^2 day atm)
Air	1615	808
Oxygen	3470	1735
Nitrogen	1180	590
Carbon dioxide	15,385	7693
Helium	2720	1360

Table 12.41 Permeability of Air vs. Temperature through Exxon Vistalon™ Ethylene–Propylene–Diene Copolymer (EPM) Rubbers[21]

Product Code	Vistalon™ 404			Vistalon™ 4608		
Temperature (°C)	23.9	65.6	93.3	23.9	65.6	93.3
Source document units, permeability coefficient (ft^3 mil/ft^2 day psi)	0.00405	0.0225	0.0637	0.00587	0.029	0.0619
Normalized units, permeability coefficient (cm^3 mm/m^2 day atm)	461	2560	7247	668	3299	7043

Test apparatus: Aminco permeability apparatus.

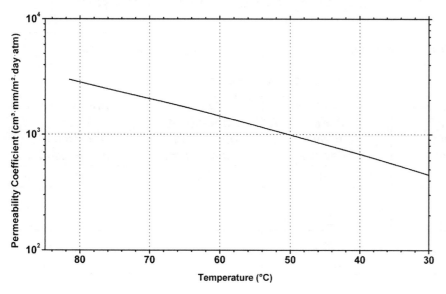

Figure 12.18 Air vs. temperature through ethylene–propylene–diene copolymer rubber.[20]

12.11 Epichlorohydrin Rubber (CO, ECO)

Epichlorohydrin rubbers include homopolymer of epichlorohydrin and a copolymer with ethylene oxide. There is also a terpolymer of epichlorohydrin, ethylene oxide, and a monomer to introduce a carbon double bond to the chain that functions as a cure site. CO is the ISO designation for the homopolymer and ECO is the ISO designation for the ethylene oxide copolymer. GECO is the designation for the terpolymer (usually allyl glycidyl ether). The monomers and polymer structures are given in Figs 12.19–12.21.

The ratios of the monomers are tailored to provide the desired properties. The epichlorohydrin monomer provides heat and ozone resistance, fuel and oil resistance, and gas permeation resistance.

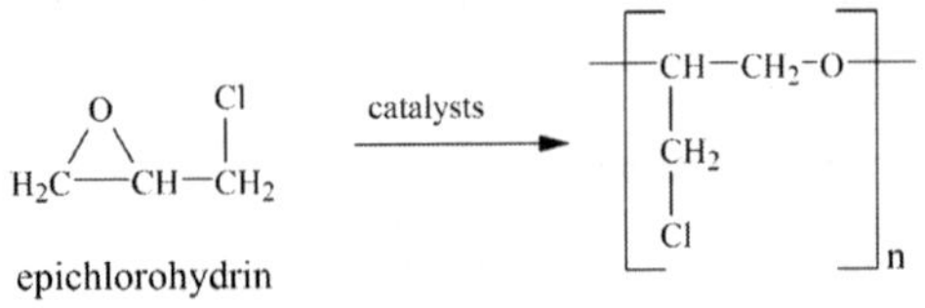

Figure 12.19 Epichlorohydrin homopolymer (CO) monomer and polymer structure.

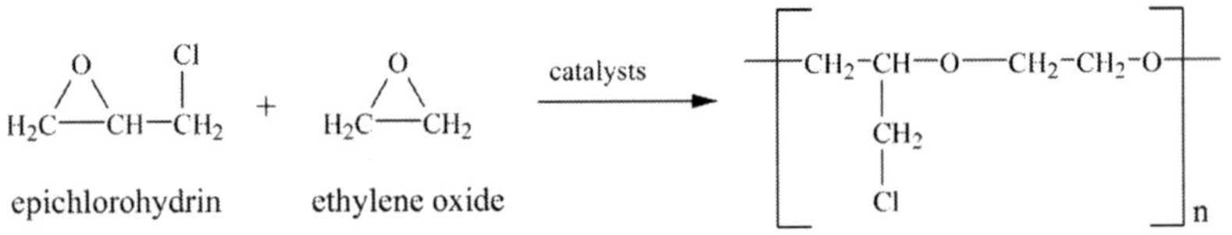

Figure 12.20 Epichlorohydrin copolymer (ECO) monomers and polymer structure.

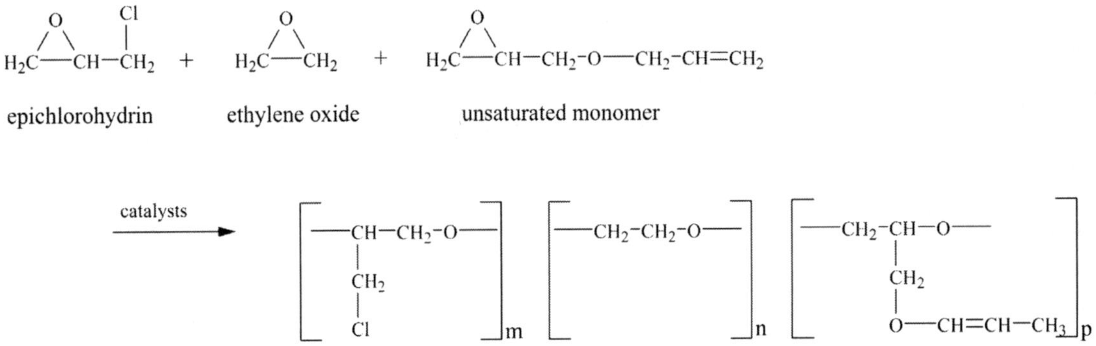

Figure 12.21 Epichlorohydrin terpolymer (GECO) monomers and polymer structures.

Table 12.42 Permeability of Various Gases at Room Temperature through Zeus Chemicals Hydrin® Epichlorohydrin Polymers[26]

Permeant Gas	Hydrin® Homopolymer		Hydrin® Copolymer	
	Permeation Coefficient			
	Source Document Units, ×10^{-8} (cm³ cm/ cm² s atm)	Normalized Units (cm³ mm/ m² day atm)	Source Document Units, ×10^{-8} (cm²/s atm)	Normalized Units (cm³ mm/ m² day atm)
Air	0.105	9	0.525	45
Nitrogen	0.048	4		
Oxygen	0.255	22		
Helium	1.97	170		
Carbon dioxide	2.6	225		

Table 12.43 Water Vapor Transmission Rate at 38 °C through Zeus Chemicals Hydrin® Epichlorohydrin Polymers[26]

Polymer Type	Vapor Transmission Rate (g/m² day)
Hydrin® homopolymer	15.5
Hydrin® copolymer	85.9

Thickness not known.

Table 12.44 Comparison of Air and Nitrogen Permeability through Different Types of Zeon Chemicals Hydrin® Epichlorohydrin Polymers[27]

Polymer Type	Permeability Coefficient (cm³ mm/m² day atm)	
	Air	Nitrogen
Homopolymer Hydrin® H	15	12
Copolymer Hydrin® C2000	43	38
Terpolymer Hydrin® T3102	26	30

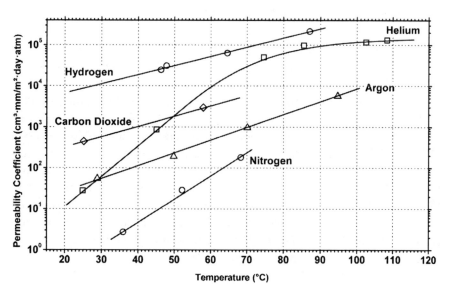

Figure 12.22 Permeation vs. temperature of gases through epichlorohydrin rubber.[28]

The ethylene monomer lowers the glass transition temperature, reduces heat resistance, and imparts static dissipative properties. The unsaturated monomer introduces a site for sulfur and peroxide curing.

Manufacturers and trade names: Hercules, Inc. Herclor, B.F. Goodrich Co., and Zeon Chemicals, Hydrin®.

Applications and uses: Fuel hoses/liners (cells), air ducts, emissions tubing, electrostatic dissipative rolls, low-temperature fuel handling curb hose, diaphragms, gaskets/O-rings, vibration dampers, dust boots, closed-cell sponge, and fabric coatings/belting (Tables 12.42–12.44).

See also Fig. 12.22.

12.12 Fluoroelastomers (FKM)

A fluoroelastomer is a special-purpose fluorocarbon-based synthetic rubber. When compared with most other elastomers, it has wide chemical resistance and superior performance, especially in high-

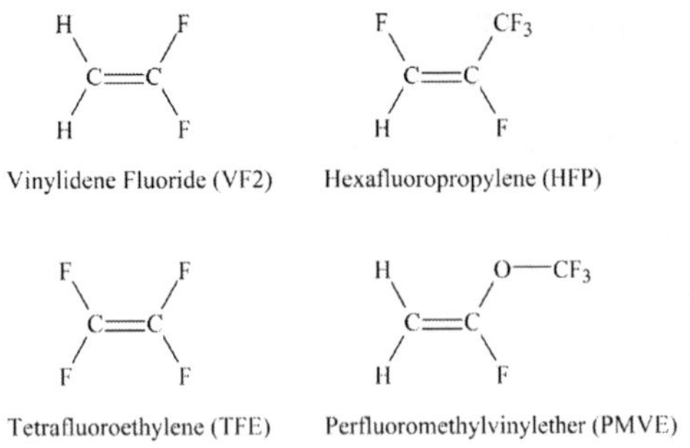

Figure 12.23 Monomers used to make fluoroelastomers.

temperature application in different media. Fluoroelastomers are categorized under the ASTM D1418 & ISO 1629 designation of FKM. The four main monomers that are used to make fluoroelastomers are shown in Fig. 12.23.

The fluorine content is an important parameter of each fluoroelastomer and is frequently reported in fact sheets. Most common grades have fluorine content that varies between 66 and 70%. Generally more fluorine means higher chemical resistance.

There are a number of types of fluoroelastomers.

- "A" Type: FKM-A is the most widely used polymer in industry today, and it is the most cost-effective polymer in relationship to performance. It has a fluorine level around 66%. This elastomer was designed in the late 1950s for the space program and today is widely used as a general-purpose FKM offering excellent fluid resistance to automotive fuels and lubricants, as well as elevated service temperatures.

- "B" Type: FKM-B is a terpolymer with an increased fluorine (68—69%) content is widely used throughout the chemical processing and power generation industries. "B" types are specified for gaskets sealing mineral acids such as sulfuric acids and other aggressive chemicals that are hauled by rail and bulk tankers. "B" types can be formulated with peroxide cure systems to resist strong acids, hot water, and steam.

- "F" Type: This terpolymer is the latest generation of "high" fluorine elastomers, with the addition of 2% more fluorine (70%). This is considered an excellent elastomer for sealing today's oxygenated automotive fuels and lubricants. "F" types can be formulated to resist concentrated aqueous inorganic acids, hot water, and steam.

- Viton® GF fluorocarbons are tetrapolymers composed of TFE, VF2, HFP, and small amounts of a cure site monomer. Presence of the cure site monomer allows peroxide curing of the compound, which is normally 70% fluorine. As the most fluid resistant of the various FKM types, Viton® GF compounds offer improved resistance to water, steam, and acids.

- Viton® GFLT fluorocarbons are similar to Viton® GF, except that perfluoromethylvinyl ether (PMVE) is used in place of HFP. The "LT" in Viton® GFLT stands for "low temperature." The combination of VF2, PMVE, TFE, and a cure site monomer is designed to retain both the superior chemical resistance and high-heat resistance of the G-series fluorocarbons. In addition, Viton® GFLT compounds (typically 67% fluorine) offer the lowest swell and the best low-temperature properties of the types discussed here Viton® GFLT can seal in a static application down to approximately −40 °F.

Table 12.45 Curing Chemistry of Fluoroelastomers

FKM Type	Monomers	Curable By	Recommended Curative
Copolymer	VF2, HFP	Amine and bisphenol	Bisphenol
Terpolymer	VF2, HFP, TFE	Amine and bisphenol	Bisphenol
Peroxide curable	VF2, HFP, TFE, CSM	Amine, bisphenol, and peroxide	Peroxide
Low temperature	VF2, HFP, TFE, PMVE, CSM	Amine, bisphenol, and peroxide	Peroxide

- FVMQ is a fluorosilicone and is covered in the chapter on polysiloxanes/silicones.
- FFKM (perfluoronated elastomers): This family of elastomers is widely known by the trade name that its inventors gave it, Kalrez®. It is essentially an elastomeric form of PTFE and retains the extreme chemical resistance at a temperature of PTFE up to 327 °C.
- AFLAS®, made by Asahi Glass Co., Ltd, is a copolymer of tetrafluoroethylene and propylene. The fluorine content is typically 57%.

The fluoroelastomers are cured by several chemical means as described in Tables 12.45 and 12.46.

Manufacturers and trade names: Dyneon Fluorel™, Solvay Solexis Tecnoflon®, DuPont Viton®, Kalrez®, Daikin Dai-el™, Asahi Glass AFLAS®.

Applications: Seals, caulks, coatings, vibration dampeners, expansion joints, gaskets, O-rings, piston seals, custom shapes, and stock rod and sheet (Tables 12.47–12.51).

See also Figs. 12.24–12.26.

Table 12.46 The Curing Chemistry Used by Solvay Solexis Tecnoflon® Fluoroelastomer Products

Grade	Fluorine Content (%)	10% Temperature of Retraction[a] (°C)	Cure Type
T 636	66	−19	Bisphenol A
L636	65	−21	Bisphenol A
PL 458	67	−24	Peroxide
PL 958	67	−24	Peroxide
PL 956	66	−26	Peroxide
PL 557	66	−29	Peroxide
PL 455	65	−30	Peroxide
PL 855	65	−30	Peroxide

[a]An industry standard for determining the ability of an elastomer to seal.

Table 12.47 Permeation of Air, Carbon Dioxide, Helium, Nitrogen, and Oxygen through DuPont Viton® Fluoroelastomer[29]

Penetrant	Air	Carbon Dioxide	Helium			Nitrogen	Oxygen
Temperature (°C)	24	30	24	121	204	24	30
Source document gas permeability (cm³ cm/ cm² s atm)	9.9×10^{-10}	5.9×10^{-8}	8.92×10^{-8}	1.74×10^{-6}	6.7×10^{-6}	5.4×10^{-10}	1.1×10^{-8}
Normalized permeability coefficient (cm³ mm/ m² day atm)	8.55	508	771	15,034	57,888	4.67	95.0

Pressure: 1 atm. Sample size: 1cm² × 1cm thick.

Table 12.48 Average Permeation Rates (Over 28 Days) of Fuel C, Ethanol 90:10 for Solvay Solexis Tecnoflon® FKM Fluoroelastomers[30]

	Permeation Rate (g mm/m² day)	
Grade	Fuel C, Ethanol 90:10	Fuel C, Ethanol 85:15
N-535	2.2	40.7
T 636	2.7	45.5
P 757	0.9	21.9
P 959	NM	4.6
P 958	3.4	16.0
PL 855	10.2	68.2

Table 12.49 Oxygen Permeation through Dyneon Fluorel™ Fluoroelastomer Grades

Grade	Type	% Fluorine	Permeability Coefficient (cm³ mm/ m² day atm)
22	Copolymer, 22% carbon	66	98
35	Copolymer, 35% carbon	66	98
FC-2110Q	Copolymer	65.9	98
FC-2110	Copolymer	65.9	98
FC-2121	Copolymer	65.9	98
FC-2123	Copolymer	65.9	98

Table 12.50 Permeation of Fuel Vapors through Various DuPont™ Elastomer Viton® Elastomers[31]

	Vapor Permeation Rate (g mm/m² day)					
Viton® Type	Fuel C	CE-10	CE-25	CE-50	CE-85	Ethanol
GLT-S	24.4	85.0	101.8	94.1	35.9	18.0
GBLT-S	20.1	52.4	64.8	56.0	20.1	9.8
GFLT-S	17.2	42.8	52.8	44.0	16.9	8.4
GF-S	4.9	11.3	13.7	8.4	3.7	1.7
F605C	5.5	14.0	16.1	9.9	4.6	1.7
VTR-9209-NPC	8.5	25.8	29.1	18.8	10.0	3.8

Table 12.51 Additional Permeation of Fuel Vapors through Various DuPont™ Elastomer Viton® Elastomers[32]

Material	Vapor Permeation Rate (g mm/m² day)			
	Fuel C at 23 °C	90% Fuel C, 10% Ethanol	85% Fuel C, 15% Methanol	Toluene at 40 °C
Fluorosilicone	455	584	635	
Viton® GLT	2.6	14	60	
Viton® AL	0.8	6.7	32	
Viton® A	0.8	7.5	36	49
Viton® GFLT	1.8	6.5	14	
Viton® B	0.7	4.1	12	
Viton® GF	0.7	1.1	3.0	7
Viton® ETP				14

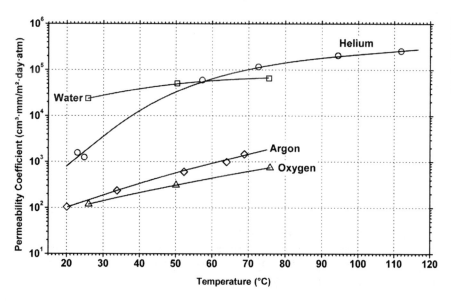

Figure 12.24 Permeation of water, helium, argon, and oxygen vs. temperature through DuPont™ advanced elastomers Viton® fluoroelastomer.[28,33]

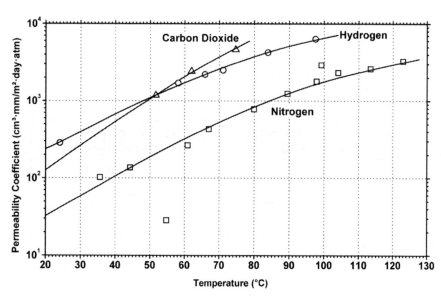

Figure 12.25 Permeation of carbon dioxide, hydrogen and nitrogen vs. temperature through DuPont™ advanced elastomers Viton® fluoroelastomer.[15,20]

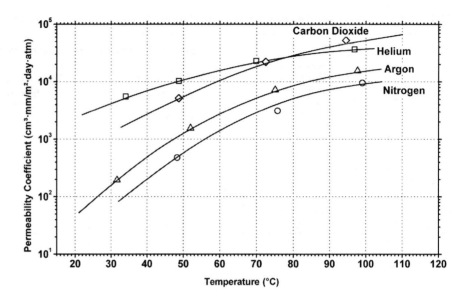

Figure 12.26 Permeation of gases vs. temperature through Asahi Glass AFLAS® fluoroelastomer.[15,20]

12.13 Natural Rubber

Natural rubber is polyisoprene. The structure of the monomer and polymer are given in Fig. 12.27. Its CAS number is 9006-04-6. Chemical and environmental resistance and mechanical properties are improved through crosslinking (vulcanizing), usually through treatment with sulfur.

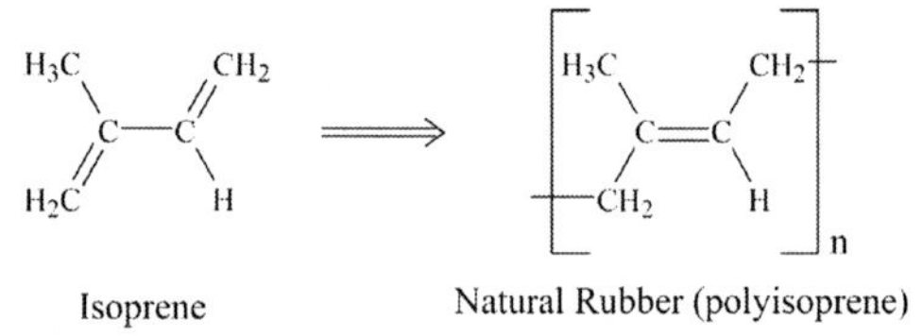

Figure 12.27 Structure of isoprene and polyisoprene.

Table 12.52 Relative Permeation of Various Gases through Natural Rubber[34]

Penetrant	Oxygen	Hydrogen	Carbon Dioxide	Nitrogen
Relative gas permeability (based on hydrogen as 100)	46	100	260	17

Table 12.53 Permeability of Gas vs. Temperature through Natural Rubber[15]

Permeant Gas	Temperature (°C)	Permeability Coefficient	
		Source Document Units (cm^2/s bar) × 10^{-9}	Normalized Units (cm^3 mm/m^2 day atm)
Air	60	250	2189
	80	400	3502
Nitrogen	60	180	1597
	80	330	2889
Carbon dioxide	60	1600	14,007
	80	2100	18,384

12: Elastomers and Rubbers

Natural rubber is more unsaturated and has fewer methyl groups than butyl rubber causing it to be 20 times more permeable to air. The presence of methyl groups generally serves to reduce the permeability of polymers.

Epoxidized natural rubber (ENR) is derived from the partial epoxidation of the natural rubber molecule, resulting in a totally new type of elastomer. The epoxide groups are randomly distributed along the natural rubber molecule.

Epoxidation results in a systematic increase in the polarity and glass transition temperature. Property changes with increasing level of epoxation include:

- an increase in damping
- a reduction in swelling in hydrocarbon oils
- a decrease in gas permeability
- an increase in silica reinforcement; improved compatibility with polar polymers like polyvinyl chloride
- reduced rolling resistance and increased wet grip

Applications and uses: Tire and other automotive (Tables 12.52–12.54).

See also Figs. 12.28–12.30.

Table 12.54 Permeation of Air vs. Temperature through Formulated Natural Rubber

	Polysar Formulation[20]			Exxon Formulation[21]		
Temperature (°C)	40	60	80	23.9	65.6	93.3
Source document units, gas permeability (ft^3 mil/ft^2 day psi)				0.00436	0.0237	0.0402
Source document units, gas permeability (cm^3 cm/cm^2 s atm)	11.8 × 10^{-8}	26.8 × 10^{-8}	43.9 × 10^{-8}			
Normalized units, permeability coefficient (cm^3 mm/m^2 day atm)	1020	2316	3793	496	2696	4574

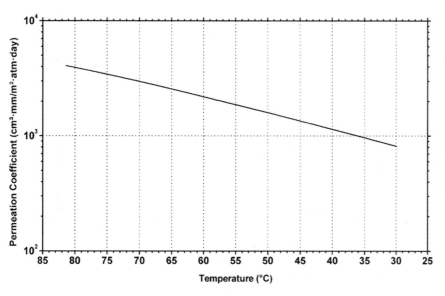

Figure 12.28 Permeation of air vs. temperature through natural rubber.[22]

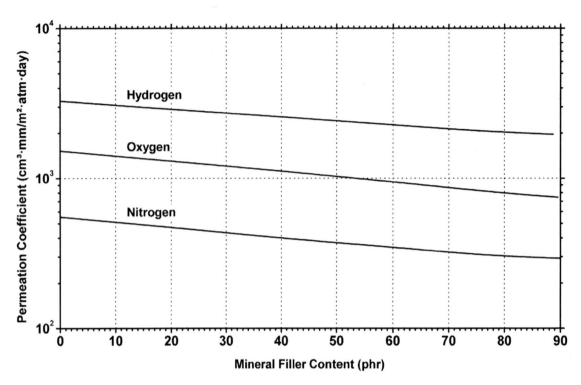

Figure 12.29 Permeation of various gases vs. mineral filler through natural rubber.[15]

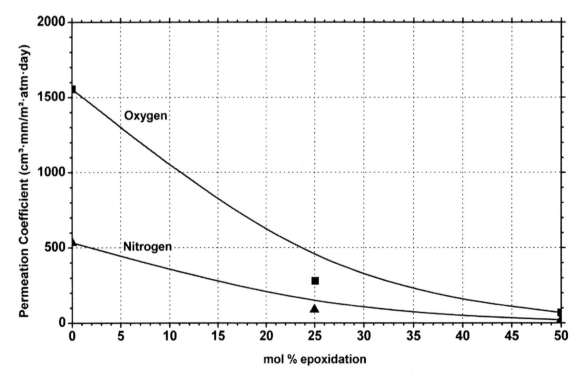

Figure 12.30 Permeation of oxygen and nitrogen vs. mol% epoxidation through natural rubber.[35]

12.14 Acrylonitrile–Butadiene Copolymer (NBR)

Acrylonitrile–butadiene copolymers (NBR) or more common nitrile rubbers are copolymers of butadiene and acrylonitrile. The monomers and polymer structure are shown in Fig. 12.31. The CAS number is 9003-18-3.

NBR is commonly considered the workhorse of the industrial and automotive rubber products industries. NBR is actually a complex family of unsaturated copolymers of acrylonitrile and butadiene. The amount of acrylonitrile in the polymer is used to manipulate the balance of NBR properties. Acrylonitrile content may range from 18 to 50%. Increasing acrylonitrile content leads to higher hardness, strength, abrasion resistance, heat resistance, and oil/fuel resistance, and lower resilience, and low-temperature flexibility.

There are several general types of NBR:

- Cold NBR—Acrylonitrile content ranges from 15 to 51%. Cold polymers are polymerized at a temperature range of 5 °C–15 °C, depending on the balance of linear-to-branched configuration desired. The lower-polymerization temperatures yield more-linear polymer chains.

- Hot NBR—Polymers are polymerized at the temperature range of 30 °C–40 °C. This process yields highly branched polymers. Branching supports good tack and a strong bond in adhesive applications. The physically entangled structure of this kind of polymer also provides a significant improvement in hot tear strength compared with a cold-polymerized counterpart. The hot polymers' natural resistance to flow makes them excellent candidates for compression molding and sponge. Other applications are thin-walled or complex extrusions where shape retention is important.

- Crosslinked Hot NBR—These are branched polymers that are further cross-linked by the addition of a difunctional monomer. These products are typically used in molded parts to provide sufficient molding forces, or back pressure, to eliminate trapped air. Another use is to provide increased dimensional stability or shape retention for extruded goods and calendered goods. This leads to more efficient extruding and vulcanization of intricate shaped parts as well as improved release from calender rolls. These NBRs also add dimensional stability, impact resistance, and flexibility for PVC modification.

- Carboxylated Nitrile (XNBR)—Addition of carboxylic acid groups to the NBR polymer's backbone significantly alters processing and cured properties. The result is a polymer matrix with significantly increased strength, measured by improved tensile, tear, modulus, and abrasion resistance. The negative effects include reduction in compression set, water resistance, resilience, and some low-temperature properties.

- Bound Antioxidant NBR—An antioxidant is polymerized into the polymer chain. The purpose is to provide additional protection for the NBR during prolonged fluid service or in cyclic fluid and air exposure. When compounding with highly reinforcing furnace carbon black the chemical reactivity between the polymer and the pigment can limit hot air aging capability. Abrasion resistance is improved when compared with conventional NBR, especially at elevated temperatures. They have also been found to exhibit excellent dynamic properties.

Manufacturers and trade names: Lanxess Perbunan®, Krynac® and Baymod® N, Girsa, Hyundai, JSR Corporation, Kumho, Nantex, Nitriflex, PetroChina, Petroflex, Polimeri Europa Europrene, Zeon Chemicals.

Applications and uses: Nonlatex gloves for the healthcare industry, automotive transmission belts, hoses, O-rings, gaskets, oil seals, V-belts, synthetic leather, printer's roller, and as cable jacketing (Tables 12.55–12.58).

See also Fig. 12.32.

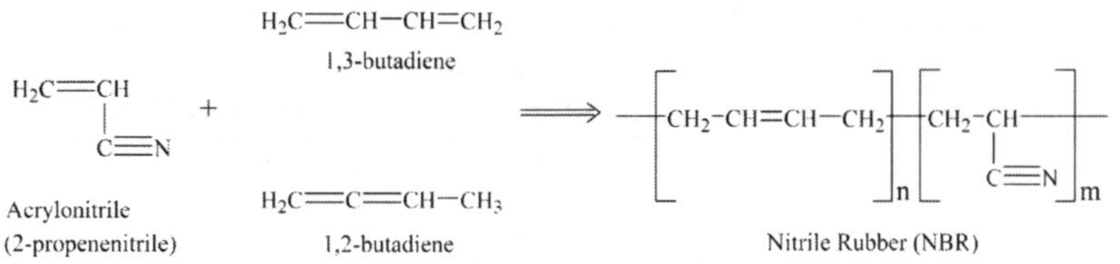

Figure 12.31 Monomers and polymer structure of NBR.

Table 12.55 Permeability of Air vs. Temperature through Bayer Lanxess Krynac® 800 Nitrile Rubber[21]

Temperature (°C)	40	60	80
Source document units, permeability coefficient (cm^3 cm/cm^2 s atm)	1.1×10^{-8}	4.1×10^{-8}	9.9×10^{-8}
Normalized units, permeability coefficient (cm^3 mm/m^2 day atm)	95	354	855

Table 12.56 Permeability of Gas vs. Temperature through Bayer Lanxess Krynac® and Perbunan® Nitrile Rubber[22]

Permeant Gas	Temperature (°C)	Permeability Coefficient	
		Source Document Units (cm^2/s bar) $\times 10^{-9}$	Normalized Units (cm^3 mm/m^2 day atm)
Air	60	25–75	219–657
	80	55–210	481–1838
Nitrogen	60	10–40	88–350
	80	25–70	219–613
Carbon dioxide	60	300–580	2626–5077
	80	480–970	4202–8492

Table 12.57 Permeability of Gases at 23 °C through High-Nitrile Content Nitrile Rubber[15]

Permeant Gas	Permeability Coefficient	
	Source Document Units [cm^3(STP)/m^2 day atm]	Normalized Units (cm^3 mm/m^2 day atm)
Air	220	110
Oxygen	145	78
Hydrogen	920	460
Nitrogen	45	23
Carbon dioxide	1165	583
Helium	950	475

Table 12.58 Permeation of Vapors at 23 °C through Low- and High-Nitrile Content Nitrile Rubber[5]

Permeant Vapor	Low-Nitrile Content NBR		High-Nitrile Content NBR	
	Vapor Transmission Rate			
	Source Document Units (g/m^2 day)	Normalized Units (g mm/m^2 day)	Source Document Units (g/m^2 day)	Normalized Units (g mm/m^2 day)
Water (0–5 days)	4.5	3.4	3.3	2.5
Water (0–5 days)	5.4	4.1	5.4	4.1

Table 12.58 (Continued)

Permeant Vapor	Low-Nitrile Content NBR		High-Nitrile Content NBR	
	Vapor Transmission Rate			
	Source Document Units (g/m² day)	Normalized Units (g mm/m² day)	Source Document Units (g/m² day)	Normalized Units (g mm/m² day)
Reference fuel B	1750	1333	390	297
Exxon unleaded gasoline	1930	1471	275	210
Diesel fuel	95	72	20	15
ASTM #3 oil	0	0	0	0

Thickness: 0.762 mm.

Figure 12.32 Air vs. temperature through Lanxess Krynac® 800 nitrile rubber.[15]

12.15 Styrene–Butadiene Rubber (SBR)

Styrene–butadiene or styrene–butadiene rubber (SBR)[2] is a synthetic rubber copolymer consisting of styrene and butadiene, its structure is shown in Fig. 12.33.

Manufacturers and trade names: Lanxess Krylene® and Krynol® and many others.

Applications and uses: Employed extensively in almost all sectors of the rubber industry. Used mainly for tires, often in blends with NR; conveyor and transmission belting, footwear soles and heels; technical goods of all kinds, for example, seals, membranes, hose, and rolls (Table 12.59).

See also Fig. 12.34.

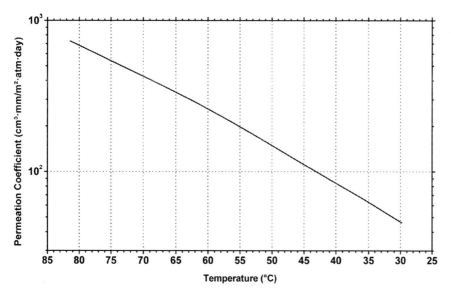

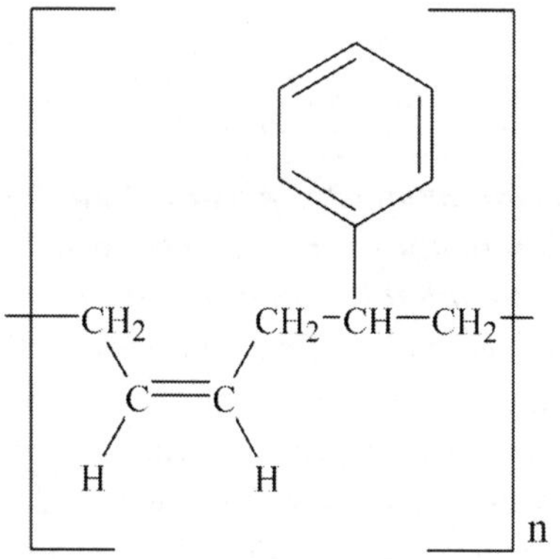

Figure 12.33 Structure of SBR.

Table 12.59 Permeability of Gas vs. Temperature through Bayer Lanxess Krylene® Styrene–Butadiene Rubber[15]

Permeant Gas	Temperature (°C)	Permeability Coefficient	
		Source Document Units (cm^2/s bar) × 10^{-9}	Normalized Units (cm^3 mm/m^2 day atm)
Air	60	150	1313
	80	260	2276
Nitrogen	60	110	963
	80	200	1751
Carbon dioxide	60	1200	10,505
	80	1500	13,132

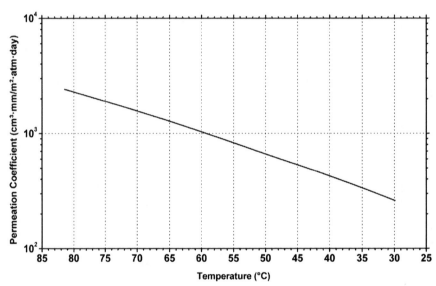

Figure 12.34 Permeation of air vs. temperature through styrene–butadiene rubber.[20]

References

1. McKeen LW. *The effect of temperature and other factors on plastics*. Plastics Design Library, William Andrew Publishing; 2008.
2. Drobny JG. *Handbook of thermoplastic elastomers*. William Andrew; 2007. p. 215–34, 191–9, 249–64, 235–47, 161–77.
3. *Estane thermoplastic polyurethane for film and sheet applications, E-30*. Noveon Inc.; 2001.
4. *Estane breathable—high moisture vapor transmission selection guide*. Lubrizol; 2007.
5. *Elastollan® material properties*. BASF; 2009.
6. *Texin and Desmopan. A guide to engineering properties*. Bayer MaterialScience; 2004.
7. *Santoprene® rubber physical properties guide*. Advanced Elastomer Systems; 2005.
8. *Hytrel® design guide—module V, H-81098*. DuPont™; 2000.
9. *Riteflex® brochure RF-001*. Ticona; 2006.
10. *Ecdel products and packaging for the medical industry*. PPM-201B Eastman; 1999.
11. *Pebax®, supplier design guide*. Atochem; 1987.
12. *Pebax® application areas*. Arkema; 2004.
13. *Pebax® breathable film*. Arkema; 2009.
14. *Styrolux product line, properties, processing, supplier design guide [B 583e/(950) 12.91]*. BASF Aktiengesellschaft; 1992.
15. *Liquid & gas permeability at room temperature, Alcryn melt-processible rubber tech notes*; 1997.
16. *Gas permeability of Vamac® ethylene acrylic elastomers*. DuPont Performance Elastomers; 2008.

17. ExxonMobil Chemical. Model formulation property sheets; 2008.
18. ExxonMobil Chemical. Exxon™ bromobutyl rubber compounding and applications manual; 2009.
19. Bromobutyl rubber optimizing key properties. Supplier marketing literature. Exxon Chemicals, 2001.
20. Polysar butyl rubbers handbook. Reference book. Miles Polysar, 1996.
21. *Elastomers Technical Information—Elastomer Permeability, supplier technical report (TI-20)*. Exxon Chemical Company; 1974.
22. *Baypren®, LXS-KA 029e, edition 2008—06*. Lanxess; 2008.
23. *LANXESS Butyl, LXS-NAKA 02us*. Lanxess; 2008.
24. *Baypren® Chloroprene, the primer all-round performer for a wide range of applications, edition 2008—06*. Lanxess; 2008.
25. *Vamac Gas Permeability Technical Information*. DuPont Performance Elastomers; 2008.
26. Hydrin® unique balance of properties. Zeon Chemicals L.P.; 4/07.
27. Seng. *Hydrin® Presentation*. Zeon Chemicals L.P.; 2006.
28. Laurenson L, Dennis N. Permeability of common elastomers for gases over a range of temperatures. *J Vacuum Science & Technology* 1985;**3**:1707—10.
29. *The engineering properties of viton fluoroelastomer. Supplier design guide (E-46315—1)*. DuPont Company; 1987.
30. *TECNOFLON®. A guide to fluoroelastomers*. Solvay Solexis; 2005.
31. Stevens RD. Fuel and permeation resistance of fluoroelastomers to ethanol blends; 2006.
32. PSP Inc. at www.pspglobal.com; 2009.
33. Reid R. Vacuum science and technology in accelerators. *Cockcroft Institute Lectures*; 2007 (Lecture 5).
34. Berins M. *Plastics engineering handbook of the society of the plastics industry, Inc*. 5th ed. New York: Society of the Plastics Industry, Van Nostrand Reinhold; 1991. p. 52—74.
35. Johnson T, Thomas S. Nitrogen/oxygen permeability of natural rubber, epoxidised natural rubber and natural rubber/epoxidised natural rubber blends. *Polymer* 1999;**40**: 3223—8.

13 Environmentally Friendly Polymers

Environmentally friendly or "green" polymers are those that are produced from *renewable resource raw materials* such as corn or that are *biodegradable or compostable*. This is a developing area in packaging materials and though there are a relatively limited number of polymers used commercially, they will certainly become more numerous and more common in the future.

Biodegradable plastics are made out of ingredients that can be metabolized by naturally occurring microorganisms in the environment. Some petroleum-based plastics will biodegrade eventually, but that process usually takes a very long time and contributes to global warming through the release of carbon dioxide.

Petroleum-based plastic is derived from oil, a limited resource. Plastic based in renewable raw materials biodegrade much faster and can be almost carbon neutral. Renewable plastic is derived from natural plant products such as corn, oats, wood, and other plants, which helps ensure the sustainability of the earth. Polylactic acid (PLA) is the most widely researched and used 100% biodegradable plastic packaging polymer currently, and is made entirely from corn-based starch. Details on PLA are included in a following section.

Cellophane™ is a polymeric cellulose film made from the cellulose from wood, cotton, hemp, or other sources. There are several modifications made to cellulose called polysaccharides (cellulose esters), which are common including cellulose acetate, nitrocellulose, carboxymethyl cellulose (CMC), and ethyl cellulose. Details on cellophane™ and its derivatives are included in several following sections.

Polycaprolactone (PCL) is biodegradable polyester that is often mixed with starch. Details on PLA are included in a following section.

Polyhydroxyalkanoates (PHAs) are naturally produced, and include poly-3-hydroxybutyrate (PHB or PH3B), polyhydroxyvalerate (PHV), and polyhydroxyhexanoate (PHH); A PHA copolymer called poly(3-hydroxybutyrate-co-3-hydroxyvalerate) (PHBV) is less stiff and tougher, and it may be used as packaging material.

Several interesting green polymers are discussed in the next few paragraphs. These are ones for which no public permeation data have been identified.

Polyanhydrides currently are used mainly in the medical device and pharmaceutical industry.[1] Figure 13.1 shows the generalized structure of an anhydride polymer and two polyanhydrides that are used to encapsulate certain drugs. The poly(bis-carboxyphenoxypropane), pCCP, is relatively slow to degrade. The poly(sebacic anhydride), pSA, is fast to degrade. Separately neither of these materials can be

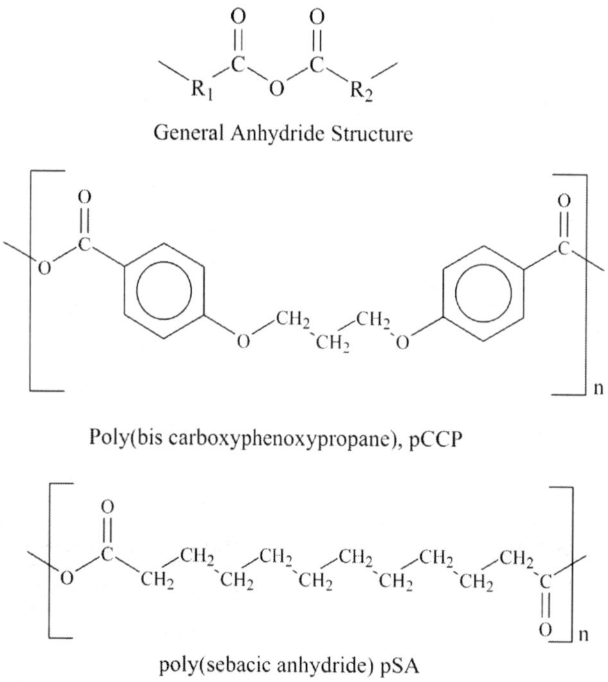

Figure 13.1 Polyanhydride chemical structures.

Permeability Properties of Plastics and Elastomers. DOI: 10.1016/B978-1-4377-3469-0.10013-X

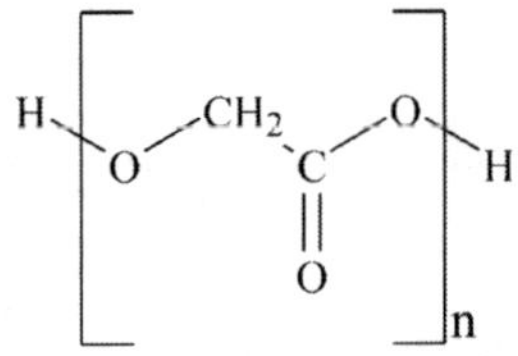

Figure 13.2 Polyglycolic acid chemical structures.

used, but if a copolymer is made in which 20% of the structure is pCCP and 80% is pSA, the overall properties meet the needs of the drug. Polyanhydrides are now being offered for general uses.

Polyglycolic acid (PGA) and its copolymers have found limited use as absorbable sutures and are being evaluated in the biomedical field, where its rapid degradation is useful. This rapid degradation has limited its use in other applications. The structure of PGA is shown in Fig. 13.2.

Interest in the "green" materials is strong as the number of commercially available materials grows. Table 13.1 lists some of the commercial materials recently available.

The following sections contain the details of several of the more common green polymers.

Table 13.1 A List of Some Environmentally Friendly Polymer Based Product Trade Name and Trademarks[2]

Trade Mark	Owner	Material
Aqua-Novon	Novon International Inc. (USA)	PCL
BAK®	Bayer AG Corporation (Germany)	Polyester amide
BioBag International AS	Polargruppen (Norway)	Mater-Bi®
Bioceta, Biocell, Biocelat	Mazucchelli, S.p.A. (Italy)	Cellulose acetate
Biofan	Gunze (Japan)	PHB/PHBV
Bioflex®	Biotec GmbH (Germany)	Starch
Biogreen	Mitsubishi Gas Chemical Co. (Japan)	PHB
Biomax®	DuPont (USA)	PBS-co-PBST
Biomer	Biomer (Germany)	Polyester, PHB
Bionolle 1000	Showa Highpolymer Co. (Japan)	PBS
Bionolle 3000	Showa Highpolymer Co. (Japan)	PBS-co-PBSA
Biopac	Biopac Ltd (UK)	Starch
BioPar®	Biop AG Biopolymer GmbH (Germany)	Starch, biodegradable synthetic polymer
Biophan®	Trespaphan GmbH (Germany)	PLA
Bioplast®	Biotec GmbH (Germany)	Starch, PLA, copolyester
Biopol™	Monsanto Co. (Italy)/Metabolix, Inc. (UK)	PHB, PHV and PHAs
Biopur®	Biotec GmbH (Germany)	Starch
Bioska	PlastiRoll Oy (Finland)	Starch/PVA
Bio-Solo	Indaco Manufacturing Ltd (Canada)	Starch, patented additives, PE
Biostarch®	Biostarch (Australia)	Maize starch
Bio-Stoll	Stoll Papierfolien (Germany)	Starch, LDPE/Ecostar, additive
Biotec®	Bioplast GmbH (Germany)	Thermoplastic starch (TPS®)

Table 13.1 (*Continued*)

Trade Mark	Owner	Material
BioRez®	Trans Furans Chemicals (Netherlands)	Furan resin
Biothene®	Biothene (UK)	Biofuels from planted soy
CAPA®	Solvay Polymers (Italy)	PCL
CelGreen PH/P-CA	Daicel Chemical Industries Ltd (Japan)	PCL/cellulose acetate
CelloTherm	UCB Films (UK)	Regular cellulose (for microwave)
Chronopol	Chronopol-Boulder, CO (USA)	PLA
Clean Green	StarchTech Inc, MN (USA)	Starch-based biopolymers
Cohpol™	VTT Chemical Technology (Finland)	Starch ester
Cornpol®	Japan Corn Starch (Japan)	Modified starch
Corterra	Shell Chemicals (USA/NL)	PTT
Degra-Novon®	Novon International Inc. (USA)	Polyolefin + additives
EarthShell®	EarthShell Corp., MD (USA)	Starch composite materials
Eastar Bio	Eastman Chemical Company (USA)	Copolyester
ECM Masterbatch Pellets	ECM Biofilms (USA)	Additives for polyolefin products
Ecoflex®	BASF Corporation (Germany)	Poly(butyleneadipate)-co-PBAT
Eco-Flow	National Starch & Chemical (USA)	Starch-based biodegradable products
Eco-Foam®	National Starch & Chemical	Foamed starch
Eco-Lam	National Starch & Chemical	Starch, PET, PP
EcoPLA®	Cargill Dow Polymers (USA)	PLA
Ecoplast	Groen Granulaat (Netherlands)	Starch
EnPol	IRe Chemical Co. Ltd (South Korea)	PBS-co-PBSA
Envirofil™	EnPac/DuPont/ConAgra (USA)	Starch/PVA
Enviromold®	Storopack Inc. (USA)	Polystyrene expanded products
EnviroPlastic®	Planet Polymer Technologies, Inc. (USA)	Cellulose acetate
EverCorn™	EverCorn, Inc. (USA)	Starch
Fasal®	Japan Corn Starch Co., Ltd (Japan); Department Agrobiotechnology, Tulln, (Austria)	50% wood wastes
FLO-PAK BIO 8®	FP International (USA)	Starch (corn or wheat)
Gohsenol	Nippon Gohsei (Japan)	PVA
GreenFill	Green Light Products Ltd (UK)	Starch/PVA
Greenpol®	SK Corporation (South Korea)	Starch, aliphatic polyester
Hydrolene®	Idroplax S.r.L. (Italy)	PVA
LACEA®	Mitsui Chemicals, Inc. (Japan)	PLA from fermented glucose
Lacty	Shimadzu Corp. (Japan)	PLA

(*Continued*)

Table 13.1 (Continued)

Trade Mark	Owner	Material
Lignopol	Borregaard Deutschland GmbH	Lignin
Loose Fill®	STOROpack (Germany)	EPS/starch
Lunare	Nippon Shokubai Co., Ltd	Polyethylenesuccinate/adipate
Mater-Bi™	Novamont S.p.A. (Italy)	Starch/cellulose derivative
Mazin	Mazin International (USA)	PLA
Mirel™	Metabolix Inc. (USA)	Corn sugar
NatureFlex™	Innovia Films (UK)	Regenerated cellulose film
NatureWorks®	Cargill Co. (USA)	PLA
Nodax™	Procter & Gamble Co. (USA)	PHB-co-PHA
Novon®	Ecostar GmbH (Germany)	Starch
Paragon	Avebe Bioplastic (Germany)	Starch
Plantic®	Plantic Technologies (Australia)	Corn-starch materials
Poly-NOVON®	Novon International	Starch additives
Polystarch	Willow Ridge Plastics, Inc. (USA)	Additives
POLYOX™	Union Carbide Corporation (USA)	Poly(ethylene oxide)
POVAL	Kuraray Povol Co., Ltd (Japan)	PVA
Pullulan	Hayashibara Biochemical (Japan)	Starch
RenaturE®	Storopack, Inc. (Germany)	Starch
ReSourceBags™	Ventus Kunststoff GmbH	Mater-Bi
Sconacell®	Buna SOW Leuna (Germany)	Starch acetate, plasticizer
Sky-green	Sunkyong Ltd (South Korea)	Aliphatic-co-polyester
Solanyl®	Rodenburg Biopolymers (Netherlands)	Starch (from potato waste)
Sorona®	DuPont Tate & Lyle (USA)	PDO
SoyOyl™	Urethane Soy Systems Co. Inc. (USA)	Soy-based products
SPI-Tek	Symphony Plastic Technologies Plc (UK)	Additives
Supol®	Supol GmbH (Germany)	Starch plant oil and sugars
TONE®	Union Carbide Corp. (USA)	PCL
Trayforma	Stora Enso Oyj (Finland)	Cellulose, food tray
Vegemat®	Vegeplast S.A.S. (France)	Starch

Abbreviations: PHBV, polyhydoxybutyrate valerate; PBS-co-PBST, polybutylene succinate, copolymer poly(butylene succinate-terephthalate); PBS-co-PBSA, polybutylene succinate copolymer polybutylene succynate adipate; PHV, polyhydroxyvalerate; LDPE, low-density polyethylene; PTT, polytrimethylene terephthalate; PBAT, poly(butylene adipate-co-terephthalate); EPS, expanded polystyrene; l PDO, poly(dioxanone).

13.1 Cellophane™

Cellophane™ is a polymeric cellulose film made from the cellulose from wood, cotton, hemp, or other sources. The raw material of choice is called dissolving pulp, which is white like cotton and contains 92–98% cellulose. The cellulose is dissolved in alkali in a process known as mercerization. It is aged several days. The mercerized pulp is treated with carbon disulfide to make an orange solution called viscose, or cellulose xanthate. The viscose solution is then extruded through a slit into a bath of dilute sulfuric acid and sodium sulfate to reconvert the viscose into cellulose. The film is then passed through several more baths, one to remove sulfur, one to bleach the film, and one to add glycerin to prevent the film from becoming brittle. Cellophane™ has a CAS number of 9005-81-6. The approximate chemical structures are shown in Fig. 13.3.

The Cellophane™ may be coated with nitrocellulose or wax to make it impermeable to water vapor. It may also be coated with polyethylene or other materials to make it heat sealable for automated wrapping machines.

Manufacturers and trademarks: Innovia Cellophane™.

Applications and uses: Cellulosic film applications include tapes and labels, photographic film, coatings for paper, glass, and plastic. Medical applications for cellulosic films include dialysis membranes (Tables 13.2–13.4).

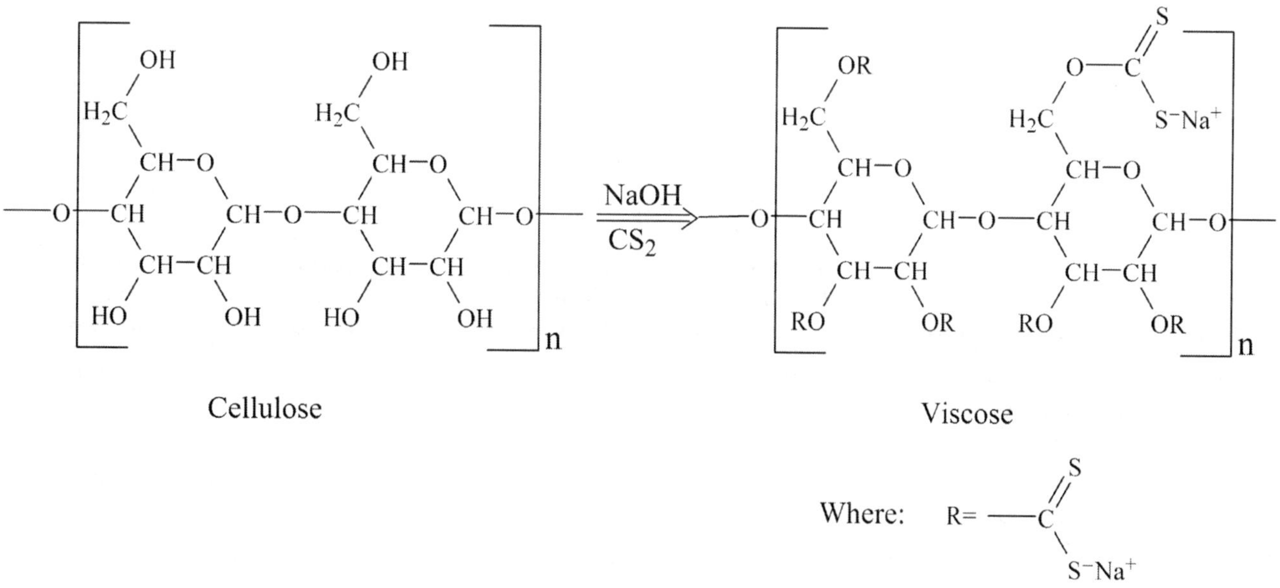

Figure 13.3 Conversion of raw cellulose to viscose.

Table 13.2 Permeability of Oxygen through Polyvinylidene Chloride (PVDC) Coated Cellophane™ Film[3]

Temperature (°C)	35	20		
Test Method	JIS Z1707	ASTM D3985		
Relative humidity (%)	0	65	85	100
Source document units Permeability coefficient (cm^3 mil/100 $in.^2$ day)	0.07	0.26	0.71	2.06
Normalized units Permeability coefficient (cm^3 mm/m^2 day atm)	0.03	0.10	0.28	0.81

Sample thickness: 0.023 mm.

Table 13.3 Permeation of Various Gases through Cellulose (Cellophane™)[4]

Penetrant	Temperature (°C)	Permeability Coefficient	
		Source Document Units ×10^{10} (cm^3 cm/cm^3 s cm Hg)	Normalized Units (cm^3 mm/m^2 day atm)
Helium	20	0.0005	0.033
Hydrogen	25	0.0065	0.427
Nitrogen	25	0.0032	0.210
Oxygen	25	0.0021	0.138
Carbon dioxide	25	0.0047	0.309
Hydrogen sulfide	45	0.0006	0.039
Sulfur dioxide	25	0.0017	0.112
Water	25	1900	12500

Table 13.4 Oxygen Gas Transmission Rate and Water Vapor Transmission Rate of Innovia Cellophane™ Films[5]

Product Code	Film Structure	Oxygen Gas Transmission Rate		Water Vapor Transmission Rate	
		Source Document Units (cm^3/100 $in.^2$ day bar)	Normalized Units (cm^3/m^2 day atm)	Source Document Units (g/100 $in.^2$ day)	Normalized Units (g/m^2 day)
DM 320	Nitrocellulose coated one side	0.19	3.0	10	183
DMS 345	Nitrocellulose coated one side	0.19	3.0	10	183
'K' HB20 (or XS)	Polyvinylidene coated both sides	0.19	3.0	0.65	12
LST 195	Nitrocellulose coated both sides	0.19	3.0	70	1284
MST/MT33	Nitrocellulose coated both sides	0.19	3.0	1.3	24
P00	Uncoated	0.19	3.0	>95	>1700
P25	Uncoated	0.19	3.0	>95	>1700

Oxygen test method: ASTM F1927, at 24 °C and 5% relative humidity. WVTR test method: ASTM E96, at 38 °C and 90% relative humidity.

13.2 Nitrocellulose

Nitrocellulose is made by treating cellulose with a mixture of sulfuric and nitric acids. This changes the hydroxyl groups (−OH) in the cellulose to nitro groups (−NO$_3$) as shown in Fig. 13.4. Nitrocellulose, also know as gun cotton and the main ingredient of smokeless gunpowder, decomposes explosively. In the early twentieth century, it was found to make an excellent film and paint. Nitrocellulose lacquer was used as a finish on guitars and saxophones for most of the twentieth century and is still used on some current applications. Manufactured by (among others) DuPont, the paint was also used on automobiles sharing the same color codes as many guitars including Fender and Gibson brands. Nitrocellulose lacquer is also used as an aircraft dope, painted onto fabric-covered aircraft to tauten and provides protection to the material. Its CAS number is 9004-70-0.

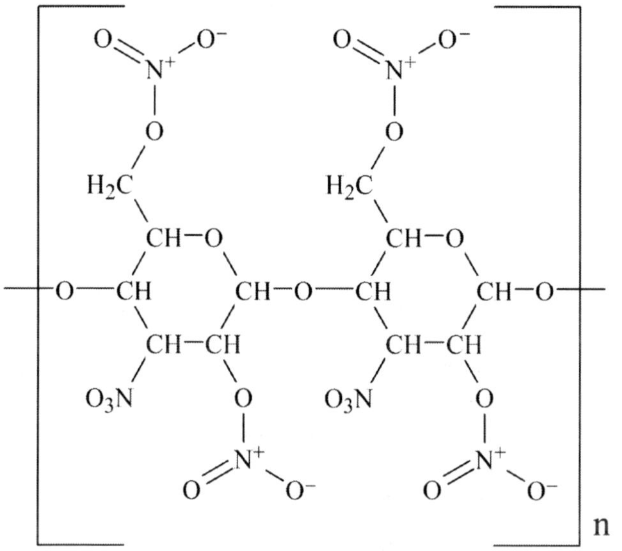

Nitrocellulose is not usually used for film applications but more commonly is part of multilayered film structures, especially those based on Cellophane™.

Manufacturers and trade names: Innovia Films Cellophane™.

Applications and uses: Food wrap (Table 13.5).

Figure 13.4 Structure of nitrocellulose.

Table 13.5 Permeation of Gases at 25 °C through Nitrocellulose Film[6]

Permeate Gas	Pressure Differential (mm Hg)	Permeability Coefficient	
		Source Document Units (cm^3 mm/cm^2 s cm Hg × 10^9)	Normalized Units (cm^3 mm/cm^2 day atm)
Helium	4.68	6.9	453
Nitrogen	5.21	0.116	7.6
Oxygen	4.995	1.95	128
Carbon dioxide	4.567	2.12	139
Sulfur dioxide	4.442	1.76	116
Ammonia	4.04	57.1	3749
Water	2.195	6295	413355
Ethane	4.92	0.063	4.1
Propane	4.57	0.0084	0.6
n-butane	4.34	~0	~0

13.3 Cellulose Acetate

Cellulose acetate is the acetate ester of cellulose. It is sometimes called Acetylated cellulose or xylonite. Its CAS number is 9004-35-7 and the approximate chemical structure is shown in Fig. 13.5.

Manufacturers and trade names: Celanese Cellulose Acetate; Eastman Chemical Company Tenite.

Applications and uses: Cellulose acetate is used as a film base in photography, as a component in some adhesives, and as a frame material for eyeglasses; it is also used as a synthetic fiber and in the manufacture of cigarette filters, found in screwdriver handles, ink pen reservoirs, x-ray films (Tables 13.6 and 13.7).

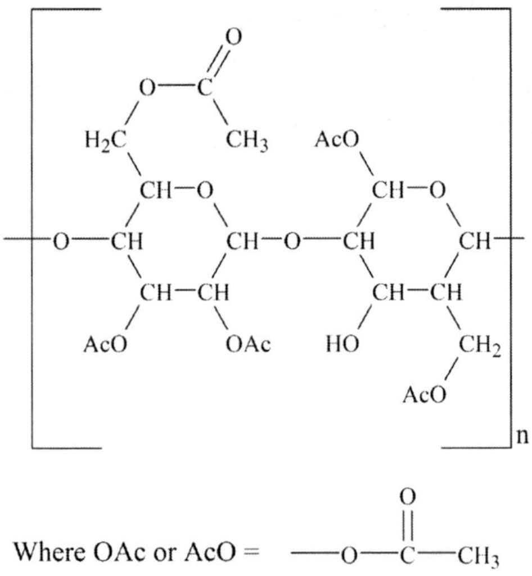

Figure 13.5 Chemical structure of cellulose acetate.

Table 13.6 Permeability of Various Gases at 35 °C through Cellulose Acetate Membranes[7]

Permeant Gas	Permeability Coefficient	
	Source Document Units (cm³ (STP)·cm/cm² s cm Hg)	Normalized Units (cm³ mm/m² day atm)
Helium	1.00×10^{-9}	656.6
Oxygen	1.30×10^{-10}	85.4
Argon	7.70×10^{-11}	50.6
Nitrogen	2.60×10^{-11}	17.1
Krypton	3.50×10^{-11}	23.0
Xenon	9.70×10^{-12}	6.4
Carbon dioxide	6.30×10^{-10}	413.7

Table 13.7 Permeability of Various Gases at 22 °C through Dense and High-Flux Cellulose Acetate[8]

Gas	Source Document Units		Normalized	
	Dense Cellulose Acetate (cm³ (STP)·cm/ cm² s cm Hg)*	High-Flux Cellulose Acetate[a] (cm³ (STP)/ cm² s cm Hg)	Dense Cellulose Acetate (cm³ mm/ m² day atm)*	High-Flux Cellulose Acetate[a] (cm³ cm/ m² day atm)
Helium	1.36×10^{-9}	2.80×10^{-5}	893	1.84×10^{6}
Neon	2.40×10^{-10}	6.00×10^{-7}	158	3.94×10^{4}
Oxygen		1.90×10^{-6}		1.25×10^{5}
Argon	3.20×10^{-11}	1.10×10^{-6}	21	7.22×10^{4}
Methane	1.40×10^{-11}	7.00×10^{-7}	9.2	4.60×10^{4}
Nitrogen	1.40×10^{-11}	6.00×10^{-7}	9.2	3.94×10^{4}
Propane	$<10^{-13}$	3.00×10^{-7}		1.97×10^{4}

See also Figs. 13.6–13.10. [a]While the membrane thickness for fully dense membrane can be measured, the nominal thickness of high-flux material cannot be used to calculate permeability from flow rates. For this reason, permeation rates, not permeabilities, are given for high-flux sample.

13.4 Ethyl Cellulose

Ethyl cellulose is similar in structure to cellulose and cellulose acetate but some of the hydroxyl (—OH) functional groups in the cellulose are replaced by the ethoxy group (—O—CH$_2$—CH$_3$). Ethyl cellulose has a CAS number of 9004-57-3 and its structure is shown in Fig. 13.11.

Manufacturers and trade names: Dow Ethocel™, Ashland Aqualon®.

Applications and uses: Pharmaceutical applications, cosmetics, nail polish, vitamin coatings, printing inks, specialty coatings, food packaging (Tables 13.8—13.10).

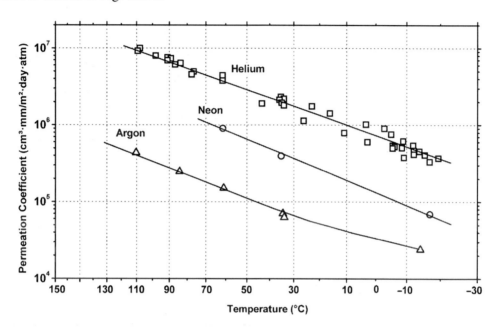

Figure 13.6 Permeation rates of noble gases at 22 °C through high-flux cellulose acetate films.[8] *The nominal thickness of high-flux material cannot be used to calculate permeability from flow rates. For this reason, permeation rates, not permeability coefficients, are given.

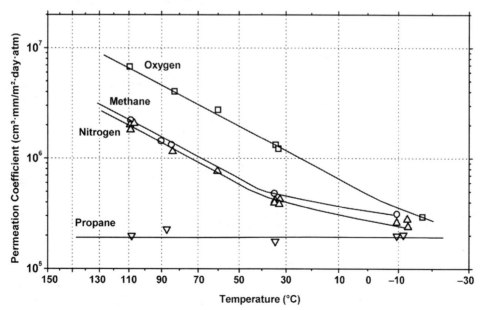

Figure 13.7 Permeation rates of common gases at 22 °C through high-flux cellulose acetate films.[8] *The nominal thickness of high-flux material cannot be used to calculate permeability from flow rates. For this reason, permeation rates, not permeability coefficients, are given.

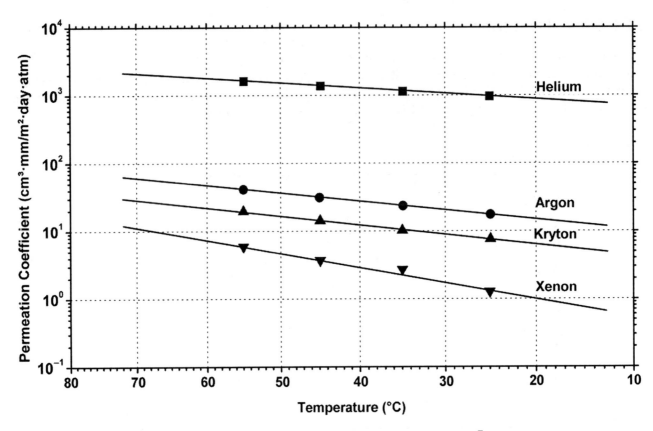

Figure 13.8 Permeation of noble gases at 35 °C through cellulose acetate films.[7]

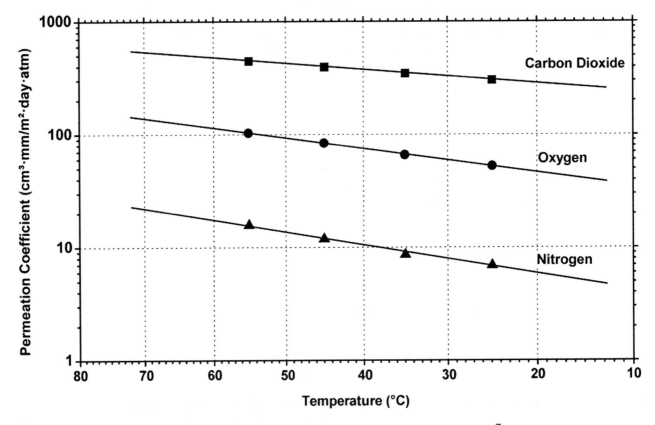

Figure 13.9 Permeation of common gases at 35 °C through cellulose acetate films.[7]

13: Environmentally Friendly Polymers

Figure 13.10 Permeation of hydrogen sulfide vs. temperature through plasticized and unplasticized cellulose acetate films.[9]

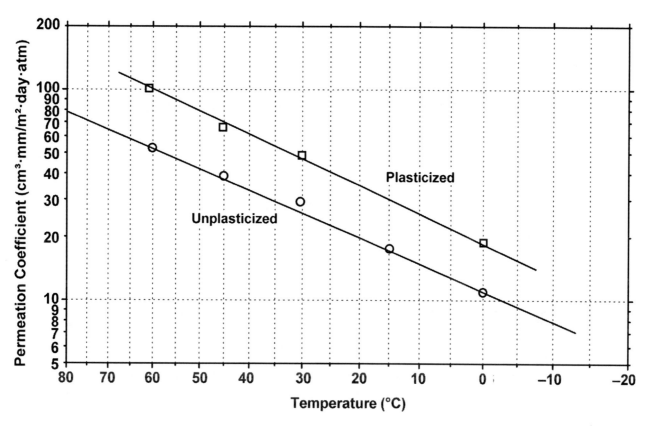

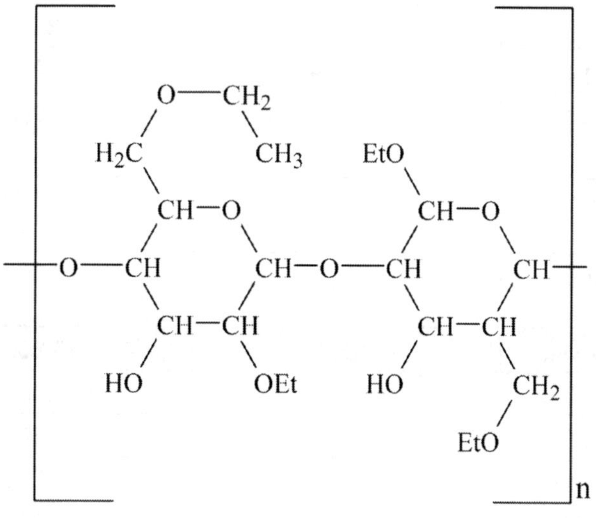

Where OEt or EtO = ──O──CH_2-CH_3

Figure 13.11 Structure of ethyl cellulose.

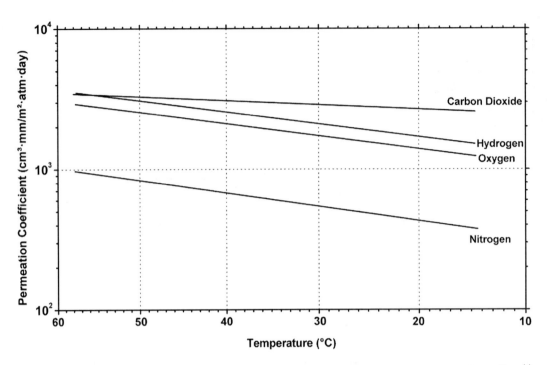

Figure 13.12 Permeation of various gases vs. temperature for Dow Ethocel™ ethyl cellulose film.[11]

Table 13.8 Permeation of Gases through Ethyl Cellulose[6]

Permeant Gas	Permeability Coefficient	
	Source Document Units × 10^9 (cm^3 mm/cm^2 s cm Hg)	Normalized Units (cm^3 mm/ m^2 day atm)
Helium	53.4	3510
Nitrogen	4.43	291
Oxygen	14.7	965
Argon	10.2	670
Carbon dioxide	113	7420
Sulfur dioxide	264	17,300
Ammonia	705	46,300
Water	8930	586,000
Ethane	9.2	604
Propane	3.7	243
n-Butane	3.87	254
n-Pentane	3.7	243
n-Hexane	7.66	503

Table 13.9 Permeation of Various Gases at 35 °C through Membranes Made from Ashland Aqualon® Ethyl Cellulose[10]

Material Grade	Ethoxy Content (%)	Permeability Coefficient				
		Source Document Units $\times 10^9$ [cm^3 (STP) cm/cm^2 s cm Hg]				
		Carbon Dioxide	Helium	Oxygen	Methane	Nitrogen
EC K-100	47.2	8.9	4.9	1.46	0.88	0.41
EC N-100	47.9	11.6	6.6	1.94	1.24	0.58
EC T-10	49.6	14.7	7.9	2.33	1.41	0.65
		Normalized Units (cm^3 mm/m^2 day atm)				
EC K-100	47.2	78	43	12.8	7.7	3.6
EC N-100	47.9	102	58	17.0	10.9	5.1
EC T-10	49.6	129	69	20.4	12.3	5.7

Pressure differential: 10 atm.

Table 13.10 Permeation Selectivity of Various Gases Pairs at 35 °C through Membranes Made from Ashland Aqualon® Ethyl Cellulose[10]

Material Grade	Ethoxy Content (%)	$\alpha(CO_2/CH_4)$	$\alpha(O_2/N_2)$	$\alpha(He/CH_4)$	$\alpha(CO_2/N_2)$	$\alpha(O_2/He)$
EC K-100	47.2	10.1	3.5	5.7	21.8	1.7
EC N-100	47.9	9.4	3.3	5.4	20.0	1.7
EC T-10	49.6	10.4	3.6	5.6	22.5	1.8

See also Fig. 13.12.

13.5 Polycaprolactone

PCL is biodegradable polyester with a low melting point of around 60 °C and a glass transition temperature of about −60 °C. PCL is prepared by ring opening polymerization of ε-caprolactone using a catalyst such as stannous octanoate. The structure of PCL is shown in Fig. 13.13.

PCL is degraded by hydrolysis of its ester linkages in physiological conditions (such as in the human body) and has therefore received a great deal of attention for use as an implantable biomaterial. In particular, it is especially interesting for the preparation of long-term implantable devices. A variety of drugs have been encapsulated within PCL beads for controlled release and targeted drug delivery. PCL is often mixed with starch to obtain a good biodegradable material at a low price.

Manufacturers and trade names: Perstorp CAPA® (previously Solvay), Dow Chemical Tone (discontinued).

Applications and uses: The mix of PCL and starch has been successfully used for making trash bags in Korea (Yukong Company) (Tables 13.11–13.14).

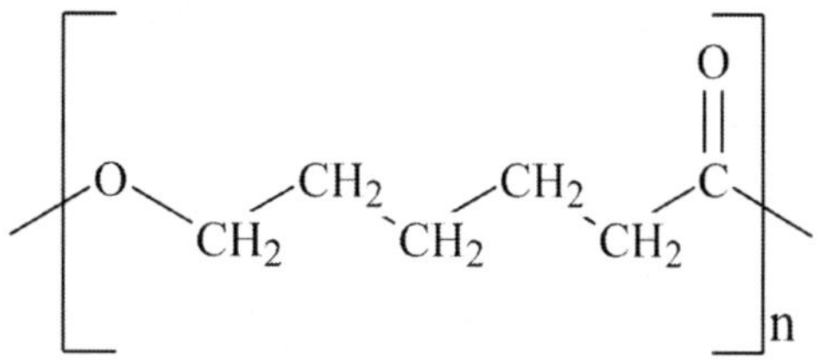

Figure 13.13 Structure of polycaprolactone.

Table 13.11 Water Vapor Permeation at 20 °C and 100% RH through Polycaprolactone Film[12]

Film Thickness (mm)	Orientation Speed (m/min)	Source Document Units, Vapor Permeation Rate	
		g mm/m² day kPaa	g mm/m² day atm[a]
1.046	2.2	64	6485
0.051	18.5	3	304

[a]Vapor permeation rates do not usually contain a pressure differential; see reference for description of nonstandard test method.

Table 13.12 Oxygen Permeation at 35 °C and 0% Relative Humidity through Polycaprolactone Film[12]

Film Thickness (mm)	Orientation Speed (m/min)	Source Document Units	Normalized Units
		Permeability Coefficient (cm³ μm/m² day kPa)	Permeability Coefficient (cm³ mm/m² day atm)
1.046	2.2	945	96
0.051	18.5	1000	101

Test equipment: Mocon Ox-Tran 2/20 Modular System.

Table 13.13 Water Vapor Permeation at 20 °C and 100% RH through Films Made from Blends of Polycaprolactone, Starch and Glycerol[12]

Film Composition: PCL/Starch/Glycerol (wt%)	Film Thickness (mm)	Orientation Speed (m/min)	Source Document Units	Normalized Units
			Vapor Permeation Rate (g mm/m² day kPa[a])	Vapor Permeation Rate (g mm/m² day atm[a])
0/60/40	0.469	0	190	19,300
2/58/40	0.204	0	100	10,100
10/54/36	0.665	0	215	21,800
10/54/36	0.195	3.8	100	10,100
20/48/32	0.208	2.2	65	6600
20/48/32	0.070	9.9	30	3040
30/42/28	0.460	0	88	8900
30/42/28	0.217	2.2	26	2600
30/42/28	0.051	8.8	11	1100
100/0/0	1.046	2.2	64	6500
100/0/0	0.051	18.5	3	300

[a]Vapor permeation rates do not usually contain a pressure differential; see reference for description of nonstandard test method.

Table 13.14 Oxygen Permeation at 35 °C and 0% Relative Humidity through Films Made from Blends of Polycaprolactone, Starch and Glycerol[12]

Film Composition: PCL/Starch/Glycerol (wt%)	Film Thickness (mm)	Orientation Speed (m/min)	Source Document Units Permeability Coefficient (cm³ µm/m² day kPa)	Normalized Units Permeability Coefficient (cm³ mm/m² day atm)
0/60/40	0.469	0	0	0
2/58/40	0.204	0	0	0
10/54/36	0.665	0	0	0
10/54/36	0.195	3.8	0	0
20/48/32	0.208	2.2	17	1.7
20/48/32	0.070	9.9	1	0.1
30/42/28	0.460	0	42	4.3
30/42/28	0.217	2.2	21	2.1
30/42/28	0.051	8.8	20	2.0
100/0/0	1.046	2.2	945	96
100/0/0	0.051	18.5	1000	101

Test equipment: Mocon Ox-Tran 2/20 Modular System.

13.6 Poly(Lactic Acid)

PLA is derived from renewable resources, such as corn starch or sugarcanes. PLA polymers are considered biodegradable and compostable. PLA is a thermoplastic, high-strength, high-modulus polymer that can be made from annually renewable sources to yield articles for use in either the industrial packaging field or the biocompatible/bioabsorbable medical device market. Bacterial fermentation is used to make lactic acid, which is then converted to the lactide dimer to remove the water molecule that would otherwise limit the ability to make high-molecular-weight polymer. The lactide dimer, after the water is removed, can be polymerized without the production of the water. This process is shown in Fig. 13.14. The PLA CAS number is 9002-97-5.

Manufacturers and trade names: FKur Bio-Flex®, Cereplast Inc. Compostables®, Mitsubishi Chemical Fozeas®, NatureEorks LLC Ingeo™, Alcan Packaging Ceramis®-PLA.

Applications and uses: Biomedical applications, such as sutures, stents, dialysis media, and drug delivery devices. It is also being evaluated as a material for tissue engineering; loose-fill packaging, compost bags, food packaging, and disposable tableware. PLA can be in the form of fibers and nonwoven textiles. Potential uses: upholstery, disposable garments, awnings, feminine hygiene products.

See also Figs. 13.15–13.19.

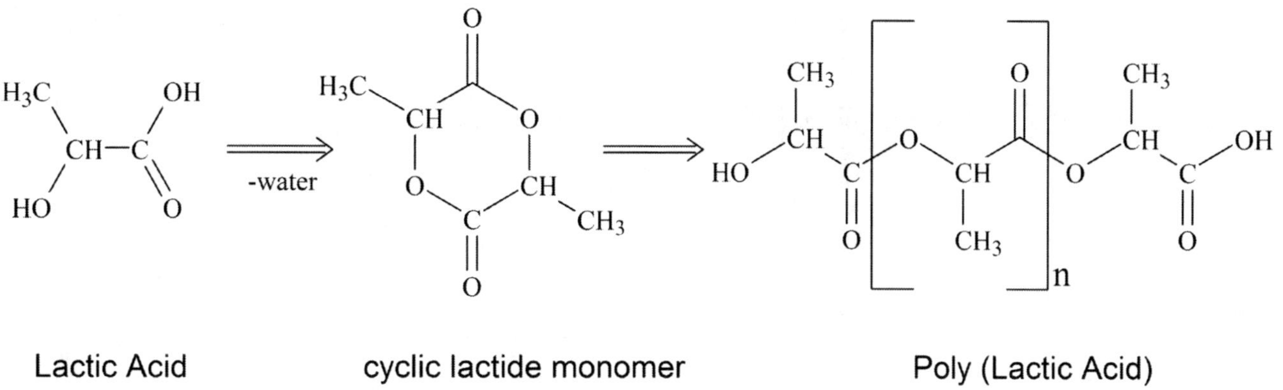

Lactic Acid   cyclic lactide monomer   Poly (Lactic Acid)

Figure 13.14 Conversion of lactic acid to polylactic acid.

Figure 13.15 Permeation coefficient of methane vs. temperature through linear polylactic acid film.[14]

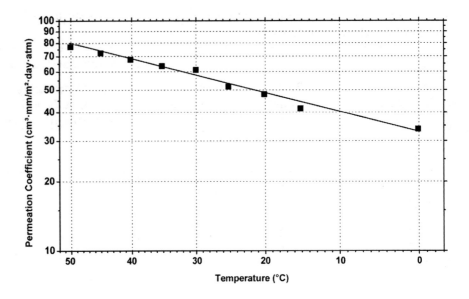

Figure 13.16 Permeation coefficient of carbon dioxide vs. temperature through linear polylactic acid film.[14]

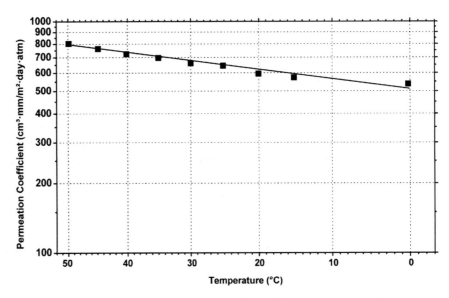

Figure 13.17 Permeation coefficient of nitrogen vs. temperature through linear polylactic acid film.[14]

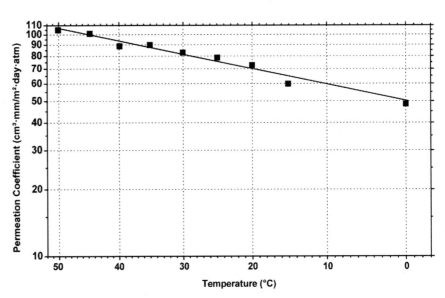

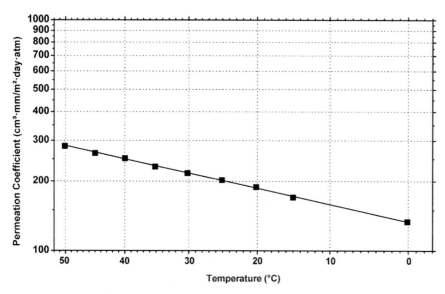

Figure 13.18 Permeation coefficient of oxygen vs. temperature through linear polylactic acid film.[14]

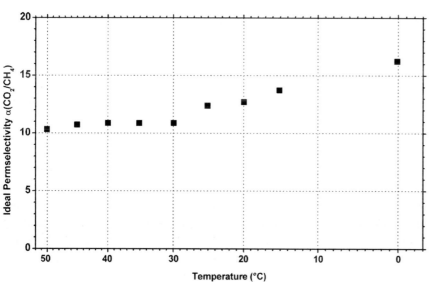

Figure 13.19 Permselectivity of carbon dioxide/methane vs. temperature through linear polylactic acid film.[14]

13.7 Poly-3-Hydroxybutyrate

PHAs are naturally produced and include PHB (or PH3B), PHV, and PHH. A PHA copolymer called PHBV is less stiff and tougher, and it may be used as a packaging material. Chemical structures of some of these polymers are shown in Fig. 13.20.

Manufacturers and trade names: FKur Bio-Flex®, Cereplast Inc. Compostables®, Mitsubishi Chemical Fozeas®, NatureWorks LLC Ingeo™, Alcan Packaging Ceramis®-PLA.

Applications and uses: Biomedical applications, such as sutures, stents, dialysis media, and drug delivery devices. It is also being evaluated as a material for tissue engineering; loose-fill packaging, compost bags, food packaging, and disposable tableware. PLA can be in the form of fibers and nonwoven textiles.

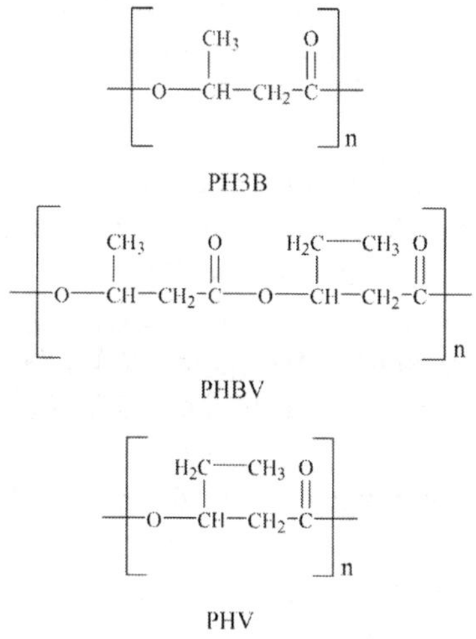

Figure 13.20 Structures of several polyhydroxyalkanoates.

Table 13.15 Permeability Coefficients for Poly-3-Hydroxybutyrate (PHB) Membranes[13]

Temperature (°C)	Methanol	Ethanol	n-Propanol	Water
Source Document Units, Permeability Coefficient Barrers				
30	–	–	–	520
50	1130	890	480	750
55	1590	900	520	990
60	1640	920	530	1050
65	2090	1230	590	1900
Normalized Units, Permeability Coefficient (cm^3 mm/m^2 day atm)				
30	–	–	–	34,100
50	74,200	58,400	31,500	49,200
55	104,000	59,100	34,100	65,000
60	108,000	60,400	34,800	68,900
65	137,000	80,800	38,700	125,000

±5% error.

Potential uses: upholstery, disposable garments, awnings, feminine hygiene products (Table 13.15).

References

1. Jain JP, Modi S, Domb AJ, Kumar N. Role of polyanhydrides as localized drug carriers. *J Control Release* 2005;**103**:541–63.
2. Chiellini E. Environmentally compatible food packaging. In: *Environmentally compatible food packaging*. Cambridge (UK): Woodhead Publishing Ltd; 2008. p. 371–95.
3. *EVAL film properties comparison. Supplier technical report*. Kuraray Co., Ltd; 2003.
4. *Affinity polyolefin plastomers. Form No. 305-01953-893 SMG*. Dow Chemical Company; 1993.
5. Innovia specification sheets. Edition USA; 2005.
6. Hsieh PY. Diffusibility and solubility of gases in ethylcellulose and nitrocellulose. *J Appl Polym Sci* 1963;**7**(5):1743–56. Available from: http://doi.wiley.com/10.1002/app.1963.070070515.
7. Nakai Y, Yoshimizu H, Tsujita Y. Enhanced gas permeability of cellulose acetate membranes under microwave irradiation. *J Memb Sci* 2005;**256**:72–7. Available from: http://linkinghub.elsevier.com/retrieve/pii/S037673880500147X.
8. Gantzel PK, Merten U. Gas separations with high-flux cellulose acetate membranes. *Ind Eng Chem Process Des Dev* 1970;**9**(2):331–2. Available from: http://pubs.acs.org/doi/abs/10.1021/i260034a028.
9. Heilman W, Tammela V, Meyer J, Stannett V, Szwarc M. Permeability of polymer films to hydrogen sulfide gas. *Ind Eng Chem* 1956;**48**:821–4.
10. Houde A, Stern S. Permeability of ethyl cellulose to light gases. Effect of ethoxy content. *J Memb Sci* 1994;**92**(1):95–101. Available from: http://linkinghub.elsevier.com/retrieve/pii/0376738894800162.
11. Brubaker DW, Kammermeyer K. Separation of gases by means of permeable membranes. Permeability of plastic membranes to gases. *Ind Eng Chem* 1952;**44**(6):1465–74. Available from: http://pubs.acs.org/cgi-bin/doilookup/?10.1021/ie50510a071.
12. Myllymäki O, Myllärinen P, Forssell P, et al. Mechanical and permeability properties of biodegradable extruded starch/polycaprolactone films. *Packag Technol Sci* 1998;**11**(6):265–74.
13. Poley LH, Silva MGD, Vargas H, Siqueira MO, Sánchez R. Water and vapor permeability at different temperatures of poly (3-hydroxybutyrate) dense membranes. *Polímeros* 2005;**15**:22–6.
14. Lehermeier H. Gas permeation properties of poly(lactic acid). *J Memb Sci* 2001;**190**:243–51.

14 Multilayered Films

This chapter will cover miscellaneous multilayered, metalized, coated, or other structured films that were not covered in the previous chapters. Gas permeation and vapor transmission coefficients may be reported for multilayered films, but strictly speaking, these are not coefficients. They are a composite of the different materials making up the multilayered films. Some of the data in this chapter are reported as permeation coefficients but they may also not include the film thickness dependence. This chapter has only a sampling of the performance of multilayered films as there are thousands of such products offered. Some multilayered structures are presented in earlier data chapters.

14.1 Metalized Films

An extremely thin metal layer (0.00005 mm) may be applied to plastic films to increase barrier properties with respect to light, water, and gases. Many plastic films can be metalized, for example, polyethylene terephthalate (PET), biaxially-oriented polypropylene (BOPP), polyvinyl chloride (PVC), and low-density polyethylene (LDPE). The process for doing this was covered in Section 3.6.7. The metalizing process takes place by evaporating aluminum in a high-vacuum chamber. Aluminum layer deposited on film is very thin (100–250 Å). The very thin layer of aluminum that condenses on film gives an excellent barrier to light, water vapor, oxygen, and other gases.

Besides advantages in barrier properties metalized films can be the economical substitute for aluminum foil in applications where aluminum foil is used to guarantee barrier effect. This extremely thin layer guarantees top barrier properties also after lamination because, unlike the foil, it is less rigid and does not crack or break. The cost of metalized film compared with aluminum foil for similar performance can be obtained by using an aluminum quantity of 130 times less at a cost of perhaps one fifth the cost of the foil.

The high-barrier properties of metalized films can also satisfy food manufacturers' needs to reduce antioxidant and preservative in their own products and also help to increase product's shelf life.

Metalized film use also helps to save energy and to increase the environmental protection for the following reasons:

- Very good utilization of aluminum and drastic reduction of unrecyclable metal waste
- Wider possibility to use single-layer film and, therefore, reduction in use of multilayer films
- Lower power consumption for production

There is also the appearance factor where the brightness of metalized film helps with the presentation of products by enhancing the particular esthetic and decorative characteristics of flexible packaging.

The quality of the metallization is important as metalized films are susceptible to pinholes (Tables 14.1 and 14.2).

14.2 Silicon Oxide Coating Technology

A coating technology that is applied with similar equipment as metalizing equipment is based on silicon oxide, called SiOx. The application process is called plasma-enhanced chemical vapor deposition (PECVD). Amcor Limited's technology for SiOx is called Ceramis®. It is deposited in equipment that is used for metalizing when an electron beam is aimed at an SiOx target and the SiOx material evaporates and is deposited onto the film. The coating is optically clear. The coating can improve barrier properties (Table 14.3).

Table 14.1 Oxygen Gas Transmission Rate and Water Vapor Transmission Rate Comparison of Generic Metalized and Unmetalized Plastic Films[1]

Product	Thickness (mm)	Oxygen Permeability (cm^3/m^2 day)	Water Vapor Transmission (g/m^2 day)
PET	0.012	140	20
PET metalized	0.012	<1.0	<1.0
Nylon	0.012	45	350
Nylon metalized	0.012	<2.5	<1.0
Oriented PP	0.020	1500	<1.5
Oriented PP metalized	0.020	<200	<0.6

Table 14.2 Oxygen Gas Transmission Rate and Water Vapor Transmission Rate of Vacmet Metalized Plastic Films[2]

Base Plastic	Film Thickness (mm)	Oxygen Gas Transmission Rate		Water Vapor Transmission Rate	
		Source Document (cm^3/100 $in.^2$ day)	Normalized (cm^3 mm/m^2 day atm)	Source Document (g/100 $in.^2$ day)	Normalized (g mm/m^2 day)
Polyester	0.024	0.070–0.080	0.028–0.032	0.045–0.061	0.018–0.024
Biaxially oriented polypropylene	0.015	3.4	0.669	0.025	0.005

Table 14.3 Oxygen Permeability and Water Transmission at 23 °C of SiOx-Coated PET Film (0.036 mm)[3]

Film	Oxygen Permeability		Water Transmission Rate (g/m^2 day)
	Source Document Units (cm^3/m^2 day bar)	Normalized Units (cm^3/m^2 day atm)	
PET	43.3	42.5	9.57
PET/SiOx	1.4	1.37	0.411

14.3 Co-continuous Lamellar Structures

Lamellar technology blends a barrier polymer into a "host" polymer creating structures with reduced permeability through the incorporation of a platelet type barrier phase. Figure 2.19 shows this kind of structure and how it increases the diffusion path through tortuosity. The lamellar barrier phase allows the barrier resin to form overlapping, discontinuous, and elongated platelets. The platelets are most often immiscible polymers such as polyvinyl alcohol (EVOH), polyamides, PVC, polyvinylidene chloride (PVDC), or platelet fillers such as talc, mica, aluminum flakes, and clay platelets.

The processing method is called lamellar injection molding (LIM). LIM technology enables low levels (2–20%) of barrier polymers to be introduced and maintained as co-continuous lamellae in molded articles. A 300-fold reduction in oxygen permeability compared to conventional blends has been achieved with minor amounts of barrier polymer (10%). Depending upon the polymer used with LIM, improvements in oxygen barrier properties generally fall between 10 and 100 times the base polymer alone.

DuPont has developed a technology that produces laminar structured films using blow molding technology, which is used for automotive parts and containers (Tables 14.4–14.11).

Table 14.4 Oxygen through Various Polymers Using Lamellar Injection Molding[4]

Film Features	Barrier Properties, High Impact, Hot Fill	Barrier Properties, High Impact, Retort	Barrier Properties	Barrier Properties, Enhanced Clarity, Hot Fill	Barrier Properties, Compatibilizer not Required, High Impact, Hot Fill, Transparent	Barrier Properties, Compatibilizer not Required, Enhanced Clarity, High Impact
Host polymer	High-density polyethylene	Polypropylene	Polyethylene terephthalate	Polystyrene	Impact polystyrene	Thermoplastic polyurethane
Barrier polymer	10% Ethylene vinyl alcohol copolymer			10% Nylon MXD6 (aromatic)		
Compatibilizer	PE-g-maleic anhydride	PP-g-maleic anhydride	Ethylene vinyl acetate	Styrene maleic anhydride copolymer		
Gas permeability	0.6				1.0	0.4
Source document units (cm^3 mil/100 $in.^2$ day)						
Permeability coefficient	0.24				0.39	0.16
Normalized (cm^3 mm/m^2 day atm)						

Table 14.5 Oxygen through Lamellar Injection Molded Nylon MXD6 at Various Amounts in Styrene Acrylonitrile Copolymer[3]

	Permeability Coefficient	
Amount of MXD6 (weight %)	Source Document (cm^3 mil/100 $in.^2$ day)	Normalized (cm^3 mm/m^2 day atm)
20%	Nondetectable	Nondetectable
10%	1.0	0.39
5%	2.3	0.91
2.5%	4.7	1.85

Table 14.6 Oxygen Permeability at 23 °C and 75% Relative Humidity through DuPont™ Nylon/HDPE Lamellar Structured Containers[3]

Lamellar Polymer	Permeability Coefficient	
	Source Document (cm³ mil/100 in.² day)	Normalized (cm³ mm/m² day atm)
10% Selar® RB 215	38.2	15.0
10% Selar® RB 300	34.0	13.4

Table 14.7 Water Vapor Transmission at 60 °C through DuPont™ Nylon/HDPE Lamellar Structured Containers[3]

Lamellar Polymer	Permeability Coefficient	
	Source Document (cm³ mil/100 in.² day)	Normalized (cm³ mm/m² day atm)
10% Selar® RB 215	1	0.39
10% Selar® RB 300	9	3.5

Table 14.8 Oxygen Permeability at 23 °C and 75% Relative Humidity through Lamellar Structured Containers[3]

Lamellar Polymer	Base Polymer	Permeability Coefficient	
		Source Document (cm³ mil/100 in.² day)	Normalized (cm³ mm/m² day atm)
15% Selar® RB 421	HDPE	1.07	0.42
15% Selar® RB 215	LDPE	90	35.4
10% Selar® RB 240	PP	36	14.2

Table 14.9 Water Vapor Transmission at 60 °C through Lamellar Structured Containers[3]

Lamellar Polymer	Base Polymer	Permeability Coefficient	
		Source Document (cm³ mil/100 in.² day)	Normalized (cm³ mm/m² day atm)
15% Selar® RB 421	HDPE	2	0.79
15% Selar® RB 215	LDPE	12	4.7
10% Selar® RB 240	PP	18	7.1

Table 14.10 Solvent Permeability through DuPont™ Selar® RB (8%)/Polyolefin Lamellar Structured Containers[3]

Solvent	Temperature (°C)	Exposure Time (Days)	Solvent Weight Loss (%)
Ethyl acetate	50	28	0.42
Isopropyl acetate	50	28	0.03
Acetone	23	180	0.48
Butyl acetate	50	28	0.08
Toluene	50	28	0.3
Xylene	50	28	0.12
Methyl isobutyl ketone	50	28	0.04
Methyl ethyl ketone	50	28	0.97
Cyclohexanone	50	28	0.02
Chlorobenzene	50	28	0.77
Hexane	50	28	0.17
Butyl alcohol	50	28	0.06
Trichloroethene	50	28	0.36
Methyl salicylate	50	28	0.01
Tetrahydrofuran	23	180	0.44
Mineral spirits	50	28	0.01
Turpentine	50	28	0.03
STP gas treatment	50	28	0.07
Paint thinner	50	28	0.06
Charcoal starter	50	28	0.04
Naptha	50	28	0.03
Kerosene	50	28	0.01
D-Limonene	50	28	0.05
Motor oils	50	28	0.02
Pine oil	50	28	0.27
Diesel fuel conditioner	50	28	0.03
Gas additive	50	28	0.03
Xylene with 25% propyl alcohol	50	28	2.84
Xylene with 50% propyl alcohol	50	28	2.61
Propyl alcohol with 25% xylene	50	28	1.44
Propyl alcohol	50	28	0.14
Xylene with 25% methyl alcohol	23	180	14.1
Xylene with 50% methyl alcohol	23	180	10.6
Methyl alcohol with 25% xylene	23	180	3.56
Methyl alcohol	23	180	0.28

Table 14.11 Solvent Permeability through DuPont™ Selar® RB (15%)/Polyolefin Lamellar Structured Containers[3]

Solvent	Temperature (°C)	Exposure Time (Days)	Solvent Weight Loss (%)
Xylene	50	28	0.11
Xylene with 25% propyl alcohol	50	28	0.16
Xylene with 50% propyl alcohol	50	28	0.11
Propyl alcohol with 25% xylene	50	28	0.02
Propyl alcohol	50	28	0.01
Xylene with 25% methyl alcohol	23	180	2.27
Xylene with 50% methyl alcohol	23	180	1.51
Methyl alcohol with 25% xylene	23	180	0.74
Methyl alcohol	23	180	0.3

14.4 Multilayered Films

See Tables 14.12–14.25. See also Figs. 14.1–14.4.

Table 14.12 Permeation of Oxygen through EVOH and PVDC Barrier Layers on Various Three-Layer Barrier Film Structures[5]

Outside (0.152 mm)	PP	PP	PET	PET	PC	PC	PS	PS
Middle layer (0.025 mm)	EVOH	PVDC	EVOH	PVDC	EVOH	PVDC	EVOH	PVDC
Inside layer (0.610 mm)	PP	PP	PP	PP	PP	PP	PP	PP
Permeability Source document units (cm³/100 in.² day atm)	0.028	0.12	0.015	0.12	0.022	0.12	0.022	0.12
Permeability Normalized units (cm³ mm/m² day atm)	0.341	1.46	0.18	1.46	0.27	1.46	0.27	1.46

65% Outside relative humidity, 100% inside relative humidity, temperature 20 °C.

Table 14.13 Permeation of Oxygen through EVOH and PVDC Barrier Layers on Various Three-Layer Barrier Film Structures[4]

Outside (0.152 mm)	HDPE	HDPE	PC	PC	Nylon	Nylon	LDPE	LDPE	PP	PP
Middle layer (0.025 mm)	EVOH	PVDC	EVOH	PVDC	EVOH	PVDC	EVOH	PVDC	EVOH	PVDC
Inside layer (0.610 mm)	PP	PP	PP	PP	PP	PP	PP	PP	PP	PP
Permeability Source document units (cm³/100 in.² day atm)	0.035	0.12	0.081	0.12	0.083	0.12	0.091	0.12	0.097	0.12
Permeability Normalized units (cm³ mm/m² day atm)	0.42	1.46	0.98	1.46	1.0	1.46	1.09	1.46	1.78	1.46

65% Outside relative humidity, 100% inside relative humidity, temperature 20 °C.

Table 14.14 Permeation of Oxygen through EVOH and PVDC Barrier Layers on Various Three-Layer Barrier Film Structures[4]

Outside (0.152 mm)	PP	PP	PET	PET	PC	PC	PS	PS
Middle layer (0.025 mm)	EVOH	PVDC	EVOH	PVDC	EVOH	PVDC	EVOH	PVDC
Inside layer (0.610 mm)	PP	PP	PP	PP	PP	PP	PP	PP
Permeability Source document units (cm³/100 in.² day atm)	0.048	0.12	0.043	0.12	0.035	0.12	0.035	0.12
Permeability Normalized units (cm³ mm/m² day atm)	0.58	1.46	0.52	1.46	0.42	1.46	0.42	1.46

75% Outside relative humidity, 100% inside relative humidity, temperature 20 °C.

Table 14.15 Permeation of Oxygen through EVOH and PVDC Barrier Layers on Various Three-Layer Barrier Film Structures[4]

Outside (0.152 mm)	HDPE	HDPE	PC	PC	Nylon	Nylon	LDPE	LDPE	PP	PP
Middle layer (0.025 mm)	EVOH	PVDC	EVOH	PVDC	EVOH	PVDC	EVOH	PVDC	EVOH	PVDC
Inside layer (0.610 mm)	PP	PP	PP	PP	PP	PP	PP	PP	PP	PP
Permeability Source document units (cm³/100 in.² day atm)	0.059	0.12	0.11	0.12	0.11	0.12	0.12	0.12	0.13	0.012
Permeability Normalized units (cm³ mm/m² day atm)	0.72	1.46	1.34	1.46	1.34	1.46	1.46	1.46	1.57	1.46

75% Outside relative humidity, 100% inside relative humidity, temperature 20 °C.

Table 14.16 Permeation of Oxygen through EVOH and PVDC Barrier Layers on Various Three-Layer Barrier Film Structures[4]

Outside (0.152 mm)	PP	PP	PET	PET	PC	PC	PS	PS
Middle layer (0.025 mm)	EVOH	PVDC	EVOH	PVDC	EVOH	PVDC	EVOH	PVDC
Inside layer (0.610 mm)	PP	PP	PP	PP	PP	PP	PP	PP
Permeability Source document units (cm³/100 in.² day atm)	0.011	0.12	0.010	0.12	0.010	0.12	0.011	0.12
Permeability Normalized units (cm³ mm/m² day atm)	0.134	1.46	0.15	1.46	0.15	1.46	0.016	1.46

65% Outside relative humidity, 10% inside relative humidity, temperature 20 °C.

Table 14.17 Permeation of Oxygen through EVOH and PVDC Barrier Layers on Various Three-Layer Barrier Film Structures[4]

Outside (0.152 mm)	PP	PP	PET	PET	PC	PC	PS	PS
Middle Layer (0.025 mm)	EVOH	PVDC	EVOH	PVDC	EVOH	PVDC	EVOH	PVDC
Inside Layer (0.610 mm)	PP	PP	PP	PP	PP	PP	PP	PP
Permeability Source document units (cm³/100 in.² day atm)	0.011	0.12	0.010	0.12	0.010	0.12	0.011	0.12
Permeability Normalized units (cm³ mm/m² day atm)	0.134	1.46	0.15	1.46	0.15	1.46	0.016	1.46

75% Outside relative humidity, 10% inside relative humidity, temperature 20 °C.

Table 14.18 Oxygen Permeability and Water Vapor Transmission through Honeywell Capran® Oxyshield® Plus (Nylon 6/EVOH/Nylon 6 Biaxially Oriented)[6]

Temperature (°C)	Relative Humidity (%)	Film Thickness (mm)	Oxygen Transmission Rate (cm³/m² day)	Water Vapor Transmission Rate (g/m² day)
25	0	0.015	1.2	
25	65	0.015	1.5	
38	100	0.015		350
25	0	0.0135	4.7	
25	65	0.0135	4.1	
38	100	0.0135		350

Table 14.19 Oxygen Permeability and Water Vapor Transmission through Honeywell Capran® Oxyshield® 1545 and 2545 (Nylon 6 Biaxially Oriented, PVDC Coated One Side)[5]

Product Code	Temperature (°C)	Relative Humidity (%)	Film Thickness (mm)	Oxygen Transmission Rate (cm³/m² day)	Water Vapor Transmission Rate (g/m² day)
1545	23	65	0.015	14	
1545	38	100	0.015		10.9–14.0
2545	25	0	0.025	14	
2545	38	100	0.025		10.9–14.0

Table 14.20 Permeation of Oxygen through Nylon/EVOH/LDPE (0.015/0.015/0.051 mm) Three-Layer Barrier Film Structures[7]

Relative humidity outside (%)	65	65	80	80	65	65	80	80
Relative humidity inside (%)	100	100	100	100	10	10	10	10
Nylon layer	Oriented	Coated	Oriented	Coated	Oriented	Coated	Oriented	Coated
Permeability	0.04	0.01	0.08	0.16	0.03	0.1	0.06	0.15
Source document units (cm³/100 in.² day atm)								
Permeability	0.05	0.14	0.1	0.2	0.04	0.14	0.08	0.21
Normalized units (cm³ mm/m² day atm)								

Temperature 20 °C.

Table 14.21 Permeation of Oxygen through EVOH/LDPE (0.015/0.051 mm) Two-Layer Barrier Film Structures[6]

Relative humidity outside (%)	65	80	65	80
Relative humidity inside (%)	100	100	10	10
Permeability	0.02	0.04	0.02	0.04
Source document units (cm³/100 in.² day atm)				
Permeability	0.02	0.04	0.02	0.04
Normalized units (cm³ mm/m² day atm)				

Temperature 20 °C.

Table 14.22 Permeation of Oxygen through PET/EVOH/LDPE (0.015/0.051 mm) Three-Layer Barrier Film Structures[6]

Relative humidity outside (%)	65	80	65	80
Relative humidity inside (%)	100	100	10	10
Permeability	0.04	0.1	0.02	0.04
Source document units (cm³/100 in.² day atm)				
Permeability	0.05	0.12	0.02	0.05
Normalized units (cm³ mm/m² day atm)				

Temperature 20 °C.

Table 14.23 Oxygen Permeability and Water Vapor Transmission through Dow Saranex™ 650, 652 Co-extruded Barrier Film (EVA/PVDC/EVA)[8]

Temperature (°C)	Relative Humidity (%)	Film Thickness (mm)	Oxygen Transmission Rate (cm³/ m² day)	Water Vapor Transmission Rate (g/m² day)
23		0.019	20	
38	90	0.019		7.0

Table 14.24 Gas Permeability and Water Vapor Transmission through Dow Saranex™ 25 Co-extruded Barrier Film (HDPE/EVA/RVDC/EVA)[9]

Permeant	Temperature (°C)	Relative Humidity (%)	Gas Permeability Coefficient (cm³ mm/ m² day atm)	Water Vapor Transmission Rate (g mm/ m² day)
Oxygen	23	10	0.4	
Carbon dioxide	23	10	1.1	
Nitrogen	23	10	0.06	
Air	23	10	0.1	
Water vapor	38	90		0.06

Film thickness: 0.051 mm.

Table 14.25 Water Vapor Transmission at 38 °C and 90% Relative Humidity through Honeywell Aclar® Multilayer Films[10]

Film Layers	Layer Thicknesses (mm)	Source Document Units — Water Vapor Transmission Rate (g/100 in.² day)	Normalized Units — Water Vapor Transmission Rate (g mm/m² day)
CTFE/ PE/PVC	0.038/0.051/0.19	0.022	0.09
CTFE/ PE/PVC	0.019/0.051/0.19	0.031	0.12
CTFE/ PVC	0.019/0.254	0.018	0.08

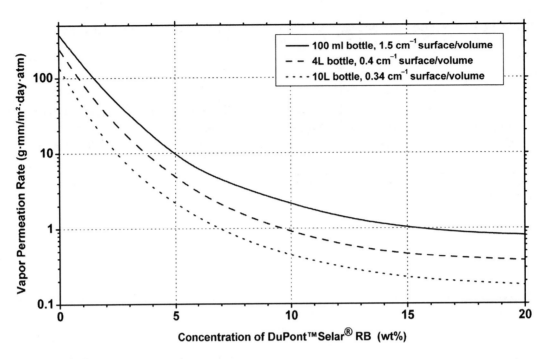

Figure 14.1 Xylene permeability through various bottle sizes vs. percentage (%) weight concentration of generic laminar multilayer structure.

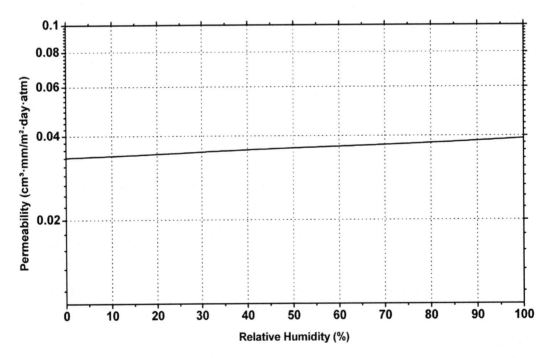

Figure 14.2 Oxygen permeation vs. relative humidity through PVDC-coated Nylon/EVOH copolymer/PE (0.02/0.02/0.51 mm) multilayer film at 20 °C, outside RH 65%.[6]

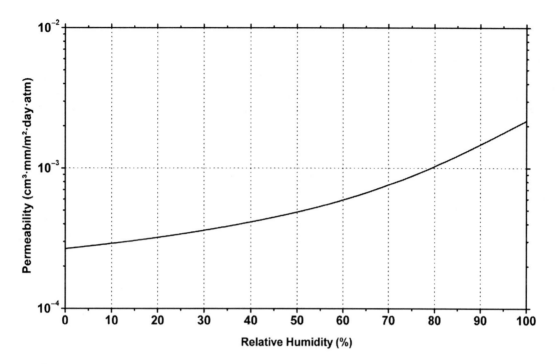

Figure 14.3 Oxygen permeation vs. relative humidity through oriented PP/EVOH copolymer/PE (0.02/0.015/0.51 mm) multilayer film at 20 °C, outside RH 65%.[6]

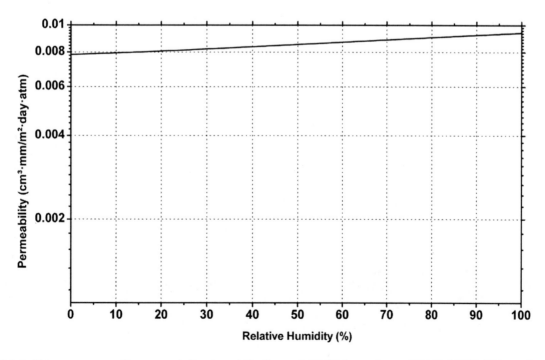

Figure 14.4 Oxygen permeation vs. relative humidity through EVOH copolymer/PE (0.015/0.051 mm) multilayer film at 20 °C, outside RH 65%.[6]

References

1. Available from: http://www.filmet.it/eng/page12.htm.
2. Vacmet Packagings (India), pvt. Ltd. *Metalized plastic films*; 2002.
3. Available from: http://www.fraunhofer.de
4. *Introducing lamellar injection molding technology—the LIM advantage (licensing bulletin). Supplier marketing literature (304-00383-493 SMG)*. Dow Chemical Company; 1993.
5. Gas barrier properties of EVAL resins—technical bulletin no. 1 10. Supplier technical report. EVAL Company of America; 2007.
6. Oxyshield® technical bulletin. 2010.
7. EVAL films the ultimate laminating film for barrier packaging applications—technical bulletin no. 160. Supplier technical report. EVAL Company of America; 2007.
8. *Saranex data sheets, Form No. 500-01185-0401 SMG*. The Dow Chemical Company; April 2001.
9. Saranex plastic film data sheets. Supplier technical report [500- (1170, 1184, 1185, 1186, 1161, 1187)- (588, 1289X, 1289)]. Dow Chemical Company; 1990.
10. *Aclar performance films. Supplier technical report (SFI-14 Rev. 9—89)*. Allied-Signal Engineered Plastics; 1989.

Appendix A: Conversion Factors

Permeability Coefficients

To Convert From	To (cm³ mm)/ (m² day atm)	Note
1×10^{-10} cm³ (STP) cm/cm² s cm Hg	65.664	
cm³ (STP) cm/cm² s cm Hg	6.5664×10^{11}	
Barrer (above)	65.664	
cm³ (STP) mm/m² s Pa	8.7545E+09	
cm³ (STP) cm/cm² s 10 torr	6.5664×10^{11}	
cm³ (STP) cm/cm² s Pa	8.75E+14	
1×10^{-10} cm³ mm/cm² s cm Hg	6.5664	
1×10^{-9} cm³ mm/cm² s cm Hg	65.664	
1×10^{-9} cm³ cm/cm² s cm Hg	656.64	
1×10^{-9} cm³ cm/cm² s bar	8.7545	
1×10^{-8} cm²/s atm	86.400	
(Same as 1×10^{-8} cm³ cm/(cm² s atm)		
1×10^{-10} cm³ cm/(cm² s atm)	0.86400	
1×10^{-8} cm³ cm/cm² s atm	86.4	
cm³ (STP) cm/cm² s atm	8.6400E+09	
cm³ (STP) mil/100 in.²/day	0.3937	
cm³ (STP) mm/m²/day	1	
cm³ μm/m² day kPa	0.10132	

(Continued)

To Convert From	To (cm³ mm)/ (m² day atm)	Note
cm³ mil/m² atm day	0.0254	
cm³ mil/m² bar day	0.025737	
cm³ mil/ 100 in.² atm day	0.3937	
cm³ mil/cm² s atm	2.1946E+07	
cm³ mil/m² atm day	0.0254	
cm³ mm/ m² mm Hg day	760	
cm³ mm/cm² atm day	10,000	
cm³ mm/cm² kPa s	8.7545E+10	
cm³ mm/m² atm day	1	
cm³ mm/ m² mm Hg day	760	
cm³ mm/m² bar day	1.0132	
cm³ mm/m² day bar	1.0132	
cm³ mm/m² Pa day	1.0132E+05	
cm³ mm/m² s atm	8.6400E+04	
cm³ mm/m² s cm Hg	6.5664E+06	
cm³ N/m² bar day	1.0133	a
cm³/100 in.² atm day	15.5	a
cm³ mm/ 100 in.² atm day	15.5	
cm³/100 in.² bar day	15.705	a
cm³/m² atm day	10	a
cm³/m² day	1	a
cm³/m² day bar	1.0133	a
cm³/m² day bar 100	0.010133	a
cm³ mil/ 100 in.² bar day	0.39892	
ft³ mil/ft² psi day	1.1377E+05	
in³ mil/100 in.² atm day	6.4516	

(Continued)

(Continued)

To Convert From	To (cm³ mm)/ (m² day atm)	Note
ml mil/m² atm day	0.0254	
mm³ mm/m² Pa day	101.32	
mm³ mm/m² s Pa	8.7545E+06	
mm³/m MPa day	0.10133	
m²/s	8.6400E+13	b
10^{-18} m³ m/m² s Pa	8.7545	

[a]Conversion factor is applicable only if the film thickness is known; multiply the converted value by the film thickness in mm.

[b]Conversion factor is applicable only if the pressure differential is known; multiply the converted value by the pressure in atm.

Vapor Permeation Rates

To Convert From	To (g mm)/ (m² day)	Note
1 × 10^{-20} kg m/m² s Pa	8.5400E-10	a
g mil/ 100 in.² atm day	0.3937	a

(Continued)

To Convert From	To (g mm)/ (m² day)	Note
g mil/ 100 in.² bar day	0.39892	a
g mil/100 in.² day	0.3937	
g mil/100 in.² h	9.4488	
g mil/ 100 in.² mm Hg day	299.21	a
g mil/day/100 in.²	0.3937	
g mil/m² atm day	0.0254	a
g mm/cm² day	10,000	
g mm/100 in.² day	15.5	
g mm/day/m²	15.5	
g mm/m² day	1	
g/100 in.² day	15.5	
g/m² day	1	
Grains/ft² h	16.74	
mg mil/in.² day	0.03937	
mg mm/m² Pa day	101.33	a
N cm³/m² bar day	0.0254	a
g mm/m² day kPa	101.32	a

[a]The unit of pressure in the original unit can be ignored.

Appendix B: Reference Fuel Compositions

ASTM Reference Fuels Defined by ASTM D471—06e1 Standard Test Method for Rubber Property—Effect of Liquids

Name of Fuel	Composition	Percentage by Volume
Reference fuel A	Isooctane	100
Reference fuel B	Isooctane	70
	Toluene	30
Reference fuel C	Isooctane	50
	Toluene	50
Reference fuel D	Isooctane	60
	Toluene	40
Reference fuel E	Toluene	100
Reference fuel F	Grade #2 diesel fuel	100
Reference fuel G	Reference fuel D	85
	Ethanol	15
Reference fuel H	Reference fuel C	85
	Ethanol	15
Reference fuel I	Reference fuel C	85
	Methanol	15
Reference fuel K	Reference fuel C	15
	Methanol	85

DIN Reference Fuels: DIN 51604-1 FAM Testing Fluid for Polymer Materials; Composition and Requirements

Name	Composition	Percentage by Volume
FAM A	Toluene	50
	Isooctane	30
	Di-isobutylene (2,4,4,-trimethyl 1-pentene)	15
	Ethanol	5
FAM B	FAM A	84.5
	Methanol	15
	Water	0.5
FAM C	FAM A	40
	Methanol	58
	Water	2

Ford Reference Fuels: AZ 105-01

Name	Composition	Percentage by Volume
Ford base fuel	Reference fuel C	80
	Methanol	15
	t-Butyl alcohol (2-methyl propan-2-ol)	5
PN 90 (peroxide number)	Ford base fuel	97.76
	70% t-butyl hydroperoxide	1.24
	Copper ion solution	1

(*Continued*)

(*Continued*)

Name	Composition	Percentage by Volume
PN 180 (peroxide number)	Ford base fuel	96.52
	70% *t*-butyl hydroperoxide	2.48
	Copper ion solution	1

Appendix C

8th Edition Permeation/Degradation Resistance Chart for Ansell Gloves

How to Read the Charts

Three categories of data are represented for each Ansell product and corresponding chemical: 1) overall degradation resistance rating; 2) permeation breakthrough time, and 3) permeation rate.

Standards for Striped-Coding

A glove-chemical combination receives VERTICAL STRIPES if either set of the following conditions is met:

- The Degradation Rating is Excellent or Good
- The Permeation Breakthrough Time is 30 minutes or greater
- The Permeation Rate is Excellent, Very Good, or Good

OR

- The Permeation Rating is not specified
- The Permeation Breakthrough Time is 240 minutes or greater
- The Degradation Rating is Excellent, or Good

A glove-chemical combination receives HORIZONTAL STRIPES if either set of the following conditions is met:

- The Degradation Rating is Poor or Not Recommended

OR

- The Degradation Rating is Degrades with Delamination (DD)
- The Permeation Breakthrough Time is less than 20 minutes

All other glove-chemical combinations receive DIAGONAL STRIPES. In other words, any glove-chemical combination not meeting either set of conditions required for Vertical Stripes, and not having a Horizontal Stripes degradation rating of either Poor or Not Recommended, receives a Diagonal Stripes rating.

Why is a product with a shorter breakthrough time sometimes given a better rating than one with a longer breakthrough time?

One glove has a breakthrough time of just 4 minutes. It is rated "very good," while another with a breakthrough time of 30 minutes is rated only "fair." Why? The reason is simple: in some cases the *rate* is more significant than the *time*.

Imagine connecting two hoses of the same length but different diameters to a faucet using a "Y" connector. When you turn on the water, what happens? Water goes through the smaller hose first because there is less space inside that needs to be filled. But when the water finally gets through the larger hose it really gushes out. In only a few minutes, the larger hose will discharge much more water than the smaller one, even though the smaller one started first.

The situation is similar with gloves. A combination of a short breakthrough time and a low permeation rate may expose a glove wearer to less chemical than a combination of a longer breakthrough time and a much higher permeation rate, if the glove is worn long enough.

Key to Permeation Rate

	Drops/hr Through a Glove (eyedropper-size drops)
E – Excellent; permeation rate of less than 0.9 µg/cm²/min.	0 to 1/2 drop
VG – Very Good; permeation rate of less than 9 µg/cm²/min.	1 to 5 drops
G – Good; permeation rate of less than 90 µg/cm²/min.	6 to 50 drops
F – Fair; permeation rate of less than 900 µg/cm²/min.	51 to 500 drops
P – Poor; permeation rate of less than 9000 µg/cm²/min.	501 to 5000 drops
NR – Not Recommended; permeation rate greater than 9000 µg/cm²/min.	5001 drops up

Key to Permeation Breakthrough

\> Greater than (time)   < Less than (time)

Key to Degradation Ratings

- E – Excellent; fluid has very little degrading effect.
- G – Good; fluid has minor degrading effect.
- F – Fair; fluid has moderate degrading effect.
- P – Poor; fluid has pronounced degrading effect.
- DD – Degrades the outer layer and delaminates it.
- NR – Not Recommended; fluid has severe degrading effect.

DD is a new degradation rating that applies to Viton/butyl gloves versus certain chemicals. It means "Degrades and Delaminates". If a chemical causes severe swelling of Viton but has little effect on butyl, the adhesion between these two rubber layers can be overcome under the relatively severe continuous liquid contact that is part of an ASTM or CEN standard permeation test. The end result of this stress is Viton "blisters" or even complete layer separation. The damage is likely to be permanent.

In cases such as these the butyl layer is providing most of the protection. But if the end use involves only the possibility of splash or intermittent contact so that the Viton layer never absorbs enough chemical to swell and delaminate, Viton/butyl gloves might still be the best choice. The ultimate decision on when to use plain butyl and when to use Viton/butyl will depend on the overall chemical mix in your facility and on the degree of exposure to each.

Specific Gloves Used for Testing

	Degradation and Permeation
Laminated LCP™ Film	Barrier® 2-100 (2.5 mil/0.06 mm)
Nitrile	Sol-Vex® 37-165 (22 mil/0.56 mm)
Neoprene Unsupported	29-865 (18 mil/0.46 mm)
Polyvinyl Alcohol Supported	PVA™
Polyvinyl Chloride Supported	Snorkel®
Natural Rubber Latex	Canners 343 (20 mil/0.51 mm)
Neoprene/Latex Blend	Chemi-Pro® 224 (27 mil/0.68 mm)
Butyl Unsupported	ChemTek® 38-320 (20 mil/0.51 mm)
Viton/Butyl Unsupported	ChemTek® 38-612 (12 mil/0.30 mm)

The first square in each column for each glove type is stripe coded to provide an overall rating for both Degradation and Permeation. The letter in each striped square is for Degradation alone.

- VERTICAL STRIPES: The glove is very well suited for application with that chemical.
- DIAGONAL STRIPES: The glove is suitable for that application under careful control of its use.
- HORIZONTAL STRIPES: Avoid use of the glove with this chemical.

SPECIAL NOTE: The chemicals in this guide highlighted in LIGHT GRAY are experimental carcinogens, according to the ninth edition of Sax' *Dangerous Properties of Industrial Materials*. Chemicals highlighted in DARK GRAY are listed as suspected carcinogens, experimental carcinogens at extremely high dosages, and other materials which pose a lesser risk of cancer.

- • – A degradation test against this chemical was not run. However, since its breakthrough time is greater than 480 minutes, the Degradation Rating is expected to be Good to Excellent.
- X – A degradation test against this chemical was not run. However, in view of degradation tests performed with similar compounds, the Degradation Rating is expected to be Good to Excellent.
- •• – A degradation test against this chemical was not run. However, in view of data obtained with similar compounds, the Degradation Rating is expected to be Fair to Poor.
- ∗ – CAUTION: This product contains natural rubber latex which may cause allergic reactions in some individuals.

8th Edition Permeation/Degradation Resistance Chart for Ansell Gloves

CHEMICAL	Laminate Film (Barrier™) Deg.	Break.	Rate	Nitrile (Sol-Vex®) Deg.	Break.	Rate	Unsupported Neoprene (29-Series) Deg.	Break.	Rate	Supported Polyvinyl Alcohol (PVA™) Deg.	Break.	Rate	Polyvinyl Chloride (Vinyl) (Snorkel®) Deg.	Break.	Rate	Natural Rubber (*Canners and Handlers™) Deg.	Break.	Rate	Neoprene/Natural Rubber Blend (*Chemi-Pro®) Deg.	Break.	Rate	Butyl Unsupported (Chemtek™ Butyl) Deg.	Break.	Rate	Viton/Butyl Unsupported (ChemTek™ Viton/Butyl) Deg.	Break.	Rate
1. Acetaldehyde	X	380	E	P	—	—	E	10	F	NR	—	—	NR	—	—	E	13	F	E	10	F	—	—	—	—	—	—
2. Acetic Acid, Glacial, 99.7%	X	150	—	G	158	—	E	390	—	NR	—	—	F	45	G	E	110	—	E	263	—	E	>480	—	DD	>480	—
3. Acetone	•	>480	E	NR	—	—	G	10	F	P	143	G	NR	<5	—	E	10	F	G	12	G	E	>480	E	DD	93	VG
4. Acetonitrile	•	>480	E	F	30	F	E	20	VG	X	150	G	NR	—	—	E	4	VG	E	13	VG	E	>480	E	DD	70	E
5. Acrylic Acid	—	—	—	G	120	—	E	395	—	NR	—	—	NR	—	—	E	80	—	E	67	—	—	—	—	—	—	—
6. Acrylonitrile	•	>480	E	—	—	—	—	—	—	•	>480	—	—	—	—	E	5	F	—	—	—	E	>480	—	E	>480	—
7. Allyl Alcohol	•	>480	E	F	140	F	E	140	VG	P	—	—	P	60	G	E	10	VG	E	20	VG	E	>480	—	E	>180	—
8. Ammonia Gas	X	19	E	•	>480	E	•	>480	—	—	—	—	—	—	—	—	—	—	X	27	E	—	—	—	—	—	—
9. Ammonium Fluoride, 40%	•	>480	E	E	>360	—	E	>480	—	NR	—	—	E	>360	—	E	>360	—	E	>360	—	—	—	—	—	—	—
10. Ammonium Hydroxide, Conc. (28-30% Ammonia)	E	30	—	E	>360	—	E	250	—	NR	—	—	E	240	—	E	90	—	E	247	—	E	>480	—	E	>480	—
11. n-Amyl Acetate	•	470	E	E	198	G	NR	—	—	G	>360	E	P	—	—	NR	—	—	P	—	—	E	128	G	F	<10	F
12. Amyl Alcohol	•	>480	E	E	>480	E	E	348	VG	G	180	G	G	12	E	E	25	VG	E	52	VG	E	>480	E	E	>480	E
13. Aniline	•	>480	E	NR	—	—	E	145	F	F	>360	E	F	62	G	E	25	VG	E	82	G	E	>480	E	E	>480	E
14. Aqua Regia	—	—	—	F	>360	—	G	>480	—	NR	—	—	G	120	—	NR	—	—	G	193	—	E	>480	—	E	>480	—
15. Benzaldehyde	•	>480	E	NR	—	—	NR	—	—	G	>360	E	NR	—	—	E	10	VG	E	27	F	E	>480	E	E	100	E
16. Benzene (Benzol)	•	>480	E	P	—	—	NR	—	—	E	>360	E	NR	—	—	NR	—	—	NR	—	—	E	20	F	E	253	VG
17. Benzotrichloride	•	>480	E	E	>480	E	NR	—	—	—	—	—	G	—	—	NR	—	—	NR	—	—	—	—	—	—	—	—
18. Benzotrifluoride	•	>480	E	E	170	G	—	—	—	—	—	—	G	<10	F	P	50	G	P	—	—	—	—	—	—	—	—
19. Bromine Water	—	—	—	E	>480	E	E	>480	E	NR	—	—	—	—	—	—	—	—	—	—	—	—	—	—	—	—	—
20. 1-Bromopropane (Propyl Bromide)	•	>480	E	••	23	F	••	<10	P	•	>480	E	••	<10	F	••	<10	P	••	<10	P	••	10	P	X	182	VG
21. 2-Bromopropionic Acid	•	>480	—	F	120	—	E	460	—	—	—	—	G	180	—	E	190	—	G	190	—	—	—	—	—	—	—
22. n-Butyl Acetate	•	>480	E	F	75	F	NR	—	—	G	>360	E	NR	—	—	NR	—	—	P	—	—	E	80	G	DD	<10	F
23. n-Butyl Alcohol	•	>480	E	E	>360	E	E	270	E	F	75	G	G	180	VG	E	35	VG	E	75	VG	E	>480	E	E	>480	E
24. Butyl Carbitol	—	—	—	E	>323	E	G	188	F	E	>480	E	E	397	VG	E	44	G	E	148	G	—	—	—	—	—	—
25. Butyl Cellosolve	•	>480	E	E	470	VG	E	180	G	X	120	G	P	60	G	E	45	G	E	48	G	E	>480	—	E	>480	—
26. gamma-Butyrolactone	•	>480	E	NR	—	—	E	245	G	E	120	VG	NR	—	—	E	60	G	E	104	F	E	>480	E	E	>480	E
27. Carbon Disulfide	•	>480	E	G	30	F	NR	—	—	E	>360	E	NR	<5	—	NR	—	—	NR	—	—	••	7	G	X	138	E
28. Carbon Tetrachloride	—	—	—	G	150	G	NR	—	—	E	>360	E	F	25	F	NR	—	—	NR	—	—	F	53	P	—	—	—
29. Cellosolve® (Ethyl Glycol Ether, 2-Ethoxyethanol)	E	>480	E	G	293	G	E	128	G	X	75	G	P	38	G	E	25	VG	E	25	VG	E	>480	E	E	465	E
30. Cellosolve Acetate® (2-Ethoxyethyl Acetate, EGEAA)	•	>480	E	F	90	G	G	40	F	X	>360	E	NR	—	—	E	10	G	E	23	G	E	>480	E	DD	105	VG
31. Chlorine Gas	•	>480	E	—	—	—	—	—	—	—	—	—	—	—	—	—	—	—	—	—	—	—	—	—	—	—	—
32. Chlorobenzene	•	>480	E	NR	—	—	NR	—	—	E	>360	E	NR	—	—	NR	—	—	NR	—	—	P	9	P	F	>480	E
33. 4-Chlorobenzotrifluoride	—	—	—	E	320	VG	F	50	F	F	—	—	F	—	—	P	—	—	P	—	—	X	75	F	X	48	F
34. 2-Chlorobenzyl Chloride	E	120	E	—	—	—	F	200	E	E	>480	E	F	65	E	F	20	F	—	—	—	E	>480	E	E	>480	E
35. Chloroform	E	20	G	NR	—	—	NR	—	—	E	>360	E	NR	—	—	NR	—	—	NR	—	—	P	5	P	X	212	VG
36. 1-Chloronaphthalene	•	>480	E	P	—	—	NR	—	—	G	>360	E	NR	—	—	NR	—	—	P	—	—	E	>480	E	E	>480	E
37. 2-Chlorotoluene	•	>480	E	G	120	G	NR	—	—	—	—	—	—	—	—	NR	—	—	NR	—	—	NR	—	—	—	—	—
38. 4-Chlorotoluene	•	>480	E	P	—	—	NR	—	—	—	—	—	P	—	—	NR	—	—	NR	—	—	••	30	F	•	>480	E

Appendix C

8th Edition Permeation/Degradation Resistance Chart for Ansell Gloves (CONTINUED)

CHEMICAL	Laminate Film (Barrier™) Deg	Perm: BT	Perm: Rate	Nitrile (Sol-Vex®) Deg	Perm: BT	Perm: Rate	Unsupported Neoprene (29-Series) Deg	Perm: BT	Perm: Rate	Supported Polyvinyl Alcohol (PVA™) Deg	Perm: BT	Perm: Rate	Polyvinyl Chloride (Vinyl) (Snorkel®) Deg	Perm: BT	Perm: Rate	Natural Rubber (*Canners and Handlers™) Deg	Perm: BT	Perm: Rate	Neoprene/Natural Rubber Blend (*Chemi-Pro®) Deg	Perm: BT	Perm: Rate	Butyl Unsupported (Chemtek™ Butyl) Deg	Perm: BT	Perm: Rate	Viton/Butyl Unsupported (ChemTek™ Viton/Butyl) Deg	Perm: BT	Perm: Rate
39. "Chromic Acid" Cleaning Solution	—	—	—	F	240	—	NR	—	—	NR	—	—	G	>360	—	NR	—	—	NR	—	—	E	>480	—	E	>480	—
40. Citric Acid, 10%	—	—	—	E	>360	—	E	>480	—	F	50	—	E	>360	E	E	>360	—	E	>480	—	—	—	—	—	—	—
41. Cyclohexane	—	—	—	•	>360	—	—	—	—	—	—	—	—	—	—	—	—	—	—	—	—	G	30	F	•	>480	—
42. Cyclohexanol	•	>480	E	E	>360	E	E	390	VG	G	>360	E	E	360	E	E	103	VG	E	47	G	E	>480	E	•	>480	E
43. Cyclohexanone	•	>480	E	F	103	G	P	23	F	E	>480	E	NR	—	—	P	—	—	P	—	—	E	>480	—	••	150	—
44. 1,5-Cyclooctadiene	•	>480	E	E	>480	E	NR	—	—	—	—	—	NR	—	—	NR	—	—	NR	—	—	P	—	—	—	—	—
45. Diacetone Alcohol	•	>480	E	G	240	E	E	208	VG	X	150	G	NR	—	—	E	43	VG	E	60	VG	E	>480	—	DD	—	—
46. Dibutyl Phthalate	—	—	—	G	>360	E	F	132	G	E	>360	E	NR	—	—	E	20	—	G	>480	E	—	—	—	—	—	—
47. 1,2-Dichloroethane (Ethylene Dichloride, EDC)	•	>480	E	NR	—	—	NR	—	—	E	>360	E	NR	—	—	P	—	—	P	—	—	—	—	—	—	—	—
48. Diethylamine	•	>480	E	F	51	F	P	—	—	NR	—	—	NR	—	—	NR	—	—	NR	—	—	F	18	—	••	19	—
49. Diisobutyl Ketone (DIBK)	•	>480	E	E	263	G	P	—	—	G	>360	E	P	—	—	P	—	—	P	—	—	E	231	G	DD	15	G
50. Dimethyl Sulfoxide (DMSO)	•	>480	E	E	240	VG	E	398	G	NR	—	—	NR	—	—	E	180	E	E	150	E	E	>480	—	DD	>480	—
51. Dimethylacetamide (DMAC)	•	>480	E	NR	—	—	NR	—	—	NR	—	—	NR	—	—	E	15	G	E	30	G	E	>480	—	DD	>480	—
52. Dimethylformamide (DMF)	•	>480	E	NR	—	—	E	45	F	NR	—	—	NR	19	—	E	25	VG	E	40	G	E	>480	E	DD	>480	E
53. Dioctyl Phthalate (DOP, DEHP)	•	>480	E	G	>360	E	G	>480	E	E	30	F	NR	—	—	P	—	—	E	>360	E	—	—	—	—	—	—
54. Di-n-Octyl Phthalate (DNOP)	—	—	—	—	—	—	—	—	—	—	—	—	—	—	—	—	—	—	—	—	—	E	>480	—	—	—	—
55. 1,4-Dioxane	•	>480	E	NR	—	—	NR	—	—	P	—	—	NR	—	—	F	5	F	F	18	F	E	>480	—	—	—	—
56. Electroless Copper Plating Solution	—	—	—	E	>360	—	E	>360	—	NR	—	—	E	>360	—	E	>360	—	—	—	—	—	—	—	—	—	—
57. Electroless Nickel Plating Solution	—	—	—	E	>360	—	E	>360	—	NR	—	—	E	>360	—	E	>360	—	—	—	—	—	—	—	—	—	—
58. Epichlorohydrin	•	>480	E	NR	—	—	P	—	—	E	300	E	NR	—	—	E	5	F	E	17	VG	E	>480	—	—	—	—
59. Ethidium Bromide, 10%	•	>480	E	•	>480	E	—	—	—	NR	—	—	—	—	—	—	—	—	—	—	—	E	>480	—	E	>480	—
60. Ethyl Acetate	•	>480	E	NR	—	—	F	10	P	F	>360	E	NR	—	—	G	5	F	F	10	F	E	196	G	DD	10	G
61. Ethyl Alcohol, Denatured, 92% Ethanol	•	>480	E	E	240	VG	E	113	VG	NR	—	—	G	60	VG	E	15	VG	E	37	VG	E	>480	E	E	>480	E
62. Ethylene Glycol	•	>480	E	E	>360	E	E	>480	E	F	120	VG	E	>360	E	E	>360	E	E	>480	E	—	—	—	—	—	—
63. Ethylene Oxide Gas	X	234	E	—	—	—	—	—	—	—	—	—	—	—	—	—	—	—	—	—	—	—	—	—	—	—	—
64. Ethyl Ether	•	>480	E	E	95	G	F	<10	F	G	>360	E	NR	—	—	NR	—	—	NR	—	—	—	—	—	—	—	—
65. Ethyl L-Lactate	E	>480	E	E	273	G	E	125	VG	E	125	G	E	15	G	E	15	VG	E	28	VG	E	>480	E	E	>480	E
66. Formaldehyde, 37% in 1/3 Methanol/Water	•	>480	E	E	>360	E	E	39	VG	P	—	—	E	100	E	E	10	G	E	32	E	E	>480	—	E	>480	—
67. Formic acid, 90%	•	>480	—	F	240	—	E	>480	—	NR	—	—	E	>360	—	E	150	—	E	>360	—	E	>480	—	—	—	—
68. Furfural	•	>480	E	NR	—	—	E	40	P	F	>360	E	NR	—	—	E	15	VG	E	43	VG	E	>480	—	G	>480	—
69. Freon TF	—	—	—	E	>360	E	E	240	E	G	>360	E	NR	—	—	NR	—	—	NR	—	—	—	—	—	—	—	—
70. Gasoline, Unleaded (Shell Premium winter blend)	•	170	E	E	>480	E	NR	—	—	G	>360	E	P	—	—	NR	—	—	NR	—	—	F	20	F	E	>480	E
71. Glutaraldehyde, 25%	—	—	—	E	>360	E	E	>480	E	P	<10	F	E	>360	E	E	210	VG	—	—	—	—	—	—	—	—	—
72. HCFC-141B	•	>480	E	E	92	F	F	33	F	P	—	—	NR	—	—	NR	—	—	NR	—	—	F	40	F	F	<10	F

8th Edition Permeation/Degradation Resistance Chart for Ansell Gloves

CONTINUED

CHEMICAL	Laminate Film (Barrier™) Deg.	Br.	Rate	Nitrile (Sol-Vex®) Deg.	Br.	Rate	Unsupported Neoprene (29-Series) Deg.	Br.	Rate	Supported Polyvinyl Alcohol (PVA™) Deg.	Br.	Rate	Polyvinyl Chloride (Vinyl) (Snorkel®) Deg.	Br.	Rate	Natural Rubber (*Canners and Handlers™) Deg.	Br.	Rate	Neoprene/Natural Rubber Blend (*Chemi-Pro®) Deg.	Br.	Rate	Butyl Unsupported (Chemtek™ Butyl) Deg.	Br.	Rate	Viton/Butyl Unsupported (ChemTek™ Viton/Butyl) Deg.	Br.	Rate
73. n-Heptane	•	>480	E	—	—	—	—	—	—	•	>480	—	NR	—	—	—	—	—	—	—	—	P	10	F	E	>480	E
74. Hexamethyldisilazine	•	>480	E	E	>360	—	E	42	—	G	>360	—	P	—	—	F	15	F	F	43	G	X	305	G	X	>480	G
75. n-Hexane	•	>480	E	E	>480	E	E	48	G	G	>360	E	NR	—	—	NR	—	—	P	—	—	P	5	F	E	>480	E
76. HFE 7100	•	>480	E	E	>480	E	E	>480	E	P	—	—	E	>480	E	E	120	E	—	—	—	—	—	—	—	—	—
77. HFE 71DE	X	164	E	F	10	F	F	<10	F	F	>480	E	NR	—	—	NR	—	—	—	—	—	—	—	—	—	—	—
78. Hydrazine, 65%	—	—	—	E	>480	—	E	386	—	NR	—	—	E	>360	—	E	150	VG	E	>360	—	E	>480	—	—	—	—
79. Hydrobromic Acid, 48%	•	>480	—	E	>360	—	E	>480	—	NR	—	—	E	>360	—	E	>360	—	E	>360	—	—	—	—	—	—	—
80. Hydrochloric Acid, 10%	—	—	—	E	>360	—	E	>480	—	NR	—	—	E	>360	—	E	>360	—	E	>360	—	—	—	—	—	—	—
81. Hydrochloric Acid, 37% (Concentrated)	•	>480	—	E	>480	—	E	>480	—	NR	—	—	E	300	—	E	290	—	E	>360	—	—	—	—	—	—	—
82. Hydrofluoric Acid, 48%	•	>480	—	E	334	—	X	>480	—	NR	—	—	X	155	—	•	>480	—	—	—	—	E	>480	—	•	>480	—
83. Hydrofluoric Acid, 95%	•	>480	E	—	—	—	X	342	VG	—	—	—	—	—	—	—	—	—	—	—	—	•	>480	E	—	—	—
84. Hydrogen Fluoride Gas	•	>480	E	X	<15	P	—	—	—	—	—	—	X	2	—	X	15	F	X	<15	F	—	—	—	—	—	—
85. Hydrogen Peroxide, 30%	—	—	—	E	>360	—	E	>480	—	NR	—	—	E	>360	—	E	>360	—	G	>360	—	—	—	—	•	>480	—
86. Hydroquinone, saturated solution	—	—	—	E	>360	E	E	108	E	NR	—	—	E	>360	E	G	>360	E	E	>360	E	—	—	—	—	—	—
87. Hypophosphorus Acid, 50%	—	—	—	E	>480	—	E	>240	—	NR	—	—	E	—	—	E	>480	—	—	—	—	—	—	—	—	—	—
88. Isobutyl Alcohol	•	>480	E	E	>360	E	E	478	E	P	—	—	F	10	VG	E	15	VG	E	52	E	E	>480	E	E	>480	E
89. Isooctane	•	>480	E	E	>360	E	E	268	VG	E	>360	E	P	—	—	NR	—	—	P	—	—	X	58	F	•	>480	E
90. Isopropyl Alcohol	•	>480	E	E	>360	E	E	110	E	NR	—	—	G	150	E	E	35	VG	E	57	E	—	—	—	—	—	—
91. Kerosene	•	>480	E	E	>360	E	E	185	G	G	>360	E	F	>360	E	NR	—	—	P	—	—	G	82	—	E	>480	—
92. Lactic Acid, 85%	•	>480	—	E	>360	—	E	>480	—	F	>360	—	E	>360	—	E	>360	—	E	>360	—	—	—	—	—	—	—
93. Lauric Acid, 36% in Ethanol	—	—	—	E	>360	—	E	>480	—	NR	—	—	F	15	—	E	>360	—	E	>360	—	—	—	—	—	—	—
94. d-Limonene	•	>480	E	E	>480	E	NR	—	—	G	>480	E	G	125	G	NR	—	—	NR	—	—	F	57	F	F	>480	E
95. Maleic Acid, saturated solution	—	—	—	E	>360	—	E	>480	—	NR	—	—	G	>360	—	E	>360	—	E	>360	—	—	—	—	—	—	—
96. Mercury	—	—	—	•	>480	E	—	—	—	—	—	—	•	>480	E	•	>480	E	—	—	—	—	—	—	—	—	—
97. Methyl Alcohol (Methanol)	•	>480	E	E	103	VG	E	73	VG	NR	—	—	G	45	G	E	12	VG	E	22	E	E	>480	—	DD	363	—
98. Methylamine, 40%	•	>480	E	E	>360	E	E	153	G	NR	—	—	E	135	VG	E	55	VG	E	100	E	E	>480	—	E	>480	—
99. Methyl Amyl Ketone (MAK)	•	>480	E	F	53	F	F	10	F	E	>360	E	NR	—	—	F	<10	F	F	<10	F	E	155	G	DD	17	F
100. Methyl-t-Butyl Ether (MTBE)	E	>480	E	E	>360	E	P	—	—	G	>360	E	NR	—	—	NR	—	—	NR	—	—	G	38	F	—	—	—
101. Methyl Cellosolve®	X	470	F	F	208	G	E	10	F	E	30	G	P	55	G	E	20	VG	—	—	—	•	>480	E	•	>480	E
102. Methylene Bromide (DBM)	•	>480	E	NR	—	—	NR	—	—	G	>360	E	NR	—	—	NR	—	—	NR	—	—	E	70	F	E	>480	E
103. Methylene Chloride (DCM)	E	20	VG	NR	—	—	NR	—	—	G	>360	E	NR	—	—	NR	—	—	NR	—	—	G	13	P	E	29	G
104. Methylene bis (4-Phenylisocyanate) (MDI)	—	—	—	—	—	—	—	—	—	—	—	—	—	—	—	•	>480	E	•	>480	E	—	—	—	—	—	—
105. Methyl Ethyl Ketone (MEK)	•	>480	E	NR	—	—	P	—	—	F	90	VG	NR	—	—	F	5	F	P	<10	F	E	183	G	DD	20	G
106. Methyl Ethyl Ketone (MEK)/Toluene, 1/1	•	>480	E	—	—	—	—	—	—	—	—	—	—	—	—	F	5	F	—	—	—	F	60	—	—	—	—
107. Methyl Iodide (Iodomethane)	•	>480	E	NR	—	—	NR	—	—	F	>360	E	NR	—	—	NR	—	—	NR	—	—	F	15	P	G	215	VG

8th Edition Permeation/Degradation Resistance Chart for Ansell Gloves (CONTINUED)

CHEMICAL	Laminate Film (Barrier™) DR	PB	PR	Nitrile (Sol-Vex®) DR	PB	PR	Unsupported Neoprene (29-Series) DR	PB	PR	Supported Polyvinyl Alcohol (PVA™) DR	PB	PR	Polyvinyl Chloride (Vinyl) (Snorkel®) DR	PB	PR	Natural Rubber (*Canners and Handlers™) DR	PB	PR	Neoprene/Natural Rubber Blend (*Chemi-Pro®) DR	PB	PR	Butyl Unsupported (Chemtek™ Butyl) DR	PB	PR	Viton/Butyl Unsupported (ChemTek™ Viton/Butyl) DR	PB	PR
108. Methyl Isobutyl Ketone (MIBK)	•	>480	E	P	45	F	NR	—	—	F	>360	E	NR	—	—	P	—	—	P	—	—	E	245	G	DD	30	G
109. Methyl Methacrylate (MMA)	•	>480	E	P	35	P	NR	—	—	G	>360	E	NR	—	—	P	—	—	NR	—	—	E	85	G	DD	10	F
110. N-Methyl-2-Pyrrolidone (NMP)	•	>480	E	NR	—	—	NR	—	—	NR	—	—	NR	—	—	E	75	VG	F	47	VG	E	>480	—	DD	—	—
111. Mineral Spirits, Rule 66	•	>480	E	E	>480	E	E	125	G	E	>360	E	F	150	VG	NR	—	—	G	23	G	—	—	—	—	—	—
112. Monoethanolamine	—	—	—	E	>360	E	E	400	E	E	>360	E	E	>480	E	E	50	E	E	57	E	—	—	—	X	>120	—
113. Morpholine	•	>480	E	NR	—	—	P	—	—	G	90	G	NR	—	—	G	20	G	E	43	G	E	>480	E	DD	235	VG
114. Naphtha, VM&P	•	>480	E	E	>360	E	G	103	G	E	420	E	F	120	VG	NR	—	—	NR	—	—	—	—	—	—	—	—
115. Nitric Acid, 10%	•	>480	—	E	>360	—	E	>480	—	NR	—	—	G	>360	—	G	>360	—	E	>360	—	—	—	—	—	—	—
116. Nitric Acid, 70% (Concentrated)	E	>480	—	NR	—	—	•	>480	—	NR	—	—	F	109	—	NR	—	—	NR	—	—	—	—	—	—	—	—
117. Nitric Acid, Red Fuming	•	>480	E	NR	—	—	NR	—	—	NR	—	—	P	—	—	P	—	—	NR	—	—	—	—	—	—	—	—
118. Nitrobenzene	•	>480	E	NR	—	—	NR	—	—	G	>360	E	NR	—	—	F	15	G	F	42	G	E	>480	—	E	>480	—
119. Nitromethane	•	>480	E	F	30	F	E	60	G	G	>360	E	P	—	—	E	10	G	E	30	E	E	>480	E	E	249	E
120. 1-Nitropropane	X	368	E	NR	—	—	F	30	G	E	>480	G	NR	—	—	E	15	G	E	25	G	E	>480	E	DD	255	E
121. 2-Nitropropane	•	>480	E	NR	—	—	F	25	F	E	>360	E	NR	—	—	E	5	G	E	30	VG	—	—	—	—	—	—
122. n-Octyl Alcohol	—	—	—	E	>360	E	E	218	E	G	>360	E	F	>360	E	E	30	VG	E	53	—	—	—	—	—	—	—
123. Oleic Acid	—	—	—	E	>360	E	F	13	VG	G	60	E	F	90	VG	F	>360	E	G	120	—	—	—	—	—	—	—
124. Oxalic Acid, saturated solution	—	—	—	E	>360	—	E	>480	—	P	—	—	E	>360	—	E	>360	—	E	>360	—	—	—	—	—	—	—
125. Pad Etch® 1 (Ashland Chemical)	—	—	—	E	>360	—	E	>360	—	F	34	—	E	>360	—	E	>360	—	E	>360	—	—	—	—	—	—	—
126. Palmitic Acid, saturated solution	—	—	—	G	30	—	E	>480	—	P	—	—	G	75	—	G	5	—	E	193	—	—	—	—	—	—	—
127. Pentachlorophenol, 5% in Mineral Spirits	—	—	—	E	>360	E	E	151	F	E	5	F	F	180	E	NR	—	—	—	—	—	—	—	—	—	—	—
128. n-Pentane	E	>480	E	E	>360	E	G	30	G	G	>360	E	NR	—	—	P	—	—	E	13	G	—	—	—	—	—	—
129. Perchloric Acid, 60%	—	—	—	E	>360	—	E	>480	—	NR	—	—	E	>360	—	F	>360	—	E	>360	—	—	—	—	—	—	—
130. Perchloroethylene (PERC)	•	>480	E	G	361	VG	NR	—	—	E	>360	E	NR	—	—	NR	—	—	NR	—	—	P	<10	F	E	>480	E
131. Phenol, 90%	•	>480	E	NR	—	—	E	353	G	F	>360	E	G	75	VG	E	90	E	E	180	E	E	>480	E	E	>480	—
132. Phosphoric Acid, 85% (Concentrated)	•	>480	—	E	>360	—	G	>360	—	NR	—	—	G	>360	—	F	>360	—	G	>360	—	—	—	—	—	—	—
133. Potassium Hydroxide, 50%	—	—	—	E	>360	—	E	>480	—	NR	—	—	E	>360	—	E	>360	—	E	>360	—	—	—	—	—	—	—
134. Propane Gas	—	—	—	•	>480	E	•	>480	E	—	—	—	X	7	VG	—	—	—	—	—	—	—	—	—	—	—	—
135. n-Propyl Acetate	—	—	—	F	20	G	P	—	—	G	120	VG	NR	—	—	P	—	—	P	—	—	E	135	G	DD	<10	F
136. n-Propyl Alcohol	E	>480	E	E	>360	E	E	323	E	P	—	—	F	90	VG	E	23	VG	E	30	E	E	>480	—	E	>480	—
137. Propylene Glycol Methyl Ether Acetate (PGMEA)	•	>480	E	E	200	F	G	37	F	E	>360	E	P	—	—	G	13	F	G	18	F	•	>480	E	X	334	E
138. Propylene Glycol Monomethyl Ether (PGME)	—	—	—	—	—	—	P	—	—	—	—	—	P	—	—	—	—	—	—	—	—	•	>480	E	•	>480	E
139. Propylene Oxide	•	>480	E	NR	—	—	NR	—	—	G	35	G	NR	—	—	P	—	—	P	—	—	X	43	F	DD	<10	F
140. Pyridine	•	>480	E	NR	—	—	NR	—	—	G	10	F	NR	—	—	F	10	F	P	10	F	•	465	E	DD	40	—
141. Rubber Solvent	—	—	—	E	>360	E	E	43	G	E	>360	E	NR	—	—	NR	—	—	NR	—	—	—	—	—	—	—	—

8th Edition Permeation/Degradation Resistance Chart for Ansell Gloves (CONTINUED)

CHEMICAL	Laminate Film (Barrier™) Deg.	Perm: Brk.	Perm: Rate	Nitrile (Sol-Vex®) Deg.	Perm: Brk.	Perm: Rate	Unsupported Neoprene (29-Series) Deg.	Perm: Brk.	Perm: Rate	Supported PVA (PVA™) Deg.	Perm: Brk.	Perm: Rate	PVC (Snorkel®) Deg.	Perm: Brk.	Perm: Rate	Natural Rubber (*Canners and Handlers™) Deg.	Perm: Brk.	Perm: Rate	Neoprene/Natural Rubber Blend (*Chemi-Pro®) Deg.	Perm: Brk.	Perm: Rate	Butyl Unsupported (Chemtek™ Butyl) Deg.	Perm: Brk.	Perm: Rate	Viton/Butyl Unsupported (ChemTek™ Viton/Butyl) Deg.	Perm: Brk.	Perm: Rate
142. Silicon Etch	•	>480	E	NR	—	—	E	>480	—	NR	—	—	F	150	—	NR	—	—	P	—	—	—	—	—	—	—	—
143. Skydrol® 500B-4	•	>480	E	NR	—	—	NR	—	—	—	—	—	NR	—	—	NR	—	—	NR	—	—	E	>480	E	DD	>480	E
144. Sodium Hydroxide, 50%	E	>480	—	E	>360	—	E	>480	—	NR	—	—	G	>480	—	E	>360	—	E	>360	—	E	>480	—	E	>480	—
145. Stoddard Solvent	•	>480	E	E	>360	E	E	139	G	E	>360	E	F	57	G	NR	—	—	G	10	G	—	—	—	—	—	—
146. Styrene	•	>480	E	NR	—	—	NR	—	—	G	>360	E	NR	—	—	NR	—	—	NR	—	—	G	26	—	E	>480	—
147. Sulfur Dichloride	—	—	—	P	>480	E	NR	—	—	—	—	—	—	—	—	NR	—	—	NR	—	—	—	—	—	—	—	—
148. Sulfuric Acid, 47% (Battery Acid)	—	—	—	E	>360	—	E	>360	—	NR	—	—	G	>360	—	E	>360	—	E	>360	—	—	—	—	—	—	—
149. Sulfuric Acid, 95-98% (Concentrated)	E	>480	E	NR	—	—	F	24	—	NR	—	—	G	26	—	NR	—	—	NR	—	—	E	>480	—	E	>480	—
150. Sulfuric Acid, 120% (Oleum)	•	>480	E	—	—	—	F	53	G	NR	—	—	••	25	G	—	—	—	—	—	—	—	—	—	—	—	—
151. Tannic Acid, 65%	—	—	—	E	>360	—	E	>480	E	P	—	—	E	>360	—	E	>360	—	E	>360	—	—	—	—	—	—	—
152. Tetrahydrofuran (THF)	•	>480	E	NR	—	—	NR	—	—	P	115	F	NR	—	—	NR	—	—	NR	—	—	F	13	F	DD	10	F
153. Toluene (Toluol)	•	>480	E	F	34	F	NR	—	—	G	>1440	E	NR	—	—	NR	—	—	NR	—	—	P	20	F	E	313	—
154. Toluene Diisocyanate (TDI)	•	>480	E	NR	—	—	NR	—	—	G	>360	E	P	—	—	G	7	G	G	65	VG	E	>480	—	E	>480	—
155. Triallylamine	•	>480	E	•	>480	E	—	—	—	—	—	—	—	—	—	—	—	—	—	—	—	—	—	—	—	—	—
156. Trichloroethylene (TCE)	•	>480	E	NR	—	—	NR	—	—	E	>360	E	NR	—	—	NR	—	—	NR	—	—	NR	—	—	DD	204	VG
157. Tricresyl Phosphate (TCP)	—	—	—	E	>360	E	F	253	F	G	>360	E	F	>360	E	E	45	E	E	>360	E	E	>480	—	E	>480	—
158. Triethanolamine (TEA)	—	—	—	E	>360	E	E	170	VG	G	>360	E	E	>360	E	G	>360	E	—	—	—	—	—	—	—	—	—
159. Turpentine	•	>480	E	E	>480	E	NR	—	—	G	>360	E	P	—	—	NR	—	—	NR	—	—	X	58	—	X	>480	E
160. Vertrel® MCA	•	>480	E	E	110	G	E	23	G	F	>360	E	G	13	F	G	<10	F	G	<10	F	X	173	VG	DD	20	G
161. Vertrel® SMT	E	10	G	P	—	—	F	<10	F	G	17	G	G	<10	E	F	<10	F	P	<10	P	••	18	F	DD	<10	F
162. Vertrel® XE	E	105	E	E	>480	E	E	47	G	F	40	VG	G	303	E	E	17	VG	E	43	VG	E	>480	E	DD	398	E
163. Vertrel® XF	E	>480	E	E	>480	E	E	>480	E	F	387	VG	E	>480	E	E	337	VG	E	204	G	E	>480	E	DD	>480	E
164. Vertrel® XM	E	>480	E	E	>480	E	E	105	E	F	10	G	P	55	G	E	23	VG	E	30	VG	—	—	—	—	—	—
165. Vinyl Acetate	•	>480	E	F	18	F	—	—	—	—	—	—	—	—	—	—	—	—	—	—	—	NR	—	—	—	—	—
166. Vinyl Chloride Gas	•	>480	E	—	—	—	—	—	—	—	—	—	—	—	—	—	—	—	—	—	—	—	—	—	—	—	—
167. Xylenes, Mixed (Xylol)	•	>480	E	G	96	F	NR	—	—	E	>360	E	NR	—	—	NR	—	—	NR	—	—	P	27	F	E	>480	E

NOTE:

This is an abridged version of Ansell's "8th Edition Chemical Resistance Guide, Permeation and Degradation Data." The complete guide is available as a pdf download at ansellpro.com. Permeation and degradation data for additional chemicals not found in the printed guide is available by using the SpecWare® link in ansellpro.com.

These recommendations are based on laboratory tests, and reflect the best judgement of Ansell in the light of data available at the time of preparation and in accordance with the current revision of ASTM F 739. They are intended to guide and inform qualified professionals engaged in assuring safety in the workplace. Because the conditions of ultimate use are beyond our control, and because we cannot run permeation tests in all possible work environments and across all combinations of chemicals and solutions, these recommendations are advisory only. The suitability of a product for a specific application must be determined by testing by the purchaser.

The data in this guide are subject to revision as additional knowledge and experience are gained. Test data herein reflect laboratory performance of partial gloves and not necessarily the complete unit. Anyone intending to use these recommendations should first verify that the glove selected is suitable for the intended use and meets all appropriate health standards. Upon written request, Ansell will provide a sample of material to aid you in making your own selection under your own individual safety requirements.

NEITHER THIS GUIDE NOR ANY OTHER STATEMENT MADE HEREIN BY OR ON BEHALF OF ANSELL SHOULD BE CONSTRUED AS A WARRANTY OF MERCHANTABILITY OR THAT ANY ANSELL GLOVE IS FIT FOR A PARTICULAR PURPOSE. ANSELL ASSUMES NO RESPONSIBILITY FOR THE SUITABILITY OR ADEQUACY OF AN END-USER'S SELECTION OF A PRODUCT FOR A SPECIFIC APPLICATION.

Skydrol is a registered trademark of Solutia Inc. Vertrel is a registered trademark of DuPont.

Index

Note: Page numbers followed by f indicate figures and t indicate tables.

2-layer barrier film structures: oxygen permeation through EVOH/LDPE, 314t
3-layer barrier film structures:
   oxygen permeation through EVOH and PVDC barrier layers, 310t, 311t, 312t, 313t
   oxygen permeation through nylon/EVOH/LDPE, 314t
   oxygen permeation through PET/EVOH/LDPE, 314t
3-layer blown film die, 42f
9-layer blown film die, 43f
10-layer stacked die, 40f

A

ABS *see* acrylonitrile-butadiene-styrene
Accelerated Keeping Test (AKT), 56, 57t
acetal copolymer, 247–250
acetal homopolymer, 247–250
Aclar Multilayer films (Honeywell), 315t
Aclar PCTFE film (Honeywell), 217t, 218t
acrylics, 145, 145f
acrylonitrile-butadiene copolymer (NBR), 281–284
   applications, 281
   monomer/polymer structure, 281f
acrylonitrile-butadiene-styrene (ABS), 77–80
   gas permeation, 78t, 79t, 80f
   INEOS ABS Lustran 246 ABS, 80f
   permselectivity vs. temperature, 80f
   Sabic Innovative Plastics Cycolac ABS, 78t
   water vapor permeation, 78t, 79t
acrylonitrile-methyl acrylate (AMA), 188–189
   applications, 188
   carbon dioxide permeation, 189f
   chemical structure, 188f
   gas permeation, 188t
   INEOS Barex Acrylonitrile-Methyl Acrylate Copolymers, 188t
   oxygen permeation, 189f
   water vapor permeation, 188t
acrylonitrile-styrene-acrylate (ASA), 80–82
   gas permeation, 81t, 82t
   water vapor permeation, 82t

addition polymerization, 21
additives, 31–37
Adflex PP (LlyondellBasell):
   carbon dioxide permeation, 158t
   oxygen permeation, 158t
adhesion tests, 17–19
Advanced Elastomer Systems Santoprene TPO:
   gas permeation, 256t
   water vapor transmission, 256t
Aegis Nylon 6 films (Honeywell), 128t
Aegis Nylon 6/66 films (Honeywell), 139t
AEM *see* ethylene acrylic elastomers
AF *see* amorphous fluoropolymer
agricultural chemicals/films, 60–65
AKT *see* Accelerated Keeping Test
alkenes, 145
alternating copolymers, 22
AMA *see* acrylonitrile-methyl acrylate
American Society for Testing and Materials *see* ASTM
amide-imide polymers, 108t, 109t
amino acids, 122f
amorphous fluoropolymer (AF), 214–215
   applications, 215
   chemical structure, 214f
   pure gas permeation, 214t
amorphous PET, 105f
amorphous plastics, 29–30
   See also individual plastics
amorphous polyamide (nylon), 122–126
   applications, 123
   carbon dioxide permeation, 123t
   chemical structure, 122f
   nitrogen permeation, 123t
   oxygen permeation, 123t, 125f, 126f
   water vapor permeation, 123t
amorphous regions, 196
Ansell gloves: permeation/degradation resistance chart, 323–328
antiblock additives, 34
antistats, 36–37
Arkema PEBAX Breathable PEBA films, 262f

329

Arkema PEBAX films:
  helium permeation, 260t
  oxygen permeation, 260t
  propane permeation, 261t
  water vapor permeation, 259t, 261t
Arkema Rilsan PA11, 132t
Arkema Voltalef PCTFE film, 219t
Arrhenius equation, 8–9
ASA *see* acrylonitrile-styrene-acrylate
Asahi Glass AFLAS Fluoroelastomer, 278f
Ashland Aqualon Ethyl Cellulose, 299t
ASTM (American Society for Testing and Materials):
  reference fuels, 321
  standards, 11t–12t
atactic polypropylene, 26
atlas cell test, 17, 18f
atmospheric gases, 59–60
automotive fuel hoses/tanks, 67–68

B

bags, 62, 60–65
ball-and-stick model: molecules, 24–25
barrier coatings *see* coatings
barrier films, 59–65
  See also films
BASF AG Lupolen V EVA, 161f
BASF AG Polystyrol 168 N GPPS film, 84t
BASF AG Terluran ABS films:
  gas permeation, 79t
  water vapor permeation, 79t
BASF AG Ultradur PBT, 93t
BASF Elastollan TPUs:
  gas permeation, 253t
  water vapor permeation, 254t
BASF Luran SAN:
  oxygen permeation, 87t
  water vapor permeation, 87t
BASF Luran S ASA films:
  gas permeation, 81t, 82t
  water vapor permeation, 82t
BASF Styrolux films:
  carbon dioxide permeation, 262t
  nitrogen permeation, 262t
  oxygen permeation, 262t
BASF Udel PSU:
  gas permeation, 239t
  water vapor permeation, 239t
BASF Ultramid A5 Nylon 66 film:
  oxygen and carbon dioxide permeation, 135t
  water vapor permeation, 136t
BASF Ultramid C35 Nylon6/66 film, 140t
Bayer Lanxess Butyl Butyl Rubber, 268t
Bayer Lanxess Krylene Styrene-Butadiene Rubber, 284t
Bayer Lanxess Krynac Nitrile Rubber:
  air permeation, 282t
  air vs. temperature, 283f
  gas permeation, 282t
Bayer MaterialScience Makrolon PC:
  carbon dioxide permeation, 96t–97t
  nitrogen permeation, 95t–96t
  oxygen permeation, 94t–95t
Bayer MaterialScience Texin TPU, 254f, 255f
beer packaging, 65
belt film casting machines, 44f
BESNO P40TL Nylon 11 (Rilsan), 132f, 133f
BET (Brunauer, Emmett, and Teller) model, 6–7, 6f
biaxial orientation, 51
biodegradable plastics, 287
  See also environmentally friendly polymers
bisphenol diamine PMDA polyetherimide, 114f
blends, 30–31
blistering, 18f, 16
blister packs, 62, 64f
block copolymers, 22
blocking, 34
blow molding, 54
blown film:
  orientation, 52, 53f
  process, 39–41
bottles, 60–65
BPADA-DDS polyetherimide sulfone, 114f
BPADA monomer structure, 114f
BPADA-PPD polyetherimide, 113f
Brampton Engineering 10-layer stacked die, 40f
branched polymers, 22–23
bromobutyl rubber, 264–266
  applications, 264–266
  crosslinking, 265f
  structural units, 264f
bromoisobutylene-isoprene copolymer, 266f
Brunauer, Emmett, and Teller (BET) model, 6–7, 6f
butyl rubber, 266
  air permeation, 267t, 268f
  applications, 266–268
  chemical structure, 266f
  Exxon Butyl Rubber, 267t
  gas permeation, 267t, 268t
  oxygen permeation, 267t

C

calendering, 41–42, 43f, 44f
Capran Oxyshield 1545 and 2545 (Honeywell), 313t
Capran Oxyshield Plus (Honeywell), 313t
Capron Nylon 6 films (Honeywell):
  gas permeation, 128t
  water vapor permeation, 128t
caps, 67
Carbon-Filled DuPont Teflon PTFE, 199t
casting film lines, 42–44, 44f
catalysts, 35

Index

Celazole PBI membrane (PBI Performance Products Inc.):
 carbon dioxide permeation, 245f
 hydrogen permeation, 244f
Celcon M25/M90/M270 POM Copolymer (Ticona), 248t
cellophane, 291–292
 applications, 291
 gas permeation, 292t
 oxygen/water vapor transmission, 292t
cellulose acetate, 294
 applications, 294
 chemical structure, 294f
 common gas permeation, 295f, 296f
 Dense and High-Flux, 294t
 gas permeation, 294t, 295f, 296f
 hydrogen sulfide permeation vs. temperature, 297f
 noble gas permeation, 295f, 296f
cellulose to viscose conversion, 291f
chain entanglement, 27–28
chemical product packaging, 60–65
Chevron Philips Marlex HDPE, 154t
Chevron Philips Ryton PPS films:
 gas permeation, 238t
 liquid vapor transmission, 238t
chlorobutyl rubber (polychloroprene), 268–270
 Exxon Chlorobutyl 1068 Chlorobutyl Rubber, 270t
chloroprene polymerization, 268f
closures, 67
coatings, 68
 corrosion protection, 10
 gravure coating, 45–46
 immersion coating, 47, 49f
 knife on roller coating, 46
 metering rod coating, 47, 48t
 permeation, 10
  testing, 15–19
 reverse roller coating, 46
 slot die coating, 47
 vacuum deposition coating, 47–48, 49f
 web coating, 44–49
COC *see* cyclic olefin copolymer
co-continuous lamellar structures, 306–310
CO, ECO *see* epichlorohydrin rubber
coefficient of friction (COF), 34
coextrusion, 39, 40
combustion modifiers, 33
composites, 31–33
compression film polypropylene, 60–65
condensation polymerization, 21, 22f
condiment bottles, 60–65
containers, 65–67
 molding, 53–56
 permeation tests, 56–57
 production, 39

conversion factors, 319–320
COPE (thermoplastic copolyester elastomers), 257–259
copolymers, 21–22
corrosion, 10
coupling agents, 36
crosslinking:
 chemical structure, 265f
 polymers, 22–23
crystalline plastics/polymers, 29–30
cyclic olefin copolymer (COC), 160–161
 applications, 160
 chemical structure, 160f
 Topas Advanced Polymers TOPAS COC, 160t
 water vapor permeation, 160t, 161t
Cycolac ABS (Sabic Innovative Plastics), 78t

D

Daikin Industries Neoflon PCTFE film, 218t
Daikin Neoflon FEP film, 207f
dairy containers, 66
Darcy's law, 1–2, 2f
Dartek Nylon 66 films (Exopack Performance Films Inc.), 135t
degree of unsaturation (DoU), 24
Delrin POM Homopolymer (DuPont), 249t
deposition coating, vacuum, 47–48, 49f
Deutsches Institut für Normung *see* DIN
device packaging, 62–65
diacids, 122f
dialysis membranes, 69–70
diamines, 122f
dimethylsilicone rubber, 235t
DIN (German Institute for Standardization) standards, 13t, 321–322
dip coating, 47, 49f
dope, 43
DoU *see* degree of unsaturation
Dow Calibre PC:
 carbon dioxide permeation, 97t
 nitrogen permeation, 97t
 oxygen permeation, 97t
Dow Chemical Attane Blown film:
 carbon dioxide permeation, 150t
 oxygen permeation, 150t
 water vapor permeation, 150t
Dow Chemical Low Acrylonitrile ABS film, 78t
Dow Chemical Low-Density Polyethylene:
 gas permeation, 151t
 solvent vapor permeation, 152t
Dow Chemical Medium Acrylonitrile Content ABS film, 78t
Dow Chemical Styron PS film, 85f
Dow Chemical Trycite Oriented PS film, 84t
Dow Chemical Tyril Low Acrylonitrile Content SAN, 87t

Dow Ethocel ethyl cellulose film, 298f
Dow Saranex 25 Co-extruded Barrier film, 315t
Dow Saranex 650, 652 Co-extruded Barrier film, 315t
Dow Saranex PVDC, 185f
Dow Saranex PVDC Multilayer films:
   oxygen permeation, 183t
   water permeation, 183t
Dow Saran PVDC films:
   gas permeation, 184t
   hydrogen sulfide permeation, 185f
   oxygen permeation, 182t
   water vapor permeation, 184t
drink containers, 66
drug encapsulation, 67
dual-mode sorption, 7, 7f
DuPont Delrin POM Homopolymer, 249t
DuPont Elastomers Vamac AEM, 263f, 264t
DuPont Elastomer Viton Elastomers, 276t, 277t
DuPont Elvax EVA:
   oxygen permeation, 162t
   water vapor permeation, 162t
DuPont Fluorocarbon FEP film:
   gas permeation, 203t
   vapor permeation, 203t–204t
DuPont Hyrtel Thermoplastic Copolyester Elastomer, 257t, 258t
DuPont Kapton film:
   gas permeation, 119t
   oxygen permeation vs. temperature, 119f
   water vapor permeation, 119t, 120t
DuPont Lamellar Structured Containers:
   oxygen permeation, 308t
   water vapor transmission, 308t
DuPont Mylar PET, 105f
DuPont Nafion, 192, 192f
DuPont nylon, 6, 130f
DuPont Nylon/HDPE Lamellar Structured Containers:
   oxygen permeation, 308t
   water vapor transmission, 308t
DuPont Performance Elastomers Neoprene Polychloroprene Rubber Vulcanizate, 269t, 270t
DuPont Selar amorphous polyamide:
   oxygen permeation vs. relative humidity, 126f
   oxygen permeation vs. temperature, 125f
DuPont Selar with nylon 6, 127f
DuPont Selar PA Amorphous Nylon, 124t
DuPont Selar RB Polyolefin Lamellar Structured Containers, 309t, 310t
DuPont SP TTR10AH9 Transparent High Gloss PVF film, 216t
DuPont Surlyn Sodium Ion Type Ionomer film:
   oxygen permeation, 191t
   water vapor permeation, 191t

DuPont Surlyn Zinc Ion Type Ionomer film:
   oxygen permeation, 190t
   water vapor permeation, 191t
DuPont Tedlar PVF, 216t
DuPont Teflon AF, 214–217, 214t
   chemical structure, 214f
   pure gas permeation coefficients, 214t
DuPont Teflon FEP copolymer:
   ammonia permeation, 205t
   gas permeation, 206f
   hydrogen permeation, 204t
   oxygen permeation, 205t
DuPont Teflon NXT, 200t, 202t
DuPont Teflon PFA film, 208t
DuPont Teflon PTFE:
   ammonia permeation, 198t
   carbon-filled, 199t
   hydrogen permeation, 197t
   nitrogen permeation, 197t, 199t
   oxygen permeation, 198t
DuPont Tefzel, 224t
DuPont Teijin films Mylar PET films:
   carbon dioxide permeation, 101t
   nitrogen permeation, 101t
   oxygen permeation, 101t–102t
   water vapor permeation, 102t
DuPont Viton Fluoroelastomer, 275t
DuPont Zytel 42 Nylon 66 film, 136t
dyes, 36
Dyneon 500 THV:
   hydrogen chloride permeation, 213t
   oxygen permeation, 213t
   water vapor permeation, 213t
Dyneon 1700 HTE:
   gas permeation, 212t
   water vapor transmission, 212t
Dyneon 6235G ETFE:
   carbon dioxide permeation, 225t
   oxygen permeation, 224t
   water vapor permeation, 225t
Dyneon 6510N PFA:
   gas permeation, 209t
   water vapor permeation, 209t
Dyneon FEP, 206t
Dyneon TF 1750 PTFE, 199t
Dyneon TFM 1700 PTFE, 201t
   gas permeation, 200t
   water vapor permeation, 201t
Dyneon TFM 1750 PTFE, 201t
Dyneon TFM PTFE, 201t

E

Eastman Eastar 5445 PCTG Copolyester film, 99t
Eastman Ecdel 9966 Thermoplastic Copolyester Elastomer, 258t

easy-peel film, 60–65
Ecdel 9966 Thermoplastic Copolyester Elastomer (Eastman), 258t
ECO see epichlorohydrin rubber
ECTFE see ethylene-chlorotrifluoroethylene copolymer
effusion, 2–3
egg cartons, 66
EIS (electro-impedance spectroscopy), 16
Elastollan TPUs (BASF):
    gas permeation, 253t
    water vapor permeation, 254t
elastomers, 31, 251–284
    definition, 251
    ethylene acrylic elastomers, 263–264
    olefinic TPEs, 256–257
    styrenic block copolymer TPEs, 262
    thermoplastic copolyester elastomers, 257–258
    thermoplastic polyether block polyamide elastomers, 259–262
    thermoplastic polyurethane elastomers, 251–255
    See also individual makes
Electometer Payne permeability cups, 10t, 15, 15f
electro-impedance spectroscopy (EIS), 16
electronegativity, 23
Elvax EVA (DuPont):
    oxygen permeation, 162t
    water vapor permeation, 162t
EMS Chemie Grivory G16 Amorphous Nylon:
    carbon dioxide and nitrogen permeation, 123t
    oxygen permeation, 123t
    water vapor permeation, 123t
EMS-Grivory Grilamid L 25 Nylon 12 Resin, 134t
EMS-Grivory Grilon BM 13 SBG, 141t
EMS-Grivory Grilon BM 20 SBG, 137t
EMS-Grivory Grilon F 34 Type 6 Nylon, 129t
EMS-Grivory Grilon F 50 Type 6 Nylon, 129t
EMS-Grivory Grilon Nylon 6/12 films, 137t, 138t
    water vapor permeation, 139t
environmentally friendly polymers, 287–304
    cellophane, 291–292
    cellulose acetate, 294
    ethyl cellulose, 295–299
    nitrocellulose, 293
    poly-3-hydroxybutyrate, 303–304
    polycaprolactone, 299–301
    polylactic acid, 301–303
    trademarks, 288t–290t
EPDM (ethylene-propylene rubbers), 270–271
epichlorohydrin rubber (CO, ECO), 272–273
    applications, 273
    gas permeation, 273f
    monomer/polymer structure, 272f
    Zeus Chemicals Hydrin Epichlorohydrin Polymers, 272t, 273t

epichlorohydrin terpolymer (GECO), 272f
EPM, EPDM (ethylene-propylene rubbers), 270–271
EPS (expanded polystyrene), 82–83
ETFE see ethylene-tetrafluoroethylene copolymer
Ethocel ethyl cellulose film (Dow), 298f
ethyl cellulose, 295–299
    applications, 295
    chemical structure, 297f
    Dow Ethocel film, 298f
    gas permeation, 298t, 298f, 299t
ethylene acrylic acid copolymer (EAA), 190
    applications, 190
    chemical structure, 190f
ethylene acrylic elastomers (AEM), 263–264
    gas permeation, 263t, 264t
    liquid permeation, 264t
    structural units, 263f
ethylene-chlorotrifluoroethylene copolymer (ECTFE), 226–230
    ammonia permeation, 227t
    applications, 226
    carbon dioxide permeation, 229f
    chemical structure, 226f
    chlorine permeation, 230f
    helium permeation, 229f
    hydrogen chloride/sulfide permeation, 230f
    nitrogen permeation, 227t, 229f
    oxygen permeation, 229f
    solvent permeation, 228t
    water vapor permeation, 228f
    See also Solvay Solexis Halar ECTFE
ethylene-propylene rubbers (EPM, EPDM), 270–271
ethylene-tetrafluoroethylene copolymer (ETFE), 223–226
    applications, 224
    carbon dioxide permeation, 225t
    chemical structure, 223f
    chlorine gas permeation, 223f
    gas permeation, 226f
    oxygen permeation, 224t
    water vapor permeation, 225t
ethylene-vinyl acetate (EVA), 161–163
    applications, 163
    oxygen permeation, 161f, 162t
    polymer structures, 161f
    water vapor permeation, 162t, 163f
ethylene-vinyl alcohol (EVOH), 163–178
    applications, 164
    chemical structure, 163f
    copolymer/PE multilayer film, 317f
    See also EVAL...; Nippon Gohsei Soarnol EVOH
EVA see ethylene-vinyl acetate
EVAL EF-XL Biaxially Oriented EVOH film:
    organic solvents, 171t
    oxygen permeation, 169t

EVAL E Series Ethylene-Vinyl Alcohol Copolymer
        (EVOH) Resins:
    carbon dioxide permeation, 176f
    gas permeation, 168t
    organic solvents, 172t
    oxygen permeation, 170t
    oxygen transmission, 166t
EVAL Ethylene-Vinyl Alcohol Copolymer (EVOH):
    fluorocarbon permeation, 166t
    Polymer Grade Series, 164t
    Resin Grades, 164t–165t
EVAL EVOH films, 174f, 175f, 176f
EVAL EVOH Grades, 168t
EVAL F Series Ethylene-Vinyl Alcohol Copolymer
        (EVOH) Resins:
    organic solvents, 172t
    oxygen permeation, 170t, 171t
    oxygen transmission rate, 165t
EVAL G Series Ethylene-Vinyl Alcohol Copolymer
        (EVOH) Resins, 167t
EVAL H Series Ethylene-Vinyl Alcohol Copolymer
        (EVOH) Resins:
    gas permeation, 169t
    oxygen permeation, 167t
EVAL K Series Ethylene-Vinyl Alcohol Copolymer
        (EVOH) Resins, 167t
EVOH *see* ethylene-vinyl alcohol
Exopack Performance films Inc. Dartek Nylon 66 films,
        135t
Exopack Scairfilm LWS-1/LWS-2 Laminating MDPE
        films, 153t
Exopack Sclairfilm LX-1 LLDPE film, 151t
expanded polystyrene (EPS), 82–83
Extem XH 1015 PEI membranes, 116t
extenders, 36
extruded polystyrene (XPS), 82–83
extrusion, 39, 47
    blow molding, 54
    blown film process, 39–41
    consumer packaging polystyrene, 60–65
Exxon Butyl Rubber, 267t
Exxon Chlorobutyl 1068 Chlorobutyl Rubber, 270t
ExxonMobil Chemical Model Formula Compounds, 265t
Exxon Vistalon Ethylene-Propylene-Diene Copolymer
        (EPM) Rubbers, 271t

F

fabric film laminates, 62–65
FEP *see* fluorinated ethylene propylene
Ferro-Advanced Polymer Alloys Alcryn AEM:
    gas permeation, 263t
    liquid permeation, 264t
fiber extruder process, 53f
fiber spinning process, 53f
Fick's law of diffusion, 3, 4
fillers, 31–33
film dies, 40f, 42f, 43f
films:
    lamination, 49–50
    post-production, 44–45
    production, 39–57
    purpose, 59
    solvent-casting process, 42–44, 44f
    See also barrier films; *individual makes of film*;
        membranes; polymer films
fire retardants, 33
FKM *see* fluoroelastomers
flame lamination, 49–50
flame retardants, 33
flavor permeation: packaging, 60
Flory-Huggins model, 6, 6f
fluorinated ethylene propylene (FEP), 203–207
    ammonia permeation, 205t
    applications, 203
    chemical structure, 203f
    fuel vapor permeation, 206t
    gas permeation, 203t, 206f, 207f
    helium permeation, 206f
    hydrogen permeation, 204t
    nitrogen permeation, 206f
    oxygen permeation, 205t
    vapor permeation, 203t–204t
fluorination, 56–57
Fluorocarbon FEP film (DuPont):
    gas permeation, 203t
    vapor permeation, 203t–204t
fluoroelastomers (FKM), 273–278
    applications, 275
    curing chemistry, 274t, 275t
    monomer structures, 274f
    Solvay Solexis Tecnoflon FKM Fluoroelastomers, 276t
fluoropolymers, 195–230
    amorphous fluoropolymer, 214–215
    ethylene-chlorotrifluoroethylene copolymer, 226–230
    ethylene-tetrafluoroethylene copolymer, 223–226
    fluorinated ethylene propylene, 203–207
    hexafluoropropylene, tetrafluoroethylene, ethylene
        terpolymer, 212
    melting point ranges, 196t
    monomer structures, 195f
    perfluoroalkoxy, 207–210
    polychlorotrifluoroethylene, 217–219
    polytetrafluoroethylene, 196–202
    polyvinyl fluoride, 215–217
    polyvinylidene fluoride, 219–222
    Teflon AF, 214–215
    tetrafluoroethylene, hexafluoropropylene, vinylidene
        fluoride terpolymer, 212–214
foam extrusion thermoforming, 62–65
food wrap film, 62–65

Ford reference fuels, 321–322
form fill and seal packaging, 60–65
fruit trays, 66
fuel composition references, 321–322
fuel hoses/tanks, 67–68

G

gases:
  effusion, 2–3
  transport through solid materials, 2–9
gas permeation:
  history, 1, 2
  test cells, 14
gas separation membranes, 71
GECO (epichlorohydrin terpolymer), 272f
general purpose polystyrene *see* polystyrene
geomembranes, 62–65
geometric isomers, 25–26
Geon 101-EP-100 PVC film (Polyone), 180f
German Institute for Standardization *see* DIN
glassy polymers, 8f
gloves, 68–69
  permeation/degradation resistance chart, 323–328
GPPS (general purpose polystyrene) *see* polystyrene
graft copolymers, 22
Graham's law, 2–3
Graham, Thomas, 1, 2
gravure coating, 45–46
'green' polymers *see* environmentally friendly polymers
Grilon *see* EMS-Grivory Grilon...
Grivory *see* EMS...

H

Halar ECTFE *see* Solvay Solexis Halar ECTFE
HDPE *see* high-density polyethylene
head-to-tail isomers, 25
heat resistant film, 62–65
heat seal resins, 62–65
heat-shrinkable polypropylene film, 62–65
Henry's law, 3, 4, 6f
hexafluoropropylene, tetrafluoroethylene, ethylene
    terpolymer (HTE), 212
  gas permeation, 212t
  water vapor transmission, 212t
high-clarity LLDPE film, 62–65
high-density polyethylene (HDPE), 153–156
  ammonia permeation, 155t
  fuel tanks, 67
  gas permeation, 154t, 156f
  hydrogen permeation, 154t
  nitrogen permeation, 155t
  NOVA Chemicals Sclair HDPE films, 154t
  oxygen permeation, 154t, 155t
high-impact polystyrene (HIPS), 22, 83, 83f
high-performance polyamide, 142–144

high-performance/temperature polymers, 233–250
HIPS *see* high-impact polystyrene
history, 1–2
hollow fiber modules, 72–73
homopolymer PTFE, 197–202
homopolymers, 21–22, 156, 197–202
Honeywell Aclar Multilayer films, 315t
Honeywell Aclar PCTFE film, 217t, 218t
Honeywell Aegis Nylon 6/66 films, 139t
Honeywell Capran Oxyshield 1545 and 2545, 313t
Honeywell Capran Oxyshield Plus, 313t
Honeywell Plastics Aegis Nylon 6 films, 128t
Honeywell Plastics Capron Nylon 6 films:
  gas permeation, 128t
  water vapor permeation, 128t
Honeywell Series R PMP films, 160t
hoses: fuel, 67–68
Hostaform POM Copolymer (Ticona), 249t, 250f
Hostaphan RN 25 Biaxially Oriented PET Release film
    (Mitsubishi):
  aroma permeation, 104t
  gas permeation, 103t
  vapor permeation, 104t
hot roll/belt lamination, 49, 50f
HTE *see* hexafluoropropylene, tetrafluoroethylene,
    ethylene terpolymer
Hydrin Epichlorohydrin Polymers (Zeus Chemicals),
    272t, 273t
hydrogen bonding, 27
Hyflon MFA (Solvay Solexis):
  oxygen and water permeation, 211f
  vapor permeation, 210t–211t
Hyflon PFA (Solvay Solexis), 210f
Hyrtel Thermoplastic Copolyester Elastomer (DuPont),
    257t, 258t

I

IBM (injection blow molding), 54
immersion coating, 47, 49f
impact copolymers, 156
impact modifiers, 35
Imperm 105 nanoclay-filled nylon-MXD6 PAA films
    (Nanocor), 143f
industrial films, 62–65
INEOS ABS Lustran 246 ABS:
  gas permeability coefficients, 80f
  permselectivity vs. temperature, 80f
INEOS Barex Acrylonitrile-Methyl Acrylate
    Copolymers:
  gas permeation, 188t
  water vapor permeation, 188t
injection blow molding (IBM), 54
Innovative Plastics *see* Sabic
Innovia Cellophane films, 292t
instron peel test, 18, 18f

inter/intramolecular attractions, 27–28
ionomers, 190–192
IPA (isophthalic acid), 142–143
isomers, 25–27
   geometric isomers, 25–26
   stereoisomers, 26–27
   structural isomers, 25
isophthalic acid (IPA), 142–143
isoprene, 278f
ISO standards, 12t–13t
isotactic polypropylene, 26

J
JIS (Japanese Industrial Standards), 13t
juice packaging, 66

K
Kapton film (DuPont):
   gas permeation, 119t
   oxygen permeation vs. temperature, 119f
   water vapor permeation, 119t, 120t
kinetic diameter, 3, 3t
knife on roller coating, 46

L
lactic acid to polylactic acid conversion, 301f
lamellar structures, 306–310
   injection molding, 307t
   oxygen permeation, 308t
   solvent permeation, 309t, 310t
   water vapor transmission, 308t
   See also DuPont Lamellar Structured Containers; DuPont Nylon/HDPE Lamellar Structured Containers; DuPont Selar RB Polyolefin Lamellar Structured Containers
lamination, 49–50
   flame lamination, 49–50
   hot roll/belt lamination, 49, 50f
Langmuir model, 6–7, 6f
Lanxess Baypren Polychloroprene Rubber Vulcanizate, 269t
Lanxess Butyl Butyl Rubber (Bayer), 268t
Lanxess Krylene Styrene-Butadiene Rubber (Bayer), 284t
Lanxess Krynac Nitrile Rubber (Bayer):
   air permeation, 282t
   air vs. temperature, 283f
   gas permeation, 282t
LCP see liquid crystalline polymers
LDPE see low-density polyethylene
linear low-density polyethylene (LLDPE), 151
linear polymers, 22–23
liquid crystalline polymers (LCP), 89–92
   applications, 89
   carbon dioxide permeation, 92t
   chemical structure, 90f
   hydrogen permeation, 91t
   oxygen permeation, 91t
   Ticona Vectra A950 LCP, 90f, 91t
   Ticona Vectra LCP films, 91t, 92t
   water vapor transmission, 92t
LISIM Technology, 52f
LLDPE (linear low-density polyethylene), 151
LlyondellBasell Adflex PP:
   carbon dioxide permeation, 158t
   oxygen permeation, 158t
LlyondellBasell polyolefins polypropylene, 157t
LlyondellBassell polyolefins polyethylene:
   gas permeation, 147f
   solvent vapor transmission, 148f
   water vapor transmission, 147f
low-density polyethylene (LDPE), 151–152
   gas permeation, 151t
   solvent vapor permeation, 152t
lubricants, 34
Lubrizol Estane Breathable TPU, 252t
Lubrizol Estane TPU:
   gas/vapor permeation, 252t
   solvent vapor permeation, 253t
Lucite Diakon Polymethyl Methacrylate:
   gas permeation, 186t
   water vapor permeation, 187t
Lupolen V EVA (BASF AG), 161f
Luran SAN (BASF):
   oxygen permeation, 87t
   water vapor permeation, 87t
Luran S ASA films (BASF):
   gas permeation, 81t, 82t
   water vapor permeation, 82t

M
machine direction orientation (MDO), 51
Makrolon PC (Bayer MaterialScience), 94t–95t, 95t–96t, 96t–97t
Marlex HDPE (Chevron Philips), 154t
MDO (machine direction orientation), 51
MDPE (medium-density polyethylene), 153–156
measurement:
   diffusivity units, 4
   permeability, 14
   solubility units, 4
   standard units, 4
   vapor transport, 14
meat trays, 66
medical molds/packaging, 62–65
medium-density polyethylene (MDPE), 153–156
melting points: fluoropolymers, 196t
membranes:
   commercial processes, 69–75
   dialysis, 69–70
   gas separation, 71

pervaporation, 70—71
processes, 69—75
production, 39—57, 52—53
purpose, 59
reverse osmosis, 70
structures, 72—75
See also films
metalized films, 305
  oxygen/water vapor transmission, 306t
  Vacmet Metalized Plastic films, 306t
metering rod coating, 47, 48t
Meyer rod coating, 47, 48t
MFA copolymer, 210—212
mica, 36
milk packaging, 66
Mirrex PVC (VPI):
  oxygen permeation, 180t
  water vapor permeation, 180t
Mitsubishi gas chemical nylon-MXD6 PAA films, 142f
Mitsubishi Polyester film Hostaphan RN 25 Biaxially Oriented PET Release film:
  aroma permeation, 104t
  gas permeation, 103t
  vapor permeation, 104t
modified atmosphere packaging, 62—65
modified PTFE, 202
moisture protection: packaging, 59
molding of containers, 53—56
molecules:
  dipole moments, 23t
  molecular weight, 28—29
  steric hindrance, 24—25
  See also polymers
monomers: molecular arrangement, 22
mulch films, 60—65
multilayered films, 9, 39, 40, 65, 65f—66f, 305—317
  lamellar structures, 306—310
  metalized films, 305
  oxygen permeation vs. relative humidity, 316f, 317f
  silicon oxide coating technology, 305—306
  xylene permeation, 316f
multiphase polymer blends, 30—31
Mylar PET (DuPont), 105f

N

Nafion (DuPont), 192, 192f
Nanocor Imperm 105 nanoclay-filled nylon-MXD6 PAA films, 143f
natural rubber, 278—280
  air permeation, 279f, 279t
  gas permeation, 280f
NBR *see* acrylonitrile-butadiene copolymer
Neoflon FEP film (Daikin), 207f
Neoflon PCTFE film (Daikin Industries), 218t
Neoprene Polychloroprene Rubber Vulcanizate (DuPont), 269t, 270t
network polymers, 22—23
nine-layer blown film die, 43f
Nippon Gohsei Soarnol EVOH:
  chloroform permeation, 174t
  kerosene permeation, 174t
  oxygen permeation vs.
    ethylene content, 177f
    film thickness, 177f
    relative humidity, 178f
    temperature, 178f
  water vapor permeation, 173t
nitrile rubber, 282t—283t, 283f
nitrocellulose, 293
  chemical structure, 293f
  gas permeation, 293t
Nollet, Jean-Antoine, 1
non-Fickian diffusion, 5
nonporous dense polymer membranes, 4
NOVA Chemicals Sclair HDPE films, 154t
nylons *see* polyamide...

O

odor permeation: packaging, 60
olefinic TPEs (TPO), 256—257
  applications, 256
  gas permeation, 256t
  water vapor transmission, 256t
OPS films, 83
optical brighteners, 35
orientation of polymers, 30, 50—52
oriented/un-oriented nylon 6, 129t

P

packaging:
  ASTM standards, 11t—12t
  atmospheric gas transmission, 59—60
  barrier films, 59—65
  categories, 59
  DIN standards, 13t
  flavor permeation, 60
  ISO standards, 12t—13t
  JIS standards, 13t
  odor permeation, 60
  oxygen transmission, 59—60
  water vapor protection, 59
PAI *see* polyamide-imide
parylene (poly(p-xylylene)), 245—247
  applications, 247
  gas permeation, 246t
  polymer molecule structures, 246f
  water vapor permeation, 247t
PB *see* polybutadiene
PBI *see* polybenzimidazole

PBI Performance Products Inc. Celazole
    PBI membrane:
  carbon dioxide permeation, 245f
  hydrogen permeation, 244f
PBT *see* polybutylene terephthalate
PC *see* polycarbonate
PCT *see* polycyclohexylene-dimethylene terephthalate
PCTFE *see* polychlorotrifluoroethylene
PDMS *see* polydimethyl siloxane
PE *see* polyethylene
PEBA *see* thermoplastic polyether block polyamide elastomers
PEBAX films *see* Arkema
PEEK *see* polyetheretherketone
PEI *see* polyetherimide
PEN *see* polyethylene naphthalate
perfluoroalkoxy (PFA), 207–211
  applications, 208
  chemical structure, 208f
  comonomers, 207t
  gas permeation, 208t, 209t, 210f
  Solvay Solexis Hyflon PFA, 210f
  water vapor permeation, 209t
perfluoropolymers *see* fluoropolymers
perfluorosulfonic acid (PFSA), 192
permeability coefficients:, 319–320
permeability cups, 10t, 15, 15f
permeants: Arrhenius equation parameters, 8t
permeation: definition, 1
permselectivity, 71
pervaporation membranes, 70–71
PES *see* polyethersulfone
PE shrink film, 62–65
PET *see* polyethylene terephthalate
PFA *see* perfluoroalkoxy
PFSA *see* perfluorosulfonic acid
PGA *see* polyglycolic acid
PH3B, 303f
PHA *see* poly-3-hydroxybutyrate
pharmaceutical blister packs, 62, 64f
PHBV: chemical structure, 303f
PHV: chemical structure, 303f
PI *see* polyimide
pigments, 36
PLA *see* polylactic acid
plastic beads, 39, 39f
plastic compositions, 30–37
plasticizers, 7f, 35
plate and frame modules, 72
PMMA *see* polymethyl methacrylate
PMP *see* polymethylpentene
polarity, 23
poly-3-hydroxybutyrate (PHA), 303–304
  applications, 303–304
  permeability coefficients, 304t

polyacrylics, 186–187
polyamide 6 (nylon 6), 126–131
  applications, 127
  carbon dioxide permeation, 130f
  chemical structure, 127f
  gas permeation, 128t, 129t
  hydrogen sulfide permeation, 130f
  oxygen permeation, 127f, 129t
  water vapor permeation, 128t
polyamide 6/12 (nylon 6/12), 137–139
  applications, 137
  chemical structure, 137f
  oxygen and carbon dioxide permeation, 137t, 138t
  water vapor permeation, 139t
polyamide 6/69 (nylon 6/69), 140–141
polyamide 11 (nylon 11), 131–133
  applications, 131
  chemical structure, 131f
  gas permeation, 132f, 132t, 133f
polyamide 12 (nylon 12), 133–134
  chemical structure, 133f
  gas permeation, 134t, 134f
  natural gas permeation vs. temperature, 134f
  water vapor permeation, 134t
polyamide 66 (nylon 66), 135–136
  applications, 135
  carbon dioxide permeation, 135t
  chemical structure, 135f
  gas permeation, 136t
  oxygen permeation, 135t
  water vapor permeation, 135t, 136t
polyamide 66/610 (nylon 66/610), 136–137
  applications, 136
  chemical structure, 136f
  gas/water vapor permeation, 137t
polyamide 666 (nylon 666 or 6/66), 139–140
polyamide-imide (PAI), 107–113
  applications, 107
  carbon dioxide permeation, 112f
  chemical structures, 107f, 108f
  gas permeation, 110t, 111f
  methane permeation, 113f
  nitrogen permeation, 111f
  oxygen permeation, 112f
polyamides (nylons), 121–144
  amino acids, 122f
  amorphous polyamide, 122–123
  diacids, 122f
  diamines, 122f
  generalized reaction, 121f
  monomers, 121t
  polyamide 6, 126–131
  polyamide 6/12, 137–139
  polyamide 6/69, 140–141

polyamide 11, 131–133
polyamide 12, 133–134
polyamide 66, 135–136
polyamide 66/610, 136–137
polyamide 666, 139–140
polyarylamide, 141–142
polyphthalamide, 142–144
polyanhydrides, 287–288, 287f
polyarylamide, 141–142
   applications, 142
   chemical structure, 141f
polybenzimidazole (PBI), 243–245
   applications, 244
   carbon dioxide permeation, 245f
   chemical structure, 244f
   hydrogen permeation, 244f
   PBI Performance Products Inc. Celazole PBI membrane, 244f, 245f
polybutadiene (PB), 158–159
   applications, 159
   nitrogen and helium permeation coefficients, 159f
   structural isomers, 158f
polybutylene terephthalate (PBT), 92–93
   applications, 93
   chemical structure, 93f
   gas/water vapor permeation, 93t
polycaprolactone, 299–301
   applications, 299
   chemical structure, 299f
   oxygen permeation, 300t, 301t
   starch and glycerol blends, 300t, 301t
   water vapor permeation, 300t
polycarbonate (PC), 93–99
   applications, 94
   carbon dioxide permeation, 96t–97t, 99f
   chemical structure, 89f, 93f, 94f
   helium permeation, 98f
   methane permeation, 98f
   nitrogen permeation, 95t–96t, 97t
   oxygen permeation, 94t–95t, 97t
polychloroprene, 268–270
polychlorotrifluoroethylene (PCTFE), 217–219
   applications, 217
   chemical structure, 217f
   gas permeation, 218t, 219t
   water vapor permeation, 217t, 218t
polycyclohexylene-dimethylene terephthalate (PCT), 99
   applications, 99
   chemical structure, 99f
polydimethyl siloxane (PDMS):
   gas permeation, 235f, 237f
   gas separation, 236f
   helium and nitrogen permeation, 236f
polyesters, 89–105
   liquid crystalline polymers, 89–92

polybutylene terephthalate, 92–93
polycarbonate, 93–99
polycyclohexylene-dimethylene terephthalate, 99
polyethylene naphthalate, 100
polyethylene terephthalate, 100–105
polyetheretherketone (PEEK), 233
   amorphous/crystalline permeation, 234t
   chemical structure, 233f
   VICTREX Morphology PEEK-Based APTIV film, 234t, 234t
   water vapor transmission, 234t
polyetherimide (PEI), 113–118
   applications, 115
   chemical structures, 113f, 114f, 114f
   gas/water vapor permeation, 115t, 116t, 118f
   hydrogen/ethane, hydrogen/methane and hydrogen/nitrogen permeation vs. pressure differential, 117f
   hydrogen/propane permeation selectivity vs. pressure differential, 117f
   Sabic Innovative Plastics Ultem 1000 PEI, 116t
   Westlake Plastics Tempalux PEI film, 115t
polyethersulfone (PES), 241–243
   applications, 241
   carbon dioxide permeation, 243f
   chemical structure, 241f
   gas permeation, 242t
   helium permeation, 242f
   methane permeation, 243f
   Sabic Innovative Plastics Ultem 1000 PES, 242t
polyethylene naphthalate (PEN), 100
   applications, 100
   chemical structure, 100f
polyethylene (PE), 145–156
   applications, 146
   crystal structure, 146f
   gas permeation, 147f, 149f
   high-density polyethylene, 153–156
   linear low-density polyethylene, 151
   low-density polyethylene, 151–152
   medium-density polyethylene, 153
   solvent vapor transmission, 148f
   sulfur dioxide permeation, 149f
   types, 146f
   ultra-low density polyethylene, 150
   unclassified polyethylene, 146–149
   water vapor transmission, 147f
polyethylene terephthalate (PET), 100–105
   applications, 101
   aroma permeation, 104t
   carbon dioxide permeation, 101t
   chemical structure, 100f
   containers, 66
   gas permeation, 105f, 103t
   hydrogen sulfide permeation, 105f

polyethylene terephthalate (PET) (*Continued*)
    Mitsubishi Polyester Film Hostaphan RN Biaxially Oriented PET Release film, 103t, 104t
    nitrogen permeation, 101t
    oxygen permeation, 101t−102t
    vapor permeation, 104t
    water vapor permeation, 102t
polyglycolic acid (PGA), 288, 288f
polyhydro-xyalkanoates, 303f
polyimide (PI), 118−120
    applications, 118
    chemical structures, 118f
    monomer structures, 118f
polyimides, 107−120
    monomer structures, 115f
    polyamide-imide, 107−113
    polyetherimide, 113−118
    polyimide, 118−120
polyisoprene, 278f
polyketones *see* polyetheretherketone
polylactic acid (PLA), 301−303
    applications, 301
    carbon dioxide/methane permselectivity vs. temperature, 303f
    carbon dioxide permeation coefficient vs. temperature, 302f
    conversion from lactic acid, 301f
    methane permeation coefficient vs. temperature, 302f
    nitrogen permeation coefficient vs. temperature, 302f
    oxygen permeation coefficient vs. temperature, 303f
polymer films:
    ASTM standards, 11t−12t
    DIN standards, 13t
    ISO standards, 12t−13t
    JIS standards, 13t
    pore-flow model, 4
    solution-diffusion model, 4, 4f
polymerization, 21
polymers:
    additives, 31−37
    Arrhenius equation parameters, 8t
    blends, 30−31
    chain entanglement, 27−28
    classification, 28−30
    definition, 21
    film-casting processes, 45t
    fluorination, 56−57
    hydrogen bonding, 27
    inter/intramolecular attractions, 27−28
    molecular weight, 28−29
    orientation, 30, 50−52
    polarity, 23
    unsaturation, 23−24
    Van der Waals forces, 27
    See also molecules

polymethyl methacrylate (PMMA), 186
    chemical structure, 186f
    gas permeation, 186t
    Lucite Diakon Polymethyl Methacrylate, 186t
    permeability data, 187f
    water vapor permeation, 187t
    Wood-Adams polymethyl methacrylate, 187f
polymethylpentene (PMP), 159−160
    applications, 159−160
    chemical structure, 159f
    water transmission, 160t
polyolefins, 145−192
    LlyondellBasell polyolefins polyethylene, 147f, 148f
    LlyondellBasell polyolefins polypropylene, 157t
    monomer structures, 145f
Polyone Geon 101-EP-100 PVC film, 180f
polyoxymethylene (POM), 247−248
polyphenylene sulfide (PPS), 234−237
    applications, 237
    chemical structure, 237f
    gas permeation, 238t
    liquid vapor transmission, 238t
polyphthalamide (PPA), 142−144, 143f
polypropylene (PP), 26, 156−158
    applications, 156
    carbon dioxide permeation, 158t
    gas permeation, 157t
    oxygen permeation, 158t
poly(p-xylylene), 245−247
polysiloxane, 233−234
    chemical structure, 234f
    helium permeation, 235f
polystyrene (PS), 82−86
    chemical structure, 82f
    gas/water vapor permeation, 84t, 85f, 86f
Polystyrol 168 N GPPS film (BASF AG), 84t
polysulfone (PSU), 237−241
    applications, 237−238
    carbon dioxide permeation, 240f, 241f
    chemical structure, 238f
    gas permeation, 239t
    helium permeation, 239f
    methane permeation, 240f
    water vapor permeation, 239t
polytetrafluoroethylene (PTFE), 196−202
    ammonia permeation, 198t
    chemical structure, 196f
    gas permeation, 200t
    hydrogen chloride permeation, 201t
    hydrogen permeation, 197t
    nitrogen permeation, 197t, 199t
    oxygen permeation, 198t, 199t

permeation comparisons, 200t
sulfur dioxide permeation, 202f
water vapor permeation, 201t
polyvinylidene chloride (PVDC), 182–186
   chemical structure, 182f
   gas permeation, 184t, 186f
   hydrogen sulfide permeation, 185f
   oxygen permeation, 182t, 183t
   oxygen transmission, 185f
   water permeation, 186f, 183t
   water vapor permeation, 184t
polyvinyl butyral (PVB), 179
   chemical structure, 179f
   gas permeation, 179t
polyvinyl chloride (PVC), 179–181
   applications, 179
   carbon dioxide permeation, 180f
   hydrogen permeation, 180f
   oxygen permeation, 180t
   oxygen permeation vs.
     plasticizer level, 181f
     temperature, 181f
   Polyone Geon 101-EP-100 PVC film, 180f
   VPI Mirrex PVC, 180t
   water vapor permeation, 180t
polyvinyl fluoride (PVF), 215–217
   applications, 215
   chemical structure, 215f
   gas permeation, 216t
   hydrogen permeation, 217f
   vapor permeation, 216t
polyvinylidene chloride (PVDC):
   applications, 182
   coated cellophane film, 291t
   coated nylon/EVOH copolymer/PE multilayer film, 316f
   Dow Saranex PVDC, 185f
   Dow Saranex PVDC Multilayer films, 183t
   Dow Saran PVDC, 182t, 184t, 185f
polyvinylidene fluoride (PVDF), 219–223
   applications, 219
   Arrhenius plot, 9f
   chemical structure, 219f
   gas permeation, 221f, 220t
   Solvay Solexis Solef 1008 PVDF, 220t
   Solvay Solexis Solef PVDF film, 221f, 222f
   water permeation, 222f
   water vapor permeation, 221f
polyvinyls, 145–192
   monomer structures, 145f
POM (polyoxymethylene), 247–250
pore-flow model, 3–9
post film formation processing, 44–45
poultry trays, 66
PP see polypropylene
PPA (polyphthalamide), 142–144
PP/EVOH copolymer/PE multilayer film, 317f
PP liners, 60–65
PPS see polyphenylene sulfide
pressure: solution-diffusion, 5–8
PS see polystyrene
PSU see polysulfone
PTFE see polytetrafluoroethylene
pull-off adhesion tests, 18–19, 19
PVB see polyvinyl butyral
PVC see polyvinyl chloride
PVDC see polyvinylidene chloride
PVDC-coated nylon/EVOH copolymer/PE multilayer film, 316f
PVDF see polyvinylidene fluoride
PVF see polyvinyl fluoride

R

random copolymers, 22, 156
reference fuel compositions, 321–322
reinforced plastics, 31–33
release agents, 33–34
renewable plastics, 287
reverse osmosis membranes, 70
reverse roller coating, 46
Rilsan BESNO P40TL Nylon, 11, 132f, 133f
Rilsan PA11 (Arkema), 132t
Riteflex Thermoplastic Copolyester Elastomer (Ticona), 257f
rotational molding/rotomolding, 55–56
rubbers, 251–284
   acrylonitrile-butadiene copolymer, 281–283
   bromobutyl rubber, 264–266
   butyl rubber, 266–268
   chlorobutyl rubber, 268–270
   epichlorohydrin rubber, 272–273
   ethylene-propylene rubbers, 270–271
   fluoroelastomers, 273–278
   natural rubber, 278–280
   styrene-butadiene rubber, 283–284
   See also individual rubbers
rubbish bags, 62–65
Ryton PPS films (Chevron Philips):
   gas permeation, 238t
   liquid vapor transmission, 238t

S

Sabic Innovative Plastics Cycolac ABS, 78t
Sabic Innovative Plastics Ultem 1000 PEI, 116t
Sabic Innovative Plastics Ultem 1000 PES, 242t
salt spray/humidity test, 16–17
SAN see styrene-acrylonitrile
sauce bottles, 60–65
SBC see styrenic block copolymer
SBR (styrene-butadiene rubber), 283–284

Scairfilm LWS-1/LWS-2 Laminating MDPE films (Exopack), 153t
Scairfilm LX-1 LLDPE film (Exopack), 151t
Sclair HDPE films (NOVA Chemicals), 154t
Sealed House for Evaporation (SHED) test, 67
Selar amorphous polyamide (DuPont), 125f, 126f
Selar with nylon 6 (DuPont), 127f
Selar PA Amorphous Nylon (DuPont), 124t
Selar RB Polyolefin Lamellar Structured Containers (DuPont), 309t, 310t
SHED (Sealed House for Evaporation) test, 67
sheeting standards, 11t–12t, 12t–13t
shipping sacks, 62–65
shrink bundling film, 62–65
silage films, 60–65
silicone rubber, 233
silicon oxide coating technology, 305
simultaneous stretching line (LISIM Technology), 52f
single-layer film: time-lag plot, 5f
SiOx-Coated PET film, 306t
SiOx coating, 305
slip additives, 34
slot die coating, 47
smoke suppressants, 33
Soarnol EVOH *see* Nippon Gohsei Soarnol EVOH
Solef 1008 PVDF (Solvay Solexis), 220t
Solef PVDF film (Solvay Solexis):
    gas permeation, 221f
    water permeation, 222f
Solexis *see* Solvay Solexis…
solid materials: transport of gases/vapors, 2–9
solubility pressure, 5–8
solution-diffusion model, 2–9
Solvay Advanced Polymers Amodel PPA, 144t
Solvay Solexis AQUIVION PFSA Membranes, 192t
Solvay Solexis Halar ECTFE:
    ammonia permeation, 227t
    carbon dioxide permeation, 229f
    chlorine permeation, 230f
    helium permeation, 229f
    hydrogen chloride/sulfide permeation, 230f
    nitrogen permeation, 227t, 229f
    oxygen permeation, 229f
    solvent permeation, 228t
    water vapor permeation, 228f
Solvay Solexis Hyflon MFA:
    oxygen and water permeation, 211f
    vapor permeation, 210t–211t
Solvay Solexis Hyflon PFA, 210f
Solvay Solexis Solef 1008 PVDF, 220t
Solvay Solexis Solef PVDF film:
    gas permeation, 221f
    water permeation, 222f
Solvay Solexis Tecnoflon FKM Fluoroelastomers, 276t
solvent-casting process, 42–44, 45t

soprene-based structural units, 264f
spectroscopy, 16
spiral wound modules, 73–75
stacked dies, 40f
standards:
    ASTM, 11t–12t
    DIN standards, 13t
    JIS standards, 13t
stationery films, 62–65
stereoisomers, 26–27
steric hindrance, 24–25
sterile packaging, 62–65
stretch blow molding, 54–55
stretch film, 62–65
structural isomers, 25
styrene: chemical structure, 77f
styrene-acrylonitrile (SAN), 77, 86–87
    chemical structure, 86f
    gas permeation, 87t
    water vapor permeation, 87t
styrene-butadiene rubber (SBR), 283–184
styrenic block copolymer (SBC) TPEs, 262
    applications, 262
    carbon dioxide permeation, 262t
    nitrogen permeation, 262t
    oxygen permeation, 262t
styrenic plastics, 77–87
Styrolux films (BASF):
    carbon dioxide permeation, 262t
    nitrogen permeation, 262t
    oxygen permeation, 262t
Styron PS film (Dow Chemical), 85f
Styron Styron PS, 83t
sugar drink containers, 66f
Surlyn Sodium Ion Type Ionomer film (DuPont):
    oxygen permeation, 191t
    water vapor permeation, 191t
Surlyn Zinc Ion Type Ionomer film (DuPont):
    oxygen permeation, 190t
    water vapor permeation, 191t
syndiotactic polypropylene, 26–27

T

Tedlar PVF (DuPont), 216t
Teflon *see* DuPont Teflon…
Tefzel (DuPont), 224t
Teijin films Mylar PET films (DuPont):
    carbon dioxide permeation, 101t
    nitrogen permeation, 101t
    oxygen permeation, 101t–102t
    water vapor permeation, 102t
Tempalux PEI film (Westlake Plastics), 115t
ten-layer stacked die, 40f
tenter frames, 52f, 51
terephthalic acid (TPA), 142–143

Terluran ABS films (BASF AG):
  gas permeation, 79t
  water vapor permeation, 79t
terpolymers, 21–22
tests:
  Accelerated Keeping Test, 56, 57t
  adhesion tests, 17–19
  atlas cell test, 17, 18f
  coatings, 15–19
  electro-impedance spectroscopy (EIS), 16
  fluorinated containers, 56–57
  gas permeation test cells, 14
  humidity test, 16–17
  instron peel test, 18f, 18
  permeation, 10–19
  pull-off adhesion tests, 18–19
  salt spray/humidity test, 16–17
  SHED test, 67
  vapor permeation cup tests, 14–15
  vapor transmission, 10–19
tetrafluoroethylene, hexafluoropropylene, vinylidene fluoride terpolymer (THV), 212–214
  applications, 213
  hydrogen chloride permeation, 213t
  oxygen permeation, 213t
  water vapor permeation, 213t
Texin TPU (Bayer MaterialScience), 254f, 255f
thermal stabilizers, 36
thermoplastic copolyester elastomers (TPE-E or COPE), 257–258
thermoplastic polyether block polyamide elastomers (PEBA), 259–262
  applications, 259
  helium permeation, 260t
  oxygen permeation, 260t
  propane permeation, 261t
  water vapor permeation, 259t, 261t
thermoplastic polyurethane elastomers (TPU), 251–255
  gas permeation, 252t, 253t, 255f
  Lubrizol Estane Breathable TPU, 252t
  Lubrizol Estane TPU, 252t, 253t
  molecular structure, 252f
  solvent vapor permeation, 253t
  water vapor permeation, 252t, 254t
  water vapor transmission, 254f
thermoplastics, 29
thermosets, 29
three-layer barrier film structures:
  oxygen permeation through EVOH and PVDC barrier layers, 310t, 311t, 312t, 313t
  oxygen permeation through nylon/EVOH/LDPE, 314t
  oxygen permeation through PET/EVOH/LDPE, 314t
three-layer blown film die, 42f
THV see tetrafluoroethylene, hexafluoropropylene, vinylidene fluoride terpolymer

Ticona Celcon M25/M90/M270 POM Copolymer, 248t
Ticona Hostaform POM Copolymer, 249t, 250f
Ticona Riteflex Thermoplastic Copolyester Elastomer, 257f
Ticona Vectra A950 LCP:
  chemical structure, 90f
  oxygen permeation, 91t
Ticona Vectra LCP films:
  carbon dioxide permeation, 92t
  hydrogen permeation, 91t
  water vapor transmission, 92t
time-lag measurement, 4, 5, 5f
Topas Advanced Polymers TOPAS COC, 160t, 161t
tortuous path effect, 32, 33f
tougheners, 35
TPA (terephthalic acid), 142–143
T-Peel Test, 18, 18f
TPE-E (thermoplastic copolyester elastomers), 257–258
TPO see olefinic TPEs
TPU see thermoplastic polyurethane elastomers
transport of gases/vapors, 2–9
transverse direction orientation, 51
trash bags, 62–65
Trycite Oriented PS film (Dow Chemical), 84t
tubular membrane modules, 73, 74f
two-layer barrier film structures: oxygen permeation through EVOH/LDPE, 314t
Tyril Low Acrylonitrile Content SAN (Dow Chemical), 87t

U

UBE 303 XA Nylon 12 Resin, 134t
UBE Industries Upilex films:
  gas permeation, 120t
  water vapor permeation, 120t
Udel PSU (BASF):
  gas permeation, 239t
  water vapor permeation, 239t
Ultem 1000 PEI (Sabic Innovative Plastics), 116t
Ultem 1000 PES (Sabic Innovative Plastics), 242t
Ultradur PBT (BASF AG), 93t
ultrafiltration, 73
ultra-low density polyethylene, 150
Ultramid C35 Nylon 6/66 film, 140t
unclassified polyethylene, 146–149
units see measurement
unsaturation, 23–24
Upilex films (UBE Industries), 120t
UV stabilizers, 35

V

Vacmet Metalized Plastic films, 306t
vacuum deposition coating, 47–48, 49f
Vamac AEM (DuPont), 264t

Van der Waals forces, 27
vapor permeation:
  cup testing, 14–15
  rates, 320
vapor transmission, 2–9
  ASTM standards, 11t–12t
  DIN standards, 13t
  ISO standards, 12t–13t
  JIS standards, 13t
  tests, 10–19
Vectra LCP films (Ticona):
  carbon dioxide permeation, 92t
  hydrogen permeation, 91t
  water vapor transmission, 92t
vegetable trays, 66
VICTREX Morphology PEEK-Based APTIV film:
  amorphous/crystalline permeation, 234t
  water vapor transmission, 234t
vinyl benzene *see* styrene
viscose: conversion from cellulose, 291f
Vistalon Ethylene-Propylene-Diene Copolymer Rubbers (Exxon), 271t
Viton Elastomers (DuPont), 276t, 277t
Viton Fluoroelastomer (DuPont), 275t
Voltalef PCTFE film (Arkema), 219t
VPI Mirrex PVC:
  oxygen permeation, 180t
  water vapor permeation, 180t
vulcanization, 265f

W

water permeation: history, 1
water vapor protection: packaging, 59
web coating, 44–49
Westlake Plastics Tempalux PEI film, 115t
wet adhesion, 10
Wood-Adams polymethyl methacrylate, 187f

X

XPS (extruded polystyrene), 82–83

Z

Zeus Chemicals Hydrin Epichlorohydrin Polymers, 272t, 273t
Zytel 42 Nylon 66 film (DuPont), 136t